3000 800028 36290
St. Louis Community College

C0-AQW-652

WITHDRAWN

Environmental Technology Handbook

Applied Energy Technology Series

James G. Speight, Ph. D., *Editor*

Khan **Conversion and Utilization of Waste Materials**
Mushrush and Speight **Petroleum Products: Instability and Incompatibility**
Speight **Environmental Technology Handbook**

IN PREPARATION

Lee **Alternative Fuels**
Wyman **Handbook on Bioethanol: Production and Utilization**

Environmental Technology Handbook

James G. Speight, Ph. D.
Western Research Institute
Laramie, Wyoming, U.S.A.

USA	Publishing Office:	Taylor & Francis 1101 Vermont Ave., N.W., Suite 200 Washington, DC 20005 Tel: (202) 289-2174 Fax: (202) 289-3665
	Distribution Center:	Taylor & Francis 1900 Frost Road, Suite 101 Bristol, PA 19007-1598 Tel: (215) 785-5800 Fax: (215) 785-5515
UK		Taylor & Francis, Ltd. 1 Gunpowder Square London EC4A Tel: 0171 583 0490 Fax: 0171 583 0581

ENVIRONMENTAL TECHNOLOGY HANDBOOK

1 2 3 4 5 6 7 8 9 0 BRBR 0 9 8 7 6

This book was set in Times Roman by James G. Speight. The editors were Christine Williams and Carol Edwards. Cover design by Michelle Fleitz. Printing and binding by Braun-Brumfield, Inc.

A CIP catalog record for this book is available from the British Library.

∞ The paper in this publication meets the requirements of the ANSI Standard Z39.48-1984 (Permanence of Paper)

Library of Congress Cataloging-in-Publication Data

Speight, J. G.

Environmental technology handbook/James G. Speight

p. cm.—(Applied energy technology series)

Includes bibliographical references and index.

1. Environmental engineering. I. Title. II. Series.

TD146.S67 1996

628—dc20 95-48034

CIP

ISBN 1-56032-315-9

Table of Contents

Preface

The history of any subject is the means by which the subject is studied, hopefully so that the errors of the past will not be repeated. In the context of this text, environmental management and environmental awareness are not new, both having been practiced in pre-Christian times. What appears to have been available and known became lost and/or forgotten during the so-called Dark Ages and remained virtually lost until recent times.

As a result of the rebirth of environmental awareness, governments in a number of nations have passed legislation to deal with waste materials. In the United States, which is used as an example throughout this text, such legislation has included the following: (1) Toxic Substances Control Act of 1976; (2) Resource Conservation and Recovery Act (RCRA) of 1976 (amended and strengthened by the Hazardous and Solid Wastes Amendments (HSWA) of 1984; and (3) Comprehensive Environmental Response, Compensation, and Liability Act (CERCLA) of 1980.

It will be a surprise to many, and perhaps no surprise to a few that environmental regulations are not new to civilization. Few people seriously discount the need for environmental regulations, but many will debate the levels at which they are proscribed - citing the cost as an unfair burden in a highly competitive economy. Regulation is of course a necessary step, but unfortunately provides no real incentive for proactive improvement. Regulation is still interpreted as a license to avoid the external costs of environmental impact. In contrast to this position, recent experience has demonstrated that systematic elimination of waste and environmental impact can provide net economic and strategic benefits: in higher quality products: more efficient operations; and in the good will of an informed public who will expect a cleaner and healthier environment. As regulations become more demanding, and the public more aware and concerned, this incentive for environmentally conscious actions will become increasingly apparent.

This book is a ready-at-hand (one-stop-shopping) guide to the many issues that are related to ecosystems as well as to pollutant mitigation and clean-up. It is an introductory overview, with a considerable degree of detail, of the various aspects of environmental technology. The book focuses on the technology of environmental issues; any chemistry in the text is used as a means of explanation of a particular point but is maintained at an elementary level.

For the purposes of the text a waste is generally referred to as *chemical* and is only classed as hazardous when the nature of the text permits. The all too general use of the descriptor *hazardous* to classify wastes is often lacking in specificity and ignores the purpose of the definition. The indiscriminate use, and twisting, of words to describe a chemical also cannot escape some criticism.

The initial chapters (Chapters 1 and 2) are an introduction to, and a description of, the various resources that can pose pollution problems. Chapters 3, 4, and 5 describe the various ecosystems. The following chapters (Chapters 6, 7, 8, 9, 10, and 11) deal with the various aspects of waste management and the final chapter (Chapter 12) covers the regulations that focus on various waste streams.

Where possible selected standard tests, as defined by the American Society for Testing and Materials (ASTM), are referenced in the text. This is an aid to the reader who may wish to consult the relevant standards and necessary tests to study their application.

The literature has been reviewed up to October 1995 but to give full references for every source used while preparing this book would require a supplementary volume. The most important sources are listed, with a preference for the most easily accessible review articles and books. This provides the reader with sources that s/he can then use to build up a more comprehensive bibliography of the subject matter.

James G. Speight, Ph.D.
Laramie, Wyoming, U.S.A.

Part I Definitions and Resources

Chapter 1

History, Definitions, and Terminology

1.0 Introduction

The *history* of any subject is the means by which the subject is studied in the hopes that the errors of the past will not be repeated. In the context of this book, environmental management is not new, having been practiced in pre-Christian times (James and Thorpe, 1994). What appears to have been available became lost and/or forgotten during the so-called Dark Ages and remained virtually lost until recent times.

Terminology is the means by which various subjects are named so that reference can be made in conversations and in writings and so that the meaning is passed on.

Definitions are the means by which scientists and engineers communicate the nature of a material to each other and to the world, through either the spoken or the written word. Thus the definition of a material can be extremely important and have a profound influence on how the technical community and the public perceive that material.

For example, water is essential to life, except in the unfortunate instance of drowning. However, a simple change in the definition of water (American Chemical Society, 1994) from *water* to *dihydrogen monoxide* conjures thoughts of dihydrogen peroxide and carbon monoxide, both dangerous chemicals, and the ramifications of the release of these chemicals to the environment. Even use of the name *oxygen dihydride* does not conjure up the familiar life-supporting liquid.

Thus precise definitions are an important aspect of any technology, especially environmental technology. It is in the environmental technology arena where many decisions are made on the basis of emotion rather than on the basis of scientific and engineering principles (Fumento, 1993). Many activities associated with the use of chemicals have a high degree of *perceived* risk (Wedin, 1994; O'Riordan, 1995). Therefore precise definitions should be the norm!

As another example, the release of wolves into Yellowstone National Park in northwestern Wyoming has been the cause of much debate. The main question appears to have been related to the ecological issue of releasing wolves that (supposedly) would feed off the herds of cattle in the area. There has been a general failure to recognize that the area has been in an ecological imbalance since the wolves were exterminated in the early decades of this century. Wolves have (by animal standards) a fairly complex hierarchy (Mowat, 1963; Lopez, 1978) and they are not the sadistic killers that certain novelists and film makers would have the general public believe. However, the interdictions of the media, a host of politicians, and a mass of uninformed wolves-rights activists and anti-wolves-rights activists created a verbal and print-media circus. Very few of the significantly uninformed masses even bothered seeking information from the people most qualified to make comment and judgment, the wildlife biologists!

A story that can evoke considerable response from both sides of the environmental issue is that of the medieval peasant. This worthy soul (whomever he may have been), while visiting the City of London, happened to pass by the street where perfumes were sold from a variety of shops. It is said that the peasant passed out from the unfamiliar odor but was revived when a shovel of the excrement was held under his nose!

This story makes an interesting anecdote but it is known that the odors of the human body that were attractive to others (certainly in medieval times and as recent as the early days of this century) are now hidden by perfumes and deodorants. It might also be surmised that it was the ammonia generated by certain types of excrement that revived the peasant. On this basis, it might be deduced that the perfume was the pollutant, being nonindigenous to that part of the city, and that the animal excrement was a nonpollutant (being indigenous to that part of the city). However, the converse can be argued just as (if not more) convincingly.

Environmental issues permeate everyday life, especially the lives of workers in various occupations where hazards can result from exposure to many external influences (Lipton and Lynch, 1994). In order to combat any threat to the environment, it is necessary to understand the nature and magnitude of the problems involved (Ray and Guzzo, 1990). It is in such situations that environmental technology has a major role to play. Environmental issues even arise when outdated laws are taken to task (*US News & World Report*, 1995). Thus the concept of what seemed to be a good idea at the time the action occurred no longer holds when the law influences the environment.

It is not the intent to subscribe to the notion that the state of the environment is acceptable, and that it always will be acceptable. However, there is the need to strike a balance between cessation of pollutant releases, defining the nature of a pollutant, and the definition of a clean environment.

2.0 Historical Aspects

Historically, the development of civilization has caused perturbation of much of the earth's ecosystem, resulting in pollution of air, land, and water. In this case, pollution is defined as the introduction into the land, water, and air systems of a chemical (or chemicals) not indigenous to these systems.

There is the general belief that environmental science and engineering are relatively new technical disciplines, but the influence of environmental effects on human (and animal) populations has been realized for centuries. In a combination of historical and geological perspectives, man's aggressive appearance on the earth, whether it is through creation or by evolution from some primordial soup (*Encyclopedia Britannica*, 1969; Miller, 1987; Krauss, 1992; Christian Bible, 1995), has also paralleled the demand for the use of the earth's resources.

The environment can be endangered by any one, or several, of a range of human activities (Mooney, 1988; Pickering and Owen, 1994). Indeed, from the first use of wood for fire as a source of warmth leading to the overwhelming modern-day use of other resources, such as coal for domestic and industrial fuel, and the demand for energy has increased the use of fossil fuels in logarithmic proportions (Tester et al., 1991).

In order to put this development into the perspective of the geological timescale of the earth, which spans some 4.5 billion (4.5×10^9) years (Figure 1.1), it is necessary to remember that the earth was virtually uninhabited for eons. For example, the Pre-Cambrian period spans approximately 3.8 billion (3.8×10^9) years, or 85% of this period. The Pre-Cambrian period is very distinct insofar as the fossil fuels do not occur in Pre-Cambrian rocks. In short, there were few, or only simplified, forms of life for much of the earth's history. Man is a comparative newcomer to the earth, the Johnny-come-lately of the earth's fauna.

The land-borne and water-borne diseases of the pre-Christian and post-Christian eras are common knowledge and are cited in many of the old texts. However, the relationship of disease to pollution is not often recognized, nor could it be expected to be recognized in those times.

Roman armies on the march recognized the need for clean, untainted drinking water. Hence the use of gall. Each Roman soldier carried in his pouch a sponge soaked in gall (water that has a small amount, ~10%, vinegar added). The vinegar killed the sources of disease that came from drinking tainted water. Records show that the water wagons following the army carried more vinegar than they did water. Water, albeit often tainted, was plentiful in Europe; vinegar was not readily available!

As the march continued, the soldier would suck on the sponge without even breaking his stride, thereby allowing the army to keep moving. He would only give up his sponge if he were dying or to a dying colleague. Such was his inclination to preserve his source of aqueous nourishment.

In fact, historically, there has been a sense that the environment was becoming tainted by the presence of large bodies of the population. For example, the location of sewage areas away from buildings, rather than next to (or even in) the building, is but one simple example of this awareness.

There is evidence of flush toilets or, at least, flowing water for sewage transport away from buildings (into the nearby river!) in several of the ancient cities (such as Nineveh, now the modern city of Mosul in Iraq) of the Fertile Crescent and in ancient Rome (James and Thorpe, 1994). Furthermore the recognition that cities needed adequate water supply and sewage systems had taken place by the Middle Ages (James and Thorpe, 1994).

The disadvantages of dumping raw sewage into rivers, upstream of villages, towns, and cities, were recognized in the pre-Christian era. Military tacticians recognized the advantage of having the army camp next to the river and upstream from the city being besieged. Of course, they knew to draw their water from even further upstream than the camp!

Tainted water was not the only recognized carrier of disease. The frequent occurrence of the plague (Cartwright and Biddiss, 1991), which was eventually

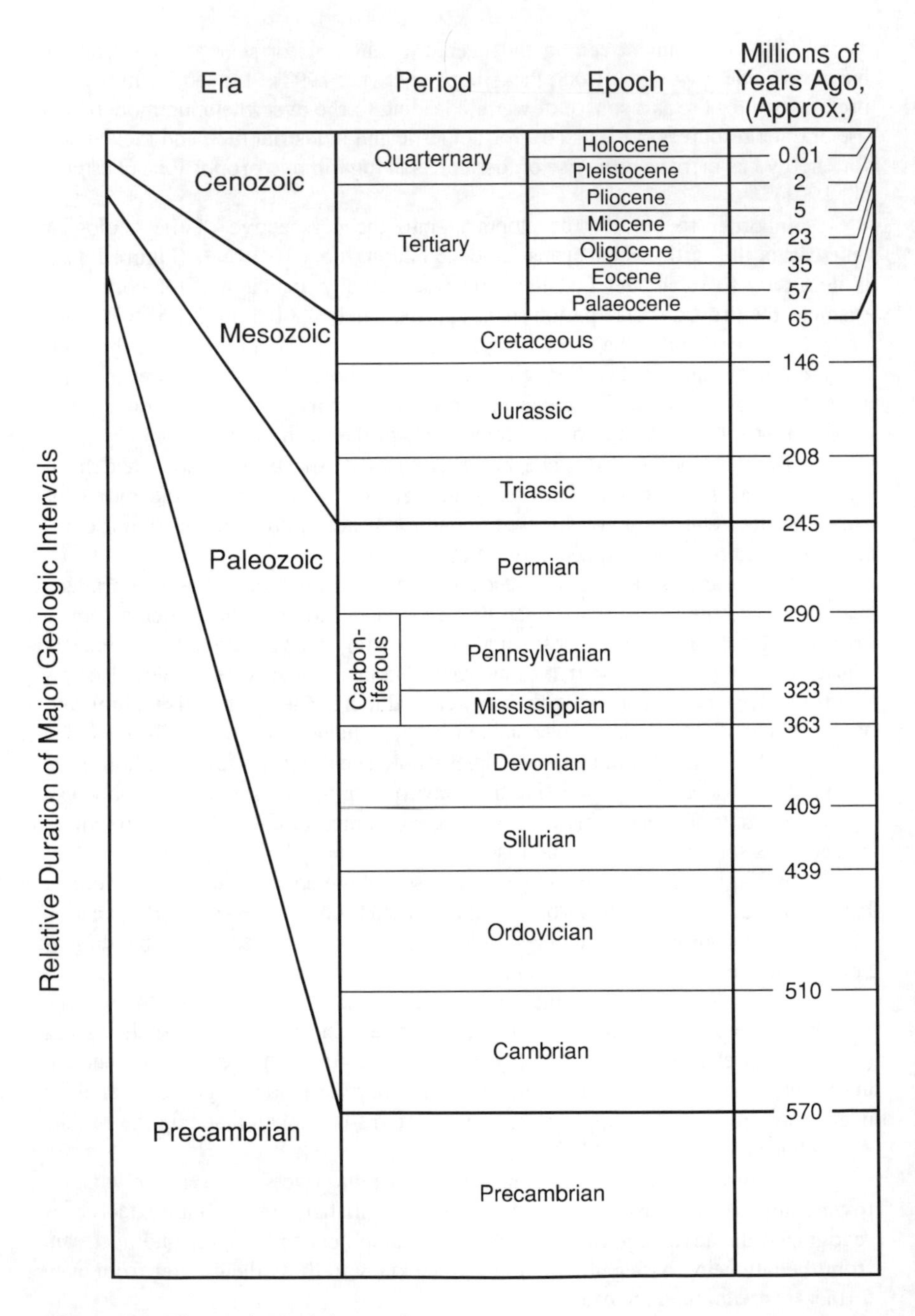

Figure 1.1: The geological time scale

related to the infestation by fleas from rats, is an example of a land-borne disease. The infestations were due to unhygienic living conditions, where piles of refuse gave the rats sustenance and, in turn, healthy rats gave the fleas their sustenance.

As industries developed during the evolution of the Industrial Age from the Middle Ages (Gimpel, 1976), they added their discharges to those of the community. When the concentration of added substances became dangerous to humans or so degraded the water that it was unfit for further use, water pollution control began. With increasing development of land areas, pollution of surface-water supplies became more critical because wastewater of an upstream community became the water supply of the downstream community.

Serious epidemics of waterborne diseases such as cholera, dysentery, and typhoid fever were caused by underground seepage from privy vaults into town wells (Rittenberg, 1987). Direct bacterial infections through water systems were the cause of disease but, of course, the role of the organism that thrived on the garbage and sewage must also be recognized.

Another form of bacteriological contamination involved the use of severed heads as ammunition for a trebuchet as a documented event in medieval warfare. These unusual forms of missiles were hurled over the walls of the besieged city of Berwick, on the English-Scottish border, in June 1333 by the English in the war against the Scots (who occupied the border city) and may have been intended for something more than the usual intimidation of the besieged persons (Turnbull, 1995). The heads, presumably some days old and subject to the heat of the summer, would also have been a source of bacteriological contamination (thus, disease) of the inner city. There are also records of the carcass of a dead horse and plague-ridden bodies being hurled over the walls and into besieged cities, presumably not for recognition by friends or relatives but for the spread of disease.

Syphilis was perhaps not necessarily a water-borne disease in the direct sense (like typhus and dysentery) but the infection of Europe from the 1490s onward seems to have been the result of the return of Columbus, by water, from the New World (Cartwright and Biddiss, 1991).

In terms of air emissions, it was well known in the Middle Ages that the air was infected with invisible spirits, some of which were benign but most of which were evil and dangerous. Whether this related to a fear of airborne disease or to a superstitious population is another question. The answer is probably the latter. Yet, it may have been the early recognition that diseases could be transmitted by air. In fact, there is one well-documented example of the recognition of air pollution in medieval times. It is recorded that a singularly important environmental event occurred in 1257 (Galloway, 1882). The event was perhaps one of the forerunners of an environmental consciousness in terms of the recognition of the adverse aspects of modern (i.e., thirteenth century) technology.

Thus it occurred that Eleanor, wife and Queen of Henry III of England, was a resident in the town of Nottingham while the King was on a military expedition into Wales. The Queen was obliged to leave Nottingham due to the noxious effects of troublesome smoke from the coal being used for heating and cooking!

Perhaps as a result of this awakening, a variety of proclamations were issued over the next several decades by Henry and by his son, Edward I. These proclamations threatened the population with the loss of various liberties, a limb or two, and even life, if the consumption of coal was not seriously decreased and, in some cases, halted.

One wonders whether Eleanor's hasty departure from Nottingham was the beginning of an environmental awakening and that the consequences of burning coal were recognized at that time. However, the proclamations seem to have had little effect (there being considerable income from the sale of coal to continental Europe) and, by the last decades of the thirteenth century, London had the dubious privilege of becoming the first city documented to suffer man-made pollution. In the years 1285 and 1288, complaints were recorded concerning the corruption of the city's air by coal fumes from the lime kilns (Galloway, 1882).

Nevertheless, the Royal positions on the pollution problem, if that is really the issue, may have solved part of the immediate problem, although it did not have any lasting effect; coal burning has continued in England from that time!

In the Middle Ages there was also pollution, which seemed to continue unabated in the fledgling industrial operations. For example, mining and smelting operations were carried out without any form of protection for the workers, as evidenced from woodcut illustrations of the period (Agricola, 1556). When these are examined in detail, perhaps one can observe the need for the Occupational Safety and Health Administration (OSHA) (see Chapter 12).

Also, in the Middle Ages, towns and cities suffered from industrial water pollution, principally due to the activities of butchers and tanners. Municipalities were always trying to move the butchers and tanners downstream, outside the precincts of the town because the industry corrupted the waters of the riverside dwellers (Fagniez, 1877).

Indeed, butchers and tanners notwithstanding, sources of drinking water have continued to be threatened by the dumping of chemical waste, which is then leached (washed) from the surface into the underground water systems (aquifers). Gaseous emissions from the combustion of fossil fuels, which were recognized early in their use (Chapter 2), as well as the emission of chlorofluorocarbons and other chemicals, can cause severe, often irreversible, damage to the atmosphere (Speight, 1993; Lipton and Lynch, 1994).

The examples of pollution from history are numerous, especially when the source and causative agent of the disease are recognized. In many cases, the causative agent could be related to living in unsanitary conditions which in turn was caused by pollution of the environment.

Because of the parallel emergence of industrial processes and of fossil fuels as major sources of energy, and the emissions that are produced, the importance of protecting the environment has become a top priority within government and industry over the last two decades (Benarde, 1989; Lipton and Lynch, 1994; Skinner, 1994). The passage of the United States' Clean Air Act in 1970 and the subsequent passage of a series of amendments, the latest in 1990 (United States Congress, 1990; Stensvaag, 1991) speaks to the recognition that the emission of noxious constituents into the atmosphere must cease. Other countries have also passed similar legislation and there are also laws to prevent pollution of the geosphere (land systems) and the aquasphere (water systems).

Columbus and his predecessors notwithstanding, the potential for damage to the environment was not recognized until recently. Any voices concerned about environmental issues were not heard with any great effect until after the 1960s. By that time, a variety of gaseous, liquid, and solid waste emissions had been liberally released to the environment without any understanding of the consequences.

The purpose of this text is to examine the effects that human development has had on the environment and also to consider the methods by which the environmental impact of this development might be mitigated. The production of minerals, the production and combustion of fossil fuels, a variety of industrial processes, and agricultural activities have all served to cause changes in the environment (Mooney, 1988). In short, the text focuses on the means by which sustainable development (Brady and Geets, 1994) might be achieved with minimum disturbance to the environment. However, in order to alleviate any confusion that might arise, it is necessary to define the relevant terms that will be used throughout the text.

3.0 Definitions and Terminology

The terminology found in the various areas of industrial technology can be extremely confusing to the uninitiated; excellent examples of the confusion that abounds are to be found in the area of fossil fuel technology (Speight, 1990, 1991, 1994). To start with an extremely relevant definition, *environmental technology* is the application of scientific and engineering principles to the study of the environment, with the goal of the improvement of the environment. Furthermore, issues related to the pollution of the environment are relative. The purity of the environment is in the eyes of the beholder!

Any organism is exposed to an *environment*, even if the environment is predominantly many members of the same organism. An example is a bacterium in a culture that is exposed to many members of the same species. Thus the environment is all external influences, abiotic (physical factors) and biotic (actions of other organisms), to which an organism is exposed. The environment affects basic life functions, growth, and reproductive success of organisms, and determines their local and geographic distribution patterns. A fundamental idea in *ecology* is that the environment changes in time and space and living organisms respond to these changes.

Since ecology is that branch of science related to the study of the relationship of organisms to their environment, an *ecosystem* is an ecological community (or living unit) considered together with the nonliving factors of its environment as a unit.

By way of brief definition, *abiotic factors* include such influences as light radiation (from the sun), ionizing radiation (cosmic rays from outer space), temperature (local and regional variations), water (seasonal and regional distributions), atmospheric gases, wind, soil (texture and composition), and catastrophic disturbances . These latter phenomena are usually unpredictable and infrequent, such as fire, hurricanes, volcanic activity, landslides, major floods, and any disturbance that drastically alters the environment and thus changes the species composition and activity patterns of the inhabitants.

Biotic factors include natural interactions (e.g., predation and parasitism) and anthropogenic stress (e.g., the effect of human activity on other organisms). Because of the abiotic and biotic factors, the environment to which an organism is subjected can affect the life functions, growth, and reproductive success of the organism and can determine the local and geographic distribution patterns of an organism.

Living organisms respond to changes in the environment by either adapting or becoming extinct. The basic principles of the concept that living organisms respond to changes in the environment were put forth by Darwin and Lamarck. The former noted the slower adaptation (evolutionary trends) of living organisms, while the latter noted the more immediate adaptation of living organisms to the environment. Both essentially espoused the concept of the survival of the fittest, alluding to the ability of an organism to live in harmony with its environment. This was assumed to indicate that the organism that competed successfully with environmental forces would survive. However, there is the alternate thought that the organism that can live in a harmonious symbiotic relationship with its environment has an equally favorable chance of survival. The influence of the environment on organisms can be viewed on a large scale (i.e., the relationship between regional climate and geographic distribution of organisms) or on a smaller scale (i.e., some highly localized conditions determine the precise location and activity of individual organisms).

Organisms may respond differently to the frequency and duration of a given environmental change. For example, if some individual organisms in a population have adaptations that allow them to survive and to reproduce under new environmental conditions, the population will continue but the genetic composition will have changed (Darwinism). On the other hand, some organisms have the ability to adapt to the environment (i.e., to adjust their physiology or morphology in response to the immediate environment) so that the new environmental conditions are less (certainly no more) stressful than the previous conditions. Such changes may not be genetic (Lamarckism).

In terms of *anthropogenic stress* (the effect of human activity on other organisms), there is the need for the identification and evaluation of the potential impacts of proposed projects, plans, programs, policies, or legislative actions upon the physical-chemical, biological, cultural, and socioeconomic components of the environment. This activity is also known as *environmental impact assessment* (EIA) and refers to the interpretation of the significance of anticipated changes related to a proposed project. The activity encourages consideration of the environment and arriving at actions that are environmentally compatible.

Identifying and evaluating the potential impact of human activities on the environment requires the identification of mitigation measures. *Mitigation* is the sequential consideration of the following measures: (1) avoiding the impact by not taking a certain action or partial action; (2) minimizing the impact by limiting the degree or magnitude of the action and its implementation; (3) rectifying the impact by repairing, rehabilitating, or restoring the affected environment; (4) reducing or eliminating the impact over time by preservation and maintenance operations during the life of the action; and (5) compensating for the impact by replacing or providing substitute resources or environments.

Nowhere is the effect of anthropogenic stress felt more than in the development of natural resources of the earth. Natural resources are various in nature and often require some definition (Chapter 2). For example, in relation to mineral resources, for which there is also descriptive nomenclature (American Society for Testing and Materials, 1995, ASTM C 294), the terms related to the available quantities of the resource must be defined. In this instance, the term *resource* refers to the total amount of the mineral that has been estimated to be ultimately available. The term *reserves* refers to well-identified resources that can be profitably extracted and utilized by means of existing technology. In many countries, fossil fuel resources are often classified as a subgroup of the total mineral resources.

In some cases, environmental pollution is a clear-cut phenomenon, whereas in others it remains a question of degree. The ejection of various materials into the environment is often cited as pollution, but there is the ejection of the so-called beneficial chemicals that can assist the air, water, and land to perform their functions. However, it must be emphasized that the ejection of chemicals into the environment, even though they are indigenous to the environment, in quantities above the naturally occurring limits can be extremely harmful. In fact, the timing and the place of a chemical release are influential in determining whether a chemical is beneficial, benign, or harmful! Thus, what may be regarded as a pollutant in one instance can be a beneficial chemical in another instance. The phosphates in fertilizers are examples of useful ("beneficial") chemicals while phosphates generated as by-products in the metallurgical and mining industries may, depending upon the specific industry, be considered pollutants (Chenier, 1992). In this case, the means by which such pollution can be prevented must be recognized (Breen and Dellarco, 1992). Thus, increased use of the earth's resources as well as the use of a variety of chemicals that are nonindigenous to the earth have put a burden on the ability of the environment to tolerate such materials.

Environmental science and engineering are disciplines involved in the study of the environment. These studies can vary from the effects of changes in the environmental conditions on the flora and fauna of a region to the more esoteric studies of animals in laboratories and can include aspects of chemistry, chemical engineering, microbiology, and hydrology as they can be applied to solve environmental problems. As an historical aside, environmental engineering (formerly known as sanitary engineering) originally developed as a subdiscipline of civil engineering.

Finally, some recognition must be made of the term carcinogen since many of the environmental effects referenced in this text can lead to cancer. *Carcinogens* are cancer-causing substances and there is a growing awareness of the presence of carcinogenic materials in the environment. A classification scheme is provided for such materials (Table 1.1) (Zakrzewski, 1991; Milman and Weisburger, 1994). The

Table 1.1: Weight-of-evidence carcinogenicity classification scheme as determined by the United States Environmental Protection Agency (Kester et al., 1994)

Group	Description
A	Human carcinogen
B1	Probable human carcinogen; limited human data are available
B2	Probable human carcinogen; carcinogen in animals but inadequate evidence in humans
C	Possible human carcinogen
D	Not classifiable as a human carcinogen
E	No carcinogenic activity in humans

number of substances with which a person comes in contact are in the tens of thousands and there is not a full understanding of the long-term effects of these substances in their possible propensity to cause genetic errors that ultimately lead to carcinogenesis. *Teratogens* are those substances that tend to cause developmental malformations.

Pollution is the introduction of indigenous (beyond the natural abundance) and nonindigenous (artificial) gaseous, liquid, and solid contaminants into an ecosystem. The atmosphere and water and land systems have the ability to cleanse themselves of many pollutants within hours or days especially when the effects of the pollutant are minimized by the natural constituents of the ecosystem. For example, the atmosphere might be considered to be self-cleaning as a result of rain. However, removal of some pollutants from the atmosphere (e.g., sulfates and nitrates) by rainfall results in the formation of *acid rain* which can cause serious environmental damage to ecosystems within the water and land systems (Johnson and Gordon, 1987; Pickering and Owen, 1994).

A *pollutant* is a substance (for simplicity most are referred to as chemicals) present in a particular location when it is not indigenous to the location or is in a greater-than-natural concentration. The substance is often the product of human activity. The pollutant, by virtue of its name, has a detrimental effect on the environment, in part or in toto. Pollutants can also be subdivided into two classes : primary and secondary.

Source → Primary pollutant → Secondary pollutant

Primary pollutants are those pollutants emitted directly from the sources. In terms of atmospheric pollutants, examples are carbon oxides, sulfur dioxide, and nitrogen oxides from combustion operations:

$$2C_{fuel} + O_2 = 2CO$$
$$C_{fuel} + O_2 = CO_2$$
$$2N_{fuel} + O_2 = 2NO$$

$$N_{fuel} + O_2 = NO_2$$
$$S_{fuel} + O_2 = SO_2$$
$$2SO_2 + O_2 = 2SO_3$$

The question of classifying nitrogen dioxide and sulfur trioxide as primary pollutants often arises, as does the origin of the nitrogen. In the former case, these higher oxides can be formed in the upper levels of the combustors. The nitrogen, from which the nitrogen oxides are formed (Chapter 2), does not originate solely from the fuel but may also originate from the air used for the combustion.

Secondary pollutants are produced by interaction of primary pollutants with another chemical or by dissociation of a primary pollutant, or other effects within a particular ecosystem. Again, using the atmosphere as an example, the formation of the constituents of acid rain is an example of the formation of secondary pollutants:

$$SO_2 + H_2O = H_2SO_3 \text{ (sulfurous acid)}$$
$$SO_3 + H_2O = H_2SO_4 \text{ (sulfuric acid)}$$
$$NO + H_2O = HNO_2 \text{ (nitrous acid)}$$
$$3NO_2 + 2H_2O = HNO_3 \text{ (nitric acid)}$$

In many cases, these secondary pollutants can have significant environmental effects, such as the formation of acid rain and smog (Chapter 10).

The source of the pollutant is as important as the pollutant itself because the source is generally the logical place to eliminate pollution. The source of the contaminant is also important because a pollutant can be converted to a contaminant, making it necessary to understand the means by which this can happen.

A *contaminant*, which is not usually classified as a pollutant unless it has some detrimental effect, can cause deviation from the normal composition of an environment.

A *receptor* is an object (animal, vegetable, or mineral) or a locale that is affected by the pollutant.

A *chemical waste* (Chapter 6) is any solid, liquid, or gaseous waste material that, if improperly managed or disposed of, may pose substantial hazards to human health and the environment. At any stage of the management process, a chemical waste may be designated, by law, a *hazardous waste* (Chapter 12). Improper disposal of these waste streams in the past has created a need for very expensive cleanup operations (Tedder and Pohland, 1993). Correct handling of these chemicals (National Research Council, 1981), as well as dispensing with many of the myths related to chemical processing (Kletz, 1990) can mitigate some of the problems that will occur when incorrect handling is the norm!

3.1 The Atmosphere

The *atmosphere* (Chapter 5) is the envelope of gases surrounding the earth and it is subdivided into regions on the basis of altitude (Linsley, 1987; Parker and

Corbitt, 1993).

The constituents of the atmosphere are primarily nitrogen (N_2), oxygen (O_2), and argon (Ar). The concentration of water vapor (H_2O) is highly variable, especially near the surface, where volume fractions can be as high as 4% in tropical regions. There are many minor constituents or trace gases, such as neon (Ne), helium (He), krypton (Kr), and xenon (Xe), that are inert, and others, such as carbon dioxide (CO_2), methane (CH_4), hydrogen (H_2), nitrous oxide (NO), and carbon monoxide (SO_2), that play important roles in radiative and biological processes.

In addition to the gaseous constituents, the atmosphere also contains suspended solid and liquid particles. Aerosols are particulate matter usually less than 1 micron (also called 1 micrometer = 1 meter x 10^{-6}) in diameter that are created by gas-to-particle reactions and are lifted from the surface by the winds. A portion of these aerosols can become centers of condensation or deposition in the growth of water and ice clouds. Cloud droplets and ice crystals are made primarily of water and contain trace amounts of particles and dissolved gases. Their diameters range up to 100 micrometers. Water or ice particles larger than about 100 microns begin to fall because of gravity and may result in precipitation at the earth's surface.

An important constituent of the atmosphere is *ozone* (Prinn, 1987). Ozone is found in trace quantities throughout the atmosphere, the largest concentrations being in the lower stratosphere between the altitudes of 9 and 18 miles (15 and 30 km). This ozone results from the dissociation by solar ultraviolet radiation of molecular oxygen in the upper atmosphere and nitrogen dioxide in the lower atmosphere. Ozone also plays an important role in the formation of photochemical smog and in the purging of trace species from the lower atmosphere (Chapters 5 and 10). The chemistry of ozone formation can be explained in relatively simple terms, although the reactions are believed to be much more complex (Chapter 5). Thus, above about 19 miles (30 km), oxygen is dissociated during the daytime by energy (*hv*) from ultraviolet light:

$$O_2 + hv = O + O$$

The oxygen atoms produced then form ozone:

$$O + O_2 + M = O_3 + M$$

where M is an arbitrary molecule required to conserve energy and momentum in the reaction.

Although present in only trace quantities, this atmospheric ozone plays a critical role for the biosphere by absorbing the ultraviolet radiation with a wavelength (λ) from 240 to 320 nanometers (nm, 1 nm = 1 meter x 10^{-9}), which would otherwise be transmitted to the surface of the earth.

This radiation is lethal to simple unicellular organisms (algae, bacteria, protozoa) and to the surface cells of higher plants and animals. It also damages the genetic material of cells (deoxyribonucleic acid, DNA) and is responsible for sunburn of human skin. In addition, the incidence of skin cancer has been statistically correlated with the observed surface intensities of the ultraviolet wavelengths from 290 to 320 nm, which are not totally absorbed by the ozone layer.

An important effect noted as a result of the changes in the constituents of the atmosphere is the tendency for the temperature close to the earth's surface to rise, a phenomenon referred to as the *greenhouse effect* (Chapter 10) (Idso, 1989). The rise in the temperature of the earth is analogous to the rise in the temperature in a greenhouse when the energy from the sun is trapped and cannot escape from the enclosed space (Figure 1.2).

Although the term greenhouse effect has generally been used for the role of the whole atmosphere (mainly water vapor and clouds) in keeping the surface of the earth warm, it has been increasingly associated (perhaps erroneously) with the contribution of carbon dioxide (Idso, 1989). However, there are various types of gases that are the result of industrial and domestic activities that can contribute to this effect and, thus, the continual rise in the earth's surface temperature (Mannion, 1991; Pickering and Owen, 1994).

The arguments (not discussions!) about the magnitude of the greenhouse effect have range back and forth for some time (Baliunas, 1994). There are those who believe that the earth is doomed to a rise in temperature and there are those who believe that we can go polluting the atmosphere without consequence. Whatever the argument, there is no doubt that the emissions, which can give rise to such an effect, must be limited (Bradley et al., 1991). The continuous pollution of the atmosphere with the so-called greenhouse gases can be of no advantage to life on earth, even if the effects of these gases are not manifested in a temperature rise but in the form of aggravating pollutants to flora and fauna.

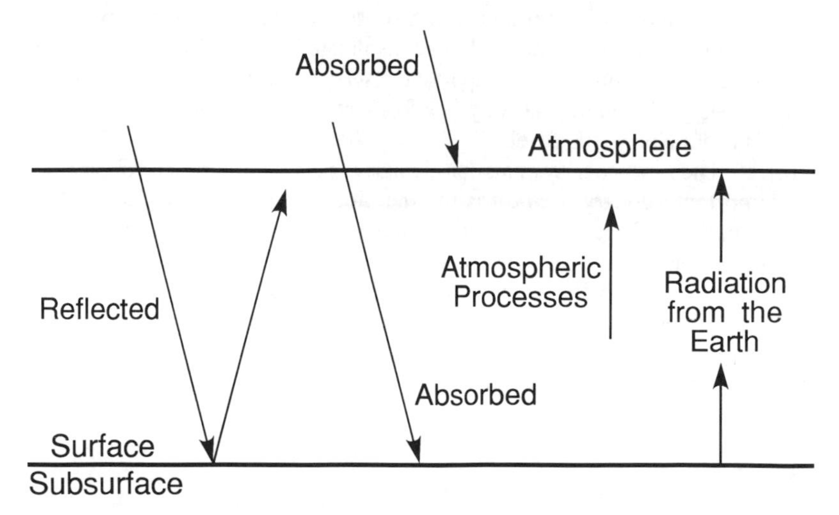

Figure 1.2: Representation of the greenhouse effect

These opposing opinions are of some concern because they may mask the reality of the situation. Analogous situations that have arisen in the last several decades are, for example, in the late 1960s and early 1970s when we were warned about an approaching ice age. In fact, it is likely that several important analysts who then warned of global cooling and imminent glaciation of northern societies are now warning of global warming (Easterbrook, 1995)! The glaciers did not arrive and now, a little more twenty years later, we are being frantically warned about a rise in temperature! We are also being advised that the rise in temperature is being accompanied by the emergence (or reemergence) of a variety of infectious diseases (Hileman, 1995). If, as has been suggested, the earth has warmed by 0.3 to 0.6°C (1°F, or less) during this century, the perception is that a rise in the temperature of the earth may lead to global catastrophes!

These types of contradictory reports and arguments add much confusion to an already difficult area of technology. What is needed at this point is a careful study of the data, the generation of new data, and less enthusiasm for catching the headlines. Obviously, one of the major challenges to the atmospheric scientist is to ensure that the assessment, through accurate measurement of the effects (Newman, 1993), of the influence of any chemicals is well understood.

3.2 Water Systems

Water systems (Chapter 4), often referred to as the aquasphere or the hydrosphere, refer to water in various forms: oceans, lakes, streams, snowpack, glaciers, polar ice caps, and water under the ground (groundwater) (Friedman, 1987; Parker and Corbitt, 1993). An important aspect of the water systems is an aquifer. An *aquifer* is a water-bearing formation in a subsurface zone that yields water to wells. An aquifer may be porous rock, unconsolidated gravel, fractured rock, or cavernous limestone. Aquifers are important reservoirs storing large amounts of water that, in theory, should be relatively free from evaporation loss or pollution. In practice, this is not always the case!

In reference to water systems (particularly lakes), the term eutrophication becomes important. *Eutrophication* is the deterioration of the esthetic and life-supporting qualities of lakes and estuaries, caused by excessive fertilization from effluents high in phosphorus, nitrogen, and organic growth substances. Algae and aquatic plants become excessive, and when they decompose, a sequence of objectional features arises. Water for consumption from such lakes must be filtered and treated. Diversions of sewage, better utilization of manure, erosion control, improved sewage treatment, and harvesting of the surplus aquatic crops alleviate the symptoms.

3.3 Land Systems

Land systems (Chapter 3) are those components that form the earth (Parker and Corbitt, 1993). In more specific terms, the *lithosphere* refers to the minerals in the earth's crust. The term *geosphere* is often more broad in coverage and refers to the complex and variable mixture of minerals, organic matter, water, and air that make up the soil.

The term *biosphere* refers to living organisms and their environments on the surface of the earth (Manahan, 1991). Included in the biosphere are all environments capable of sustaining life above, on, and beneath the earth's surface as well as in the oceans. Consequently, the biosphere includes virtually all of the water systems (the aquasphere, the hydrosphere) as well as portions of the atmosphere and the upper lithosphere (land systems). In addition, there are relationships between the atmosphere, the water systems, the land systems, and the biosphere. These relationships are physical and are also caused by environmental forces (Figure 1.3). More than four centuries ago, Paracelsus (Philippus Aureolus Theophrastus Bombast von Hohenheim 1493-1541) taught that the macrocosm (the heavens) and the microcosm (the earth and all its creatures) were linked together and that the macrocosm directed the growth and development of the microcosm (Huser, 1589).

In present day relationships between the atmosphere, the water systems, and the land systems, *acid rain* (Chapter 10) is the precipitation phenomenon that incorporates anthropogenic acids (i.e., those acids that are the result of human activities) and other acidic materials. The deposition of acidic materials into the water systems and onto the land occurs in both wet and dry forms as rain, snow, fog, dry particles, and gases.

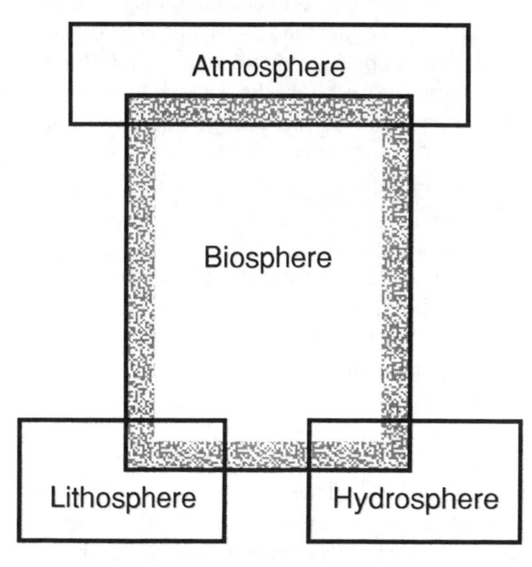

Figure 1.3: Relationship between the atmosphere, the lithosphere (land systems), and the hydrosphere (water systems)

The effect of acid rain (acid deposition) on a particular ecosystem depends largely on the sensitivity of the ecosystem to the acid deposition (Johnson and Gordon, 1987). There is also the ability of the ecosystem to neutralize the acid as well as the concentration and composition of acid reaction products and the amount of acid added to the system.

Trace element is a term that refers to those elements that occur at very low levels in a given system. The somewhat ambiguous term probably arose from the inadequacy of earlier analytical techniques, i.e., before modern methods such as atomic absorption, plasma emission, neutron-activation analysis, gas chromatography, and mass spectrometry extended the limits of detection to the very low levels currently attainable. In many early investigations, it was possible to detect only the presence of an element, as it was said to be present at a trace level. A reasonable definition of a trace element is one that occurs at a level of a few parts per million or less. The term *trace substance* is a more general one applied to both elements and chemical compounds.

There are a variety of trace elements encountered in natural waters, some of which are the nutrients required for animal and plant life. Of these, many are essential at low levels but toxic at higher levels, a point that must be kept in mind in judging whether a particular element is beneficial or detrimental. Some of these elements, such as lead or mercury, have such toxicological and environmental significance that they are discussed in detail in separate sections.

Some of the *heavy metals* are among the most harmful of the elemental pollutants. These metals include essential elements like iron as well as toxic metals like lead (Pb), cadmium (Cd), and mercury (Hg). Most of them have a tremendous affinity for sulfur and attack sulfur bonds in enzymes, thus immobilizing the enzymes. Protein carboxylic acid ($-CO_2H$) and amino ($-NH_2$) groups are also chemically bound by heavy metals. Cadmium, copper, lead and mercury ions bind to cell membranes, hindering transport processes through the cell wall. Heavy metals may also precipitate phosphate compounds or catalyze their decomposition.

4.0 Ecological Cycles

There are several ecological cycles that also need some definition, as each one plays a role in the interrelationship of the various ecological systems as a whole (Prinn, 1987; Clark, 1989; Frosch and Gallopoulos, 1989; Graedel and Crutzen, 1989; Maurits la Riviere, 1989; Schneider, 1989).

Briefly, *ecological cycles* are those cycles involving land systems, water systems, and the atmosphere and which are important to life. The cycle might involve oxygen, nitrogen, carbon dioxide, or some other element that is fundamental to life or to the environment in general. These cycles can be perturbed by anthropogenic stress as well as by natural occurrences.

For example, the oxygen cycle (Figure 1.4) illustrates the role of oxygen in the land, water and air either as oxygen per se or combined with carbon as carbon dioxide and as carbonates. The cycle also includes industrial activities (such as fossil fuel

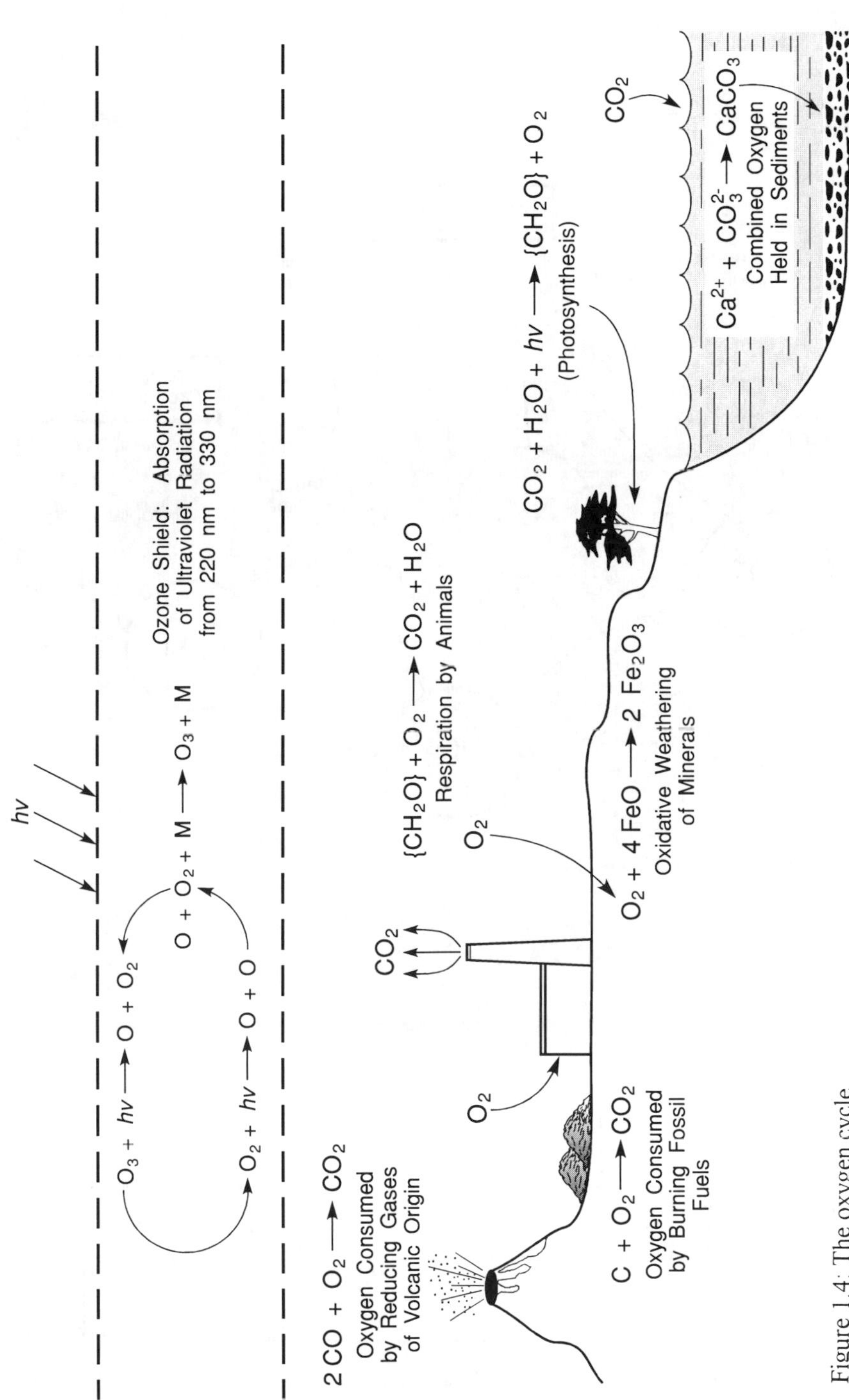

Figure 1.4: The oxygen cycle

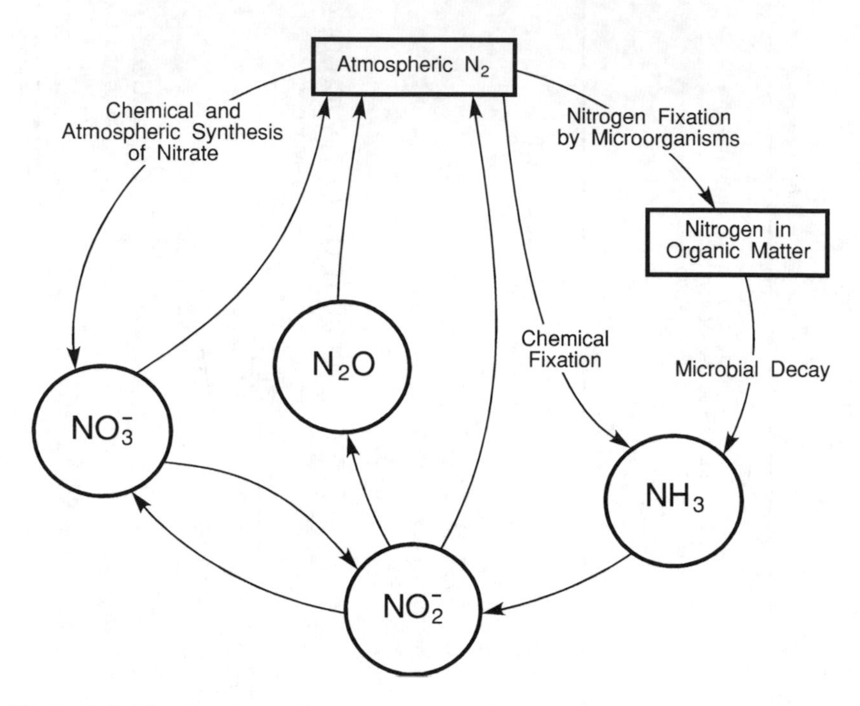

Figure 1.5: The nitrogen cycle

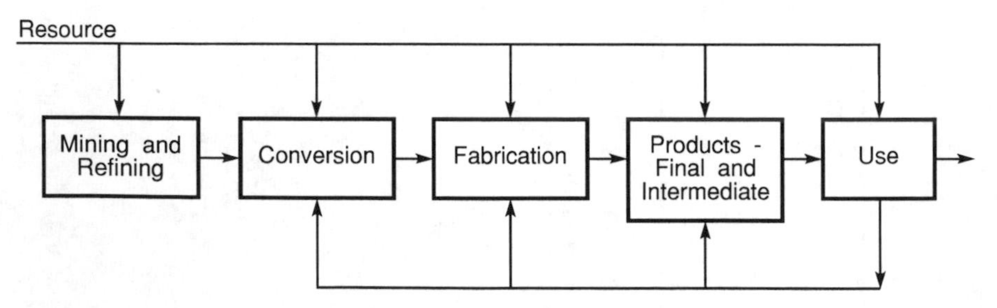

Figure 1.6: The industrial cycle

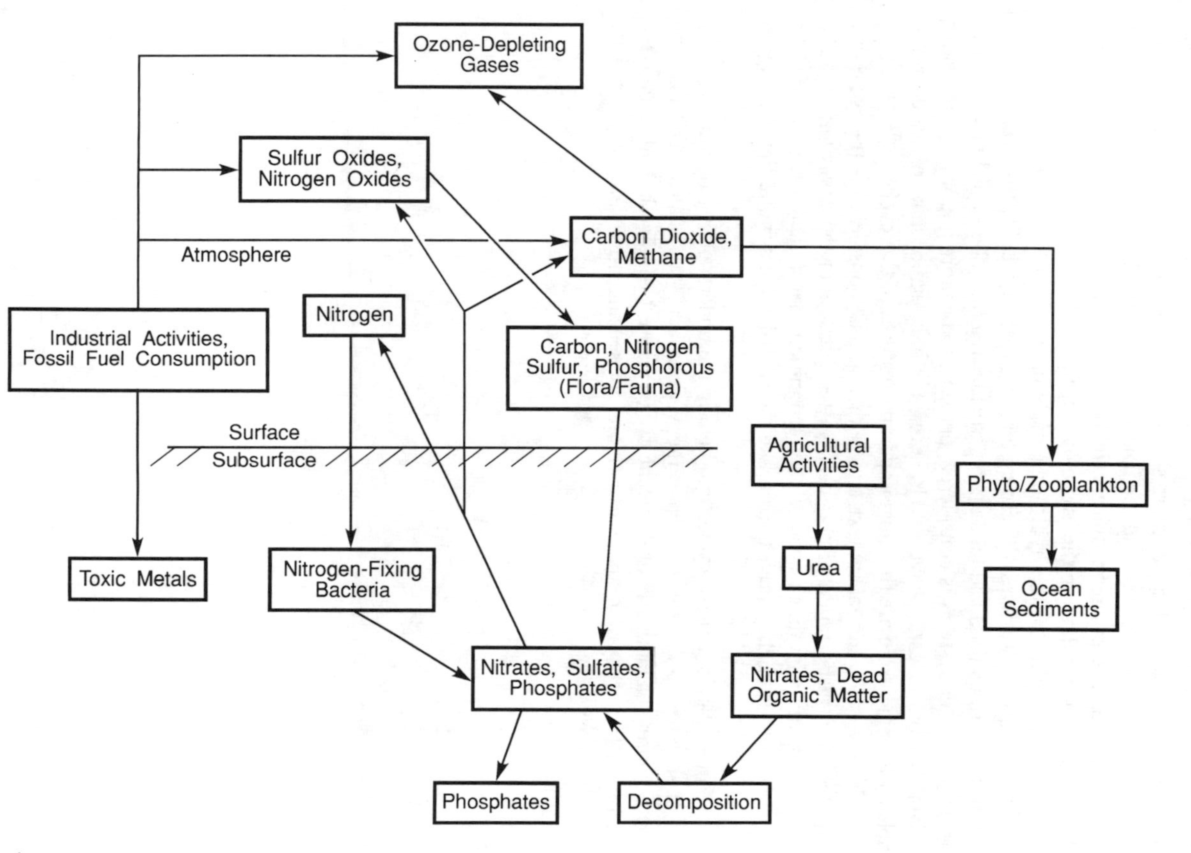

Figure 1.7: Pollutant entry into the environment

burning) and natural activities (represented by volcanic activity). The nitrogen cycle (Figure 1.5) relates not only to nitrogen in the atmosphere but also to nitrogen fixation in the soil and its use in plant growth.

The water cycle (also referred to as the hydrologic cycle) (Walsh, 1987) (Chapter 4) is the means by which water is transported throughout several ecosystems. In theory, there should be a harmonious balance between the water movements on the earth. Indeed, there should be a harmonious balance between the atmosphere, the water systems, and the land systems (Gore, 1993), but the existence of contrasting desert areas and tropical rain forests speaks to the apparent imbalance in these cycles, particularly in the hydrologic cycle.

A subcategory of the hydrologic cycle involves the dispersion of water in various groundwater systems. These systems are responsible for the availability of water to the various flora and fauna in a particular region. The availability of water from groundwater systems is considered, perhaps, to be more important than water availability from lakes and rivers. However, the interrelationships of water in the whole hydrologic cycle are the most important features of this cycle.

Finally, in the current context, a cycle that must be considered is the cycle that illustrates the transportation of chemical wastes throughout the environment. The main feature of a simplified cycle is the recognition that a chemical pollutant can interact with the land, the water, the air, and the animals/plants found therein (Figure 1.7).

Understanding these cycles can help in the understanding of the means by which a chemical pollutant can enter the various ecosystems as a result of being produced during an industrial or domestic cycle (Figure 1.6), thereby influencing the behavior of the ecosystem and, furthermore, influencing floral and faunal activity within the ecosystem.

5.0 References

Agricola, Georgius (Bauer, Georg). 1556. *De Re Metallica*. Froben, Basel, Switzerland.

American Chemical Society. 1994. Dihydrogen Monoxide Hit Hard by Internet Message. *Chemical and Engineering News*. American Chemical Society, Washington, D.C.

American Society for Testing and Materials. 1995. ASTM C 294. Descriptive Nomenclature of Constituents of Natural Mineral Aggregates. *Annual Book of ASTM Standards*. Vol. 04.02. American Society for Testing and Materials, Philadelphia, Pennsylvania.

Baliunas, S. 1994. *Coal Voice* 17(1): 34.

Benarde, M.A. 1989. *Our Precarious Habitat: Fifteen Years Later*. John Wiley & Sons Inc., New York.

Bradley, R.A., Watts, E.C., and Williams, E.R. (eds.). 1991. *Limiting Net Greenhouse Gas Emissions in the United States. Vol. I: Energy Technologies. Vol. II: Energy Responses*. United States Department of Energy, Washington, D.C..

Brady, G.L., and Geets, P.C.F. 1994. *International Journal of Sustainable Development and World Ecology* 1(3): 189.

Breen, J.J., and Dellarco, M.J. (eds.). 1992. *Pollution Prevention in Industrial Processes*. Symposium Series No. 508. American Chemical Society, Washington, D.C.

Cartwright, F.F., and Biddiss, M.D. 1991. *Disease and History*. Dorset Press, New York.

Chenier, P.J. 1992. *Survey of Industrial Chemistry*. 2nd ed. VCH Publishers Inc., New York.

Christian Bible. 1995. New International Version. Book of Genesis. Chapters 1 and 2. Zondervan Publishing House, Grand Rapids, Michigan.

Clark, W.C. 1989. *Scientific American* 261(3): 46.

Easterbrook, G. 1995. *A Moment on the Earth: The Coming Age of Environmental Optimism*. Viking Press, New York.

Encyclopedia Britannica. 1969. Vol. 6, p. 709. William Benton, Chicago.

Fagniez, G. 1877. Etudes sur l'industrie et la classe industrielle a Paris aux XIIIe and XIVe siecles (Studies on industry and the industrial class in Paris during the 13th and 14th centuries) , p. 22. Paris, France.

Friedman, H.L. 1987. In *Encyclopedia of Science and Technology*. 6th ed. Vol. 19, p. 171. Edited by S.P. Parker. McGraw-Hill, New York.

Frosch, R.A., and Gallopoulos, N.E. 1989. *Scientific American* 261(3): 144.

Fumento, M. 1993. *Science Under Siege: Balancing Technology and the Environment*. p.372. William Morrow and Company Inc., New York.

Galloway, R.L. 1882. *A History of Coal Mining in Great Britain*. Macmillan & Co., London, England.

Gimpel, J. 1976. *The Medieval Machine: The Industrial Revolution of the Middle Ages*. Chap. 4. Holt, Reinhart and Winston, New York.

Gore, A. 1993. *Earth in the Balance: Ecology and the Human Spirit*. Penguin Books USA, New York.

Graedel, T.E., and Crutzen, P.J. 1989. *Scientific American* 261(3): 58.

Hileman, B. 1995. *Chemical and Engineering News* 73(40): 19.

Huser. (ed.). 1589. *Paracelcus: Opera Omnia*. Huser, Basel, Switzerland.

Idso, S.B. 1989. *Carbon Dioxide and Global Change: Earth in Transition*. S.B. Idso, Tempe, Arizona. .

James, P., and Thorpe, N. 1994. *Ancient Inventions*. Ballantine Books, New York.

Johnson, R.W., and Gordon, G.E. (eds.). 1987. *The Chemistry of Acid Rain: Sources and Atmospheric Processes*. Symposium Series No. 349. American Chemical Society, Washington, D.C.

Kester, J.E., Hattemer-Frey, H.A., and Krieger, G.R. 1994. In *Environmental Science and Technology Handbook*. p. 37. Edited by K.W. Ayers et al. Government Institutes Inc., Rockville, Maryland.

Kletz, T.A. 1990. *Improving Chemical Engineering Practices*. 2nd ed. Hemisphere Publishers, Washington, D.C.

Krauss, L.M. 1992. In *Mysteries of Life and the Universe*. p. 47. Edited by W.H. Shore. Harcourt Brace and Co., New York.

Linsley, R.K. 1987. In *Encyclopedia of Science and Technology*. 6th ed. Vol. 2, p. 158. Edited by S.P. Parker. McGraw-Hill, New York.

Lipton, S., and Lynch, J. 1994. *Handbook of Health Hazard Control in the Chemical Process Industry*. John Wiley & Sons Inc., New York.

Lopez, B.H. 1978. *Of Wolves and Men*. Charles Scribner and Sons, New York.

Manahan, S.E. 1991. *Environmental Chemistry*. Lewis Publishers, Chelsea, Michigan.

Mannion, A.M. 1991. *Global Environmental Change: A Natural and Cultural Environmental History*. Longman Scientific and Technical Publishers, Harlow, Essex, England.

Maurits la Riviere, J.W. 1989. *Scientific American* 261(3): 80.

Miller, S.L. 1987. In *Encyclopedia of Science and Technology*. 6th ed. Vol. 10, p. 45. Edited by S.P. Parker. McGraw-Hill, New York.

Milman, H.A. and Weisburger, E.K. 1994. *Handbook of Carcinogen Testing. 2nd ed.* Noyes Data Corp., Park Ridge, New Jersey.

Mooney, H. 1988. *Towards an Understanding of Global Change*. National Academy Press, Washington, D.C.

Mowat, F. 1963. *Never Cry Wolf*. Little, Brown and Co., New York.

National Research Council. 1981. *Prudent Practices for Handling Hazardous Chemicals in Laboratories*. National Academy Press, Washington, D.C.

Newman, L. (ed.). 1993. *Measurement Challenges in Atmospheric Chemistry*. Advances in Chemistry Series No. 232. American Chemical Society, Washington, D.C.

O'Riordan, T. (ed.). 1995. *Perceiving Environmental Risks*. Academic Press Inc., San Diego, California.

Parker, S.P., and Corbitt, R.A. (eds.). 1993. *Encyclopedia of Environmental Science and Engineering*. McGraw-Hill, New York.

Pickering, K.T., and Owen, L.A.. 1994. *Global Environmental Issues*. Routledge

Publishers, New York.
Prinn, R.G. 1987. In *Encyclopedia of Science and Technology*. 6th ed. Vol. 2, p. 171, 185. Edited by S.P. Parker. McGraw-Hill, New York.
Ray, D.L. and Guzzo, L. 1990. *Trashing the Planet: How Science Can Help Us Deal with Acid Rain, Depletion of the Ozone, and Nuclear Waste (Among Other Things)*. Regnery Gateway, Washington, D.C.
Rittenberg, S.C. 1987. In *Encyclopedia of Science and Technology*. 6th ed. Vol. 19, p. 317. Edited by S.P. Parker. McGraw-Hill, New York.
Schneider, S.H. 1989. *Scientific American* 261(3): 70.
Skinner, J.P. 1994. *Today's Chemist* 3(3): 40.
Speight, J.G. 1990. *Fuel Science and Technology Handbook*. Marcel Dekker, Inc., New York.
Speight, J.G. 1991. *The Chemistry and Technology of Petroleum.* 2nd ed. Marcel Dekker Inc., New York.
Speight, J.G. 1993. *Gas Processing: Environmental Aspects and Methods*. Butterworth Heinemann, Oxford, England.
Speight, J.G. 1994. *The Chemistry and Technology of Coal.* 2nd ed. Marcel Dekker Inc., New York.
Stensvaag, J-M. 1991. *Clean Air Act Amendments: Law and Practice*. John Wiley & Sons Inc., New York.
Tedder, D.W., and Pohland, F.G. (eds.). 1993. *Emerging Technologies in Hazardous Waste Management III*. Symposium Series No. 518. American Chemical Society, Washington, D.C.
Tester, J.W., Wood, D.O., and Ferrari, N.A. (eds.). 1991. *Energy and the Environment in the 21st Century*. MIT Press, Cambridge, Massachusetts.
Turnbull, S. 1995. *The Book of the Medieval Knight*. Arms and Armour Press, London, England.
United States Congress. 1990. Public Law 101-549. An Act to Amend the Clean Air Act to Provide for Attainment and Maintenance of Health Protective National Ambient Air Quality Standards, and for Other Purposes. November 15.
US News & World Report. 1995. March 13. p.34.
Walsh, J.J. 1987. In *Encyclopedia of Science and Technology*. 6th ed. Vol. 19, p. 317. Edited by S.P. Parker. McGraw-Hill, New York.
Wedin, R.E. 1994. *Today's Chemist*. 3(3): 12.
Zakrzewski, S.F. 1991. *Principles of Environmental Toxicology*. ACS Professional Reference Book. American Chemical Society, Washington, D.C.

Chapter 2

Resources and Resource Utilization

1.0 Introduction

Resources are those usable materials that occur naturally within the earth. They are often referred to as *natural resources* and include the *fossil fuels* (coal, petroleum, and natural gas), *mineral resources*, *wood*, and other *renewable energy resources* (Figure 2.1) (Grenon, 1976; Jahn and Strauss, 1983; Wood and Baldwin, 1985; Cassedy and Grossman, 1990; Johansson et al., 1993; Dovers, 1994; Pickering and Owen, 1994). Energy sources can be divided into *primary energy sources* (Table 2.1) and *secondary energy sources* (Table 2.2). Ores that yield radioactive materials on processing are also included under the general term of *minerals* (Meyerhoff and Meyerhoff, 1987).

In addition, there are also questions relating to the policies concerning resource use (Cooper, 1994). The issues from such questions, although not the subject of this text, must be considered when resource development is planned. The development of resources is usually concerned with the production or use of energy (Table 2.3). *Energy* is a common word, used daily, often several hundred times per day by scientists and technologists. The word arises from the Greek word *energeia*, which is itself a combination of the words *en* (in) and *ergon* (work) and was first used in the nineteenth century as a term to describe *the ability to (do) work*. From that time the word has changed (although the term *erg* is used today as a unit of energy). More modern meanings have extended the word to describe the output of fuel sources.

However, in terms of energy production, it is the fossil fuel resources that are by far the most important, in that they constitute (using the United States as an example, Figure 2.2) the majority of the energy sources in the major industrialized nations (Dryden, 1975; EIA, 1994; Bisio and Boots, 1995). Indeed, the relationship of energy source development and the resulting effects on the environment are well known (Tester et al., 1991; Dovers, 1994; Pickering and Owen, 1994).

The use of natural resources has continued in an as-needed (often haphazard, often uncontrolled) manner for several centuries and perhaps even millennia and into prehistoric times, if the use of wood for fires and the use of specific ores to produce tools are included (Singer et al., 1958).

Industrial operations (Mooney, 1988; Austin, 1984) that produce any product from a natural resource must make every attempt to ensure that natural resources such as air, land, and water remain unpolluted. The environmental aspects of such an operation need to be carefully, and continually, addressed (Chenier, 1992).

While the focus of this text is on the environmental aspects of resource utilization, it is necessary to understand the means by which pollution can occur. An

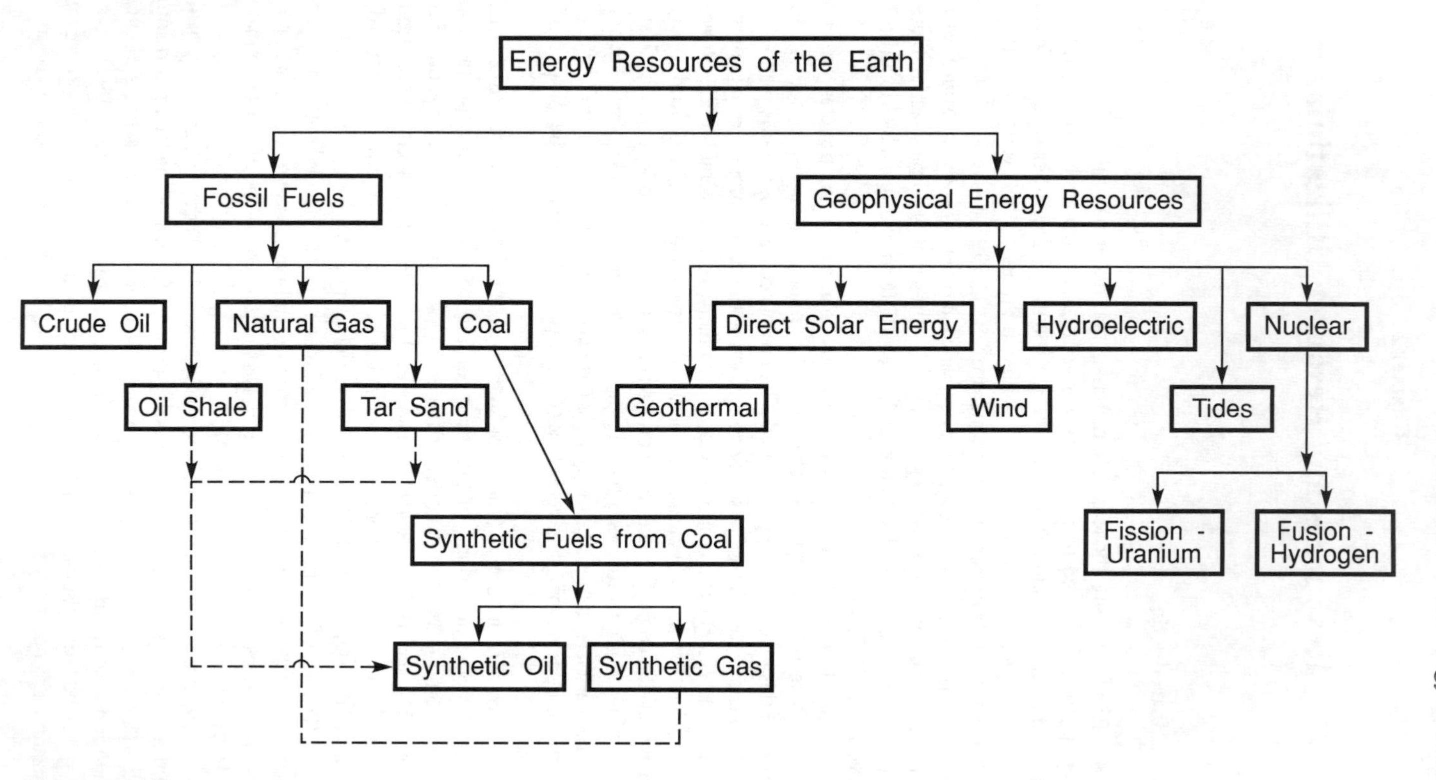

Figure 2.1: Energy resources

Table 2.1: Primary energy sources

Classification	Energy type	Source
Nonrenewable energy	Fossil	coal
		peat
		crude petroleum
		natural gas
	Nuclear	uranium
		thorium
		deuterium
		lithium
		beryllium
Renewable energy	Solar	solar-thermal conversion
		photoelectric energy
		photochemical conversion
		stored solar heat
	Hydro	river-reservoir energy
	Tidal	tidal energy
	Wind	windmill energy
	Oceans	ocean heat
		ocean current
		wave energy
	Geothermal	natural steam
		hot water
		hot dry rocks
	Biomass	wood and other vegetation

important aspect is an understanding of the various cycles that exist in the biosphere and atmosphere (Chapter 1).

For the present purposes, it is more pertinent to consider the use of natural resources since the eighteenth and the time of the onset of the industrial revolution, which have predominantly focused on the production of energy (Gray et al., 1991). It is necessary to understand that natural resources, the commodity, and the environment are intimately related:

Resource → Commodity → Environment

Perturbations in one area usually cause perturbations in one, or both, of the other two. For example, the availability of many metals depends upon the quantity of energy used and the amount of environmental damage tolerated in the extraction of low-grade ores.

Mining and processing of resources offer the world valuable commodities

Table 2.2: Secondary energy sources

Energy type	Source
Electric	electric power generation
	fuel cells
Nuclear	tritium
	plutonium
Fossil (coal-derived)	coke
	char
	tar
	blast furnace gas
	water gas and carbureted water gas
	producer gas
	town gas
	briquettes
	coal slurries
	coal gasification
	coal methanol
Fossil (petroleum-derived)	gasoline
	kerosene
	coke
	shale oil
	synthetic crude oil (oil sand)
	fuel oils
	liquefied natural gas (LNG)
	liquefied petroleum gas (LPG)
	propane
	butane
Biomass	wood waste and bark
	bagasse
	hulls (grain, rice, cottonseed)
	nut shells
	sugar beets
	tobacco stems
	citrus rinds
	corncobs
	garbage and trash
	methane gas
	alcohol (ethanol, methanol)

Table 2.3: General classification of minerals and their uses.

Materials	Uses
Fuels	
Coal	Fuel, electricity, gas, chemicals
Petroleum	Gasoline, heating, chemicals, plastics
Natural gas	Fuel, chemical
Uranium	Nuclear power, explosives
Ferrous metals	
Chromium	Alloys, chemicals
Cobalt	Alloys, carbides
Columbium	Alloys
Iron	Steel, cast iron
Mangancse	Batteries
Molybdenum	Alloys
Nickel	Alloys, coinage
Tungsten	Alloys
Vanadium	Alloys
Nonferrous metals	
Copper	Electrical conductors, coinage
Lead	Batteries, construction
Tin	Tinplate, solder
Zinc	Galvanizing, chemicals
Light metals	
Aluminum	Transportation, building materials
Beryllium	Alloys
Magnesium	Building materials, refractories
Titanium	Pigments, construction
Zirconium	Alloys, chemicals, refractories
Precious metals	
Gold	Jewelry, coinage
Platinum	Chemistry, catalysts
Silver	Photography, electronics, jewelry
Industrial	
Clays	Ceramics, filters, adsorbents
Feldspar	Ceramics, fluxes
Fluorspar	Fluxes
Phosphates	Fertilizers, chemicals
Sodium chloride	Chemical, glass, metallurgy
Sulfur	Fertilizers, acid, metallurgy, paper, foods, textiles

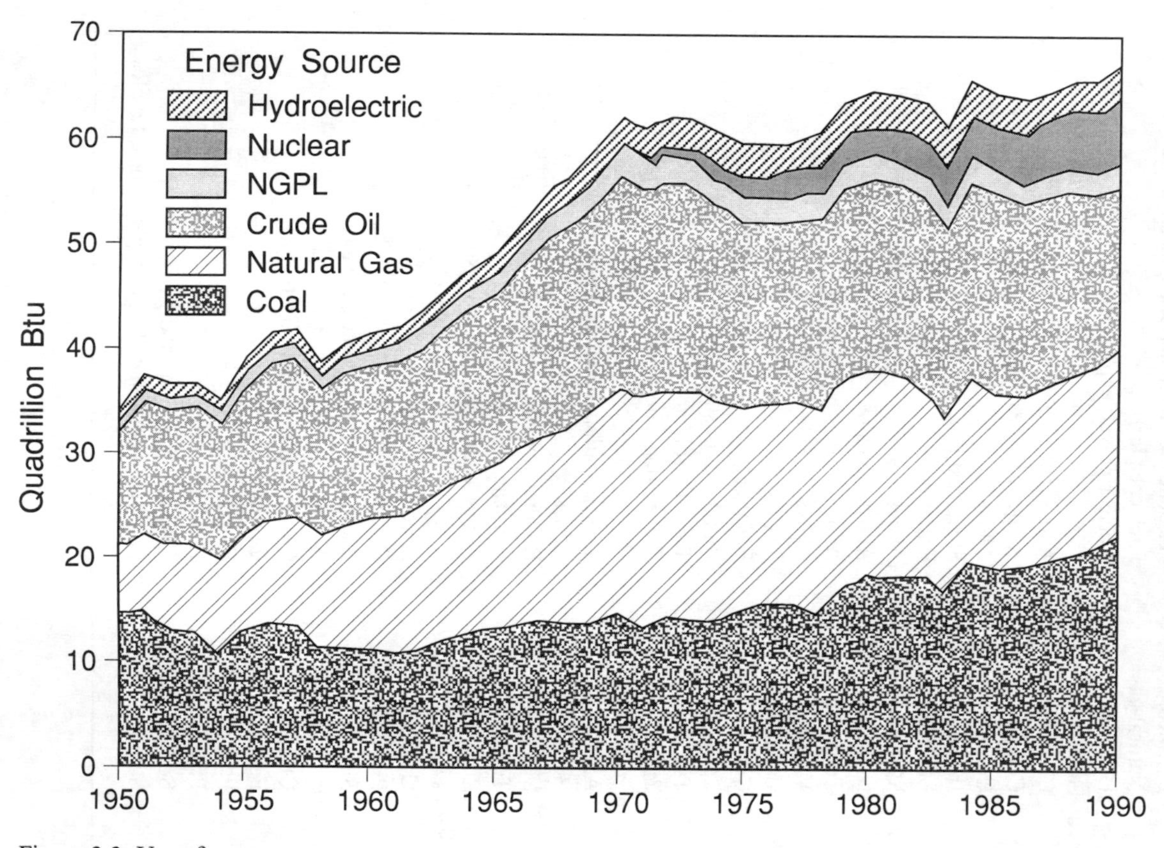

Figure 2.2: Use of energy sources

suitable for a variety of industrial and domestic uses as well as a variety of contaminants (Table 2.4). Of particular interest are those contaminants that are organic in nature since they play a major role in chemical process design and environmental design (Lyman et al., 1990). However, production and use may also involve disturbance of land, air pollution from dust and smelter emissions, and water pollution from disrupted aquifers. This problem is aggravated by the fact that the general trend in mining involves utilization of less rich ores, almost analogous to the acceptance by refineries of lower quality crude oils (Speight, 1991). However before launching into a description of the various natural resources, it is necessary to understand how the reserves of these natural resources are defined.

A *resource* is the entire commodity that exists in the sediments and strata whereas the *reserves* represent that fraction of a commodity that can be recovered economically. However, the use of the term *reserves* as being descriptive of the resource is subject to much speculation. In fact, it is subject to word variations! For example, reserves are classed as *proved*, *probable*, and *possible* (Figure 2.3).

Proved reserves are those reserves of the mineral that have been positively identified as recoverable with current technology. *Probable reserves* are those reserves of the mineral that are nearly certain but about which a slight doubt exists. *Possible reserves* are those reserves where there is an even greater degree of uncertainty about recovery but about which there is some information. An additional term *potential reserves* is also used on occasion; these reserves are based upon geological information about the types of sediments where such resources are likely to occur and they are considered to represent an educated guess. Then there are the so-called *undiscovered reserves*, which are little more than figments of the imagination! Thus, in reality, and in spite of the use of self-righteous word-smithing, the proved reserves may be a very small part of the total hypothetical and/or speculative amounts of a resource.

At some time in the future, certain resources may become reserves. Such a reclassification can arise as a result of improvements in recovery techniques which may either make the resource accessible or bring about a lowering of the recovery costs and render winning of the resource an economical proposition. In addition, other uses may also be found for a commodity, and the increased demand may result in an increase in price. Alternatively, a large deposit may become exhausted and unable to produce any more of the resource thus forcing production to focus on a resource that is lower grade but has a higher recovery cost.

Finally, it must be remembered that the definition of an economical resource will change over time. For example in the 1880s, copper (for the new-born electrical industry) was extracted from ores that contained 5% copper; lesser copper-containing ores were considered uneconomical. Currently, copper is extracted from ores that contain only 0.3 percent copper, at a price relative to other goods and services that was lower in 1990 than it was in the late nineteenth century (Cooper, 1994).

In addition, what was worthless material has become an economic resource. "Dirt" is a natural resource now that the tailings (waste) of some old mines are being worked for the third time. Similarly, Pennsylvania rock oil was considered a sticky pollutant of streams until Benjamin Silliman, a professor of chemistry at Yale College,

Table 2.4: Potential contaminants from various industrial operations

Industry	Examples of sites	Contaminants
Chemical	Acid/alkali works Dyeworks Fertilizer, pesticides and pharmaceutical handling sites Paint works Wood treatment plants	Acids; alkalis; solvents; phenols; organic compounds
Petrochemical	Oil refineries Tank farms Fuel storage depots Tar distilleries	Hydrocarbons; acids; alkalis
Metal	Iron and steel works Foundries, smelters Electroplating, anodizing and galvanizing works Engineering works Shipbuilding/shipbreaking Scrap reduction plants	Metals; Fe, Cu, Ni, Cr, Zn, Cd, and Pb; asbestos
Energy	Gas works Power stations	Phenols, cyanides; sulfur compounds
Transport	Garages, vehicle and maintenance workshops Railway depots	Combustibles; Asbestos
Mineral extraction	Mines and spoil heaps	Metals
Land restoration (including waste sites)	Pits and quarries Filled sites	Gases Leachates
Water supply and sewage treatment	Waterworks Sewage treatment plants	Metals (sludges) Microorganisms
Miscellaneous	Docks, wharfs and quays Tanneries	Metals; organics methane
	Rubber works	Toxic flammable or explosive substances

discovered how to convert it into kerosene (a valuable lamp oil) in the 1850s, thus laying the basis for the modern petroleum industry. A nuisance was thus converted into a highly sought natural resource. Copper and petroleum, the examples used here, are not exceptional: the real price of most natural resources has declined over the past century.

2.0 Fossil Fuel Resources

Fossil fuels are those fuels, namely coal, petroleum (including heavy oil and bitumen), natural gas, and oil shale produced by the decay of plant remains over geological time (Speight, 1990). They are carbon-based and represent a vast source of energy. In fact, at the present time, the majority of the energy consumed by humans is produced from the fossil fuels, coal, petroleum, and natural gas while smaller amounts of energy come from nuclear and hydroelectric sources (Figure 2.2). As an aside, the nuclear power industry (Zebroski and Levenson, 1976; Rahn, 1987) is truly an industry where the future is uncertain or, at least, at the crossroads (Lowinger and Hinman, 1994). As a result, fossil fuels are projected to be the major sources of energy for the next fifty years.

Planning for the future use of these materials requires (as it does for any resource) an estimate of the amounts of fossil fuels that remain. However, the estimates of the fossil fuels available for future use differ considerably. The issue lies with the use of so-called *undiscovered* (or *yet-to-be discovered*) *reserves* of the fossil fuels. Suppositions that there are reserves of fossil fuels yet-to-be-discovered are open to much debate and many questions!

Resource availability is a matter of understanding the amount of a resource that is available, recovery of the resources with the maximal efficiency (minimal cost), and the application of recovery methods that will cause minimal damage to (disturbance of) the environment. Estimates of the quantities of recoverable fossil fuels in the world vary and are subject to much speculation because of the loose definition of the term resource (see also Chapter 1) (Speight, 1990, 1991).

Fossil fuels provide the major source of energy for industrial and domestic consumption. They also provide and, in many cases, the means (energy) by which other resources can be recovered and converted to products for consumer use. Therefore fossil fuels will impact not only the evolution of science and engineering but also the environment. Some actions will be necessary to ensure reductions in emissions from the use of such fuels (Walker and Wirl, 1994).

Fossil fuels are *not* the only source of pollutants, but they are a major source of energy. As such, their use has been cause for considerable criticism. However, it is not the fuel that is at fault but the manner in which the fuels are used! Thus it is necessary to reconsider the strategies for fossil fuel use whereupon it may be discovered that fossil fuels can be used in an environmentally acceptable manner. Hence, it is necessary to understand the various types of fossil fuels that generate energy and the evolution of their use at the present time. This will also aid in understanding the means by which fossil fuels can produce noxious emissions and the

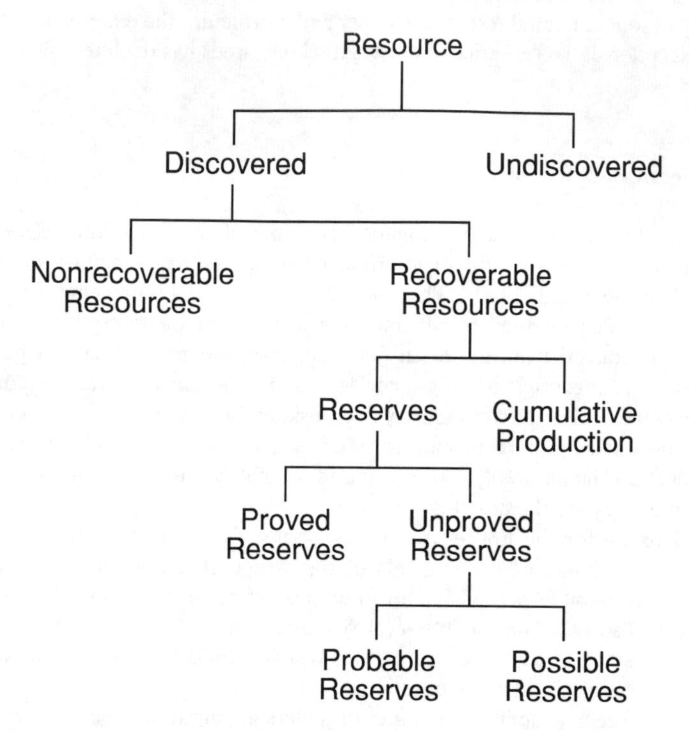

Figure 2.3: Resource classification

means by which the emissions can be reduced, if not eliminated completely.

In summary, understanding the nature of fossil fuels and their recovery can be equated to the recovery of any mineral and its use as a commodity. There is also the need to understand the influence of the various processes on the ecological cycles (Chapter 1) as well as, in many cases, the potential for the production of acid rain and the potential for global warming, or any similar effect (Kasibhatla et al., 1993; Easterbrook, 1995).

2.1 Coal

Coal is an abundant fossil fuel and is classified by rank (degree of metamorphosis/progressive alteration) as a natural series from lignite to anthracite (American Society for Testing and Materials, 1995, ASTM D 388). Coal forms a major part of the earth's fossil fuel resources (Berkowitz, 1979; Hessley et al., 1986; Hessley, 1990; Walker, 1993), the amount available being subject to the method of estimation and to the definition of the resources (Energy Information Administration, 1988, 1989, 1991a, 1991b, 1992). In fact, coal has been referred to as the keystone to a variety of energy scenarios (Schmidt and Hill, 1976).

Coal is, perhaps, the most familiar of the fossil fuels not necessarily because of its usage throughout the preceding centuries (Galloway, 1882; James and Thorpe, 1994), but more because of its extremely common usage during the nineteenth century. Coal was, in large part, the fuel that allowed the industrial revolution to proceed. And because of this, as well as the continuing popularity of coal, it is extremely important in the present context because of the associated, and very necessary, cleanup of the emissions when it is burned.

Coal has probably been known and used for a considerable time, but the records are somewhat less than complete. There are frequent references to coal in the Christian Bible (Cruden, 1930) but, all in all, the recorded use of coal in antiquity is not well documented (Galloway, 1882; James and Thorpe, 1994).

Over the centuries, coal use increased substantially. In fact, in the late sixteenth century, an increasing shortage of wood in Europe resulted in coal becoming an exploitable resource in England, France, Germany, and Belgium. In the middle to late eighteenth century, the use of coal increased dramatically in Britain with the successful development of coke smelters and the ensuing use of the coke to produce steam power. By the early nineteenth century , coal suppled most of Britain's energy requirements; the use of gas from coal (*town gas*) for lighting was also established.

In contrast, in the United States, wood was more plentiful, and the population density was much lower than in Europe. Any coal required for energy was imported from Britain and Nova Scotia. After the Revolutionary War (1775-1781), coal entered the picture as an increasing source of energy. As an example, the state of Virginia supplied coal to New York City.

However, attempts to open the market to accept coal as a fuel were generally ineffective in the United States, and progress was extremely slow. It was not until the period from 1850 to 1885 that coal use in the United States increased, spurred by the emerging railroad industry, both for the manufacture of steel rails and as a fuel for locomotives.

Coal occurs in various forms, defined by rank or type (Speight, 1994); it is not only a solid hydrocarbon material with the ability to produce energy on demand but it also has the potential to produce considerable quantities of carbon dioxide, nitrogen oxides, and sulfur oxides as a result of combustion and the resultant oxidation of the elements in coal (Manowitz and Lipfert, 1990; Afonso et al., 1993; Chagger et al., 1993; Davidson, 1993; Speight, 1993; Davidson, 1994).

Coal production is predicted to increase 1% per year to meet the domestic and foreign demand. Currently, coal consumption for electricity generation accounts for more than 56%, of the total electricity generation and this is expected to rise to nearly 90% by the year 2010 (Burlington Northern Railroad, 1995; Energy Information Administration, 1995). As a result of the increased use of coal for power generation, electricity generators are seen as being responsible for one-third of the carbon emissions (Energy Information Administration, 1995). There are other aspects of coal use that requires monitoring as well. Coal combustion, for example, in coal-fired power plants, is also responsible for the production of bottom ash and fly ash (Williamson and Wigley, 1994). Both types of ash have the ability to promote erosion and corrosion of equipment as well as pollution of the environment, and the leachability of trace elements from the ash is cause for environmental concern (Palmer et al., 1995).

2.2 Petroleum

Liquid fossil fuels are usually given one or more of the names *oil*, *crude oil*, *petroleum*, and *heavy oil*. Unlike coal, there is no universally acceptable system by which petroleum can be classified. Classification of petroleum, using a physical property, has been attempted (Speight, 1991), but much of the classification nomenclature tends to be focused on general descriptions. For example, heavy oil is so-called because of its tendency to contain lesser amounts of the distillable fractions and have a density (0.90-1.00) close to that of water (1.00). Thus at the temperature of density measurement, heavy oil is lighter than water. On the other hand, the near-solid bitumen, often referred to as natural asphalt, is heavier than water (density 1.00-1.05). The term natural asphalt is actually incorrect, since asphalt is a product of refinery processing and may be an altered, rather than natural, material (Speight, 1990, 1991). In summary, more complete descriptions of petroleum would be necessary in any universally acceptable classification scheme.

The recorded use of petroleum and the nonvolatile derivative of petroleum, asphalt, have been known for about 6000 years (Agricola, 1556; Abraham, 1945; Speight, 1991; James and Thorpe, 1994). There are references to oil seepages, the product being frequently referred to as pitch (and slime in the Bible) in the ancient world of the Greeks and Persians (Herodotus, 447 B.C.). Alexander the Great is reputed to have discovered crude oil on the banks of the River Oxus (Abraham, 1945).

There are also records of the use of mixtures of pitch and sulfur as a weapon of war during the Battle of Palatea, Greece, in the year 429 B.C. (Forbes, 1959). There are references to the use of a liquid material, naft (presumably the volatile fraction of petroleum which we now call naphtha and which is used as a solvent or as a precursor to gasoline), as an incendiary material during various battles of the pre-Christian era (James and Thorpe, 1994). This is the so-called Greek fire, a precursor and distant cousin to napalm.

With the onset of the industrial revolution in the seventeenth century, the use of fossil fuels as sources of energy became well established. First coal was used which carried the revolution through the nineteenth century, and then petroleum came into use. It is petroleum that has fueled the military vehicles of many nations becoming a prime commodity in the twentieth century and, in the thoughts of some, the prime cause of all wars since the Franco-Prussian disagreements of 1870-1871 (Yergin, 1992).

After World War II, there was a phenomenal rise in the use of petroleum and its various products, mostly through the development of the internal combustion engine (the automobile) and the accompanying expansion of industrial operations. At the same time, the growth of electricity use during the last four decades is related to the rise in the use of household consumer products such as refrigerators and electric stoves.

Currently, petroleum refining is a complex sequence of events that result in the production of a variety of products (Figure 2.4). In fact, petroleum refining might be considered as a collection of individual, yet related processes that are each capable of

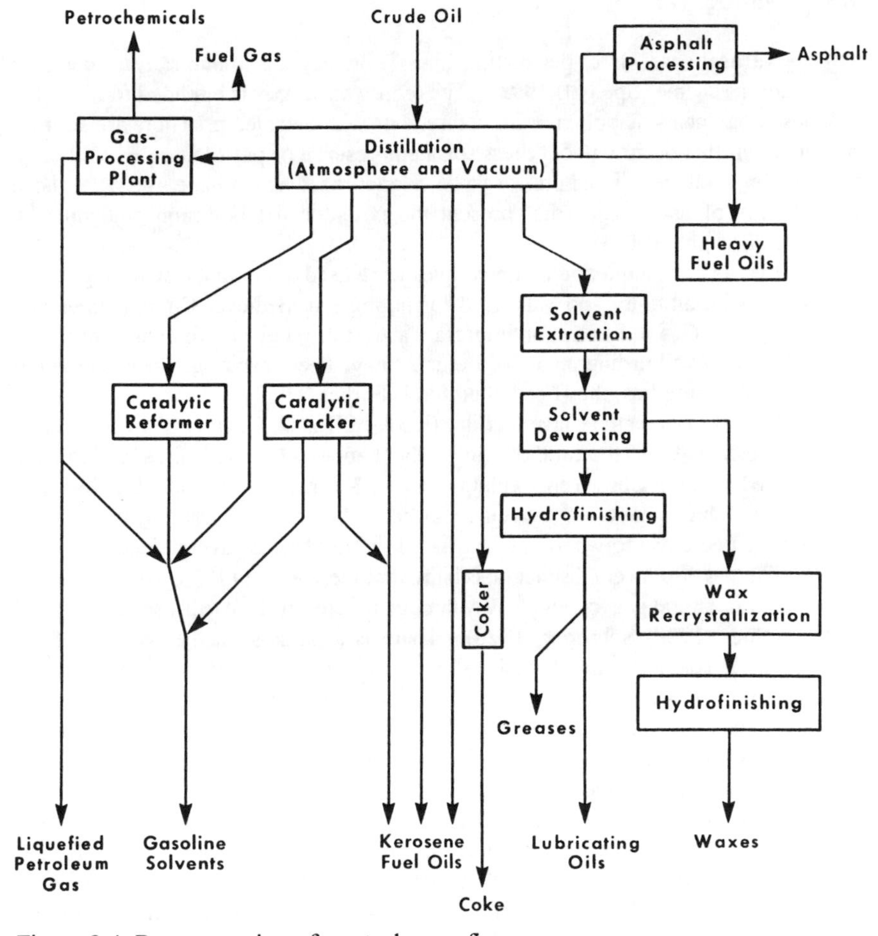

Figure 2.4: Representation of a petroleum refinery

producing effluents.

Petroleum refining as it is currently known will continue at least for the next three decades. In spite of the various political differences that have caused fluctuations in petroleum imports, it is predicted (EIA, 1995) that petroleum imports will reach 59% of petroleum consumption in the United States by the year 2010. It is also predicted (EIA, 1995) that use of petroleum for the transportation sector will increase as increases in travel offset increased efficiency. As a consequence of this increase in use, petroleum will be the largest single source of carbon emissions from fuel.

2.3 Natural Gas

Natural gas, like petroleum, does not have a universally acceptable classification scheme (Speight, 1993). The term *natural gas* is intended to mean the gaseous components that often occur in reservoirs with petroleum. There are sources of natural gas that occur without the associated presence of petroleum. Even though the gas may, to all intents and purposes, be characterized as methane, there are those constituents of natural gas that present the potential for pollution and must be removed (Speight, 1993).

Just as petroleum and its derivatives were used in antiquity, natural gas was also known in antiquity and was used for lighting in medieval China (James and Thorpe, 1994). Gas wells were an important aspect of religious life in ancient Persia and exhibitions of burning pillars of fire must have been inspiring, since, as the old name varishnak implies, they needed no food (Forbes, 1958).

Note is made in the Christian Bible (Book of Daniel, Chapter 3) of the eternal fires as they existed in the time of King Nebuchadnezzar. These fires, which have been quoted as varying between 9 cubits (12 feet, 3.7 meters) and 40 cubits (60 feet, 18.3 meters) high, have since been reported as being a self-igniting natural gas seepage located close to where the Tigris and the Euphrates Rivers meet.

There is also a very distinct possibility that the *voices of the gods*, as noted in old texts, was natural gas forcing its way through fissures in the earth's surface (Scheil and Gauthier, 1909; Schroder, 1920). There is a passage in the records of the Assyrian King Tukulti Ninurta (~885 B.C.) in which the voices of the gods are heard arising from rocks near Hit, the site of the bitumen deposits that became so well known as a building mastic. Similar sounds attributed to be the voices of the gods are noted to have also occurred in the region around Kirkuk.

In classical times these wells were often flared (Lockhart, 1939; Forbes, 1958, 1964), environmental aspects notwithstanding. They are depicted as burning near Apollo's shrine, on coins of Apollonia, and near the present Selenizza (Albania) (Forbes, 1958). Plutarch mentions that Alexander the Great saw burning gas wells near to Ecbatana. Plutarch describes these as a gulf of fire streaming from an inexhaustible source!

Historical records indicate that the use of natural gas, other than for religious purposes, dates back to about 250 A.D. when it was used as a fuel in China. The gas was obtained from shallow wells and was distributed through a piping system constructed from hollow bamboo stems. Gas wells were also known in Europe in the Middle Ages and were reputed to eject oil, such as the phenomenon observed at the site near Mineo in Sicily (Forbes, 1958). Natural gas was used on a small scale for heating and lighting in northern Italy during the early seventeenth century.

In a more modern context, there is the record of a burning spring in 1775 near Charleston, West Virginia, as well as on land owned by George Washington (Lincoln, 1785). The fire may have been caused by the ignition of naphtha seepages, but there is the very strong possibility a gas seepage that was ignited by the flaming torch held by one of the investigators. There is also the first record of a natural gas well being drilled (to a depth of 27 feet, 8.2 meters) in the United States near a burning spring at Fredonia, New York, in 1821 (Speight, 1993, and references cited therein).

In the years following this discovery, natural gas use was restricted to the local environs since the available technology for storage and transportation (bamboo pipes notwithstanding!) was not well developed. Also, at that time, natural gas had little or no commercial value. In fact, in the 1930s when petroleum refining was commencing an expansion in technology that continues today, gas was not considered to be a major fuel source and was only produced as an unwanted by-product of crude oil production. The principal gaseous fuel source at that time (i.e., the 1930s) was the gas produced by the surface gasification of coal (Linden et al., 1976). In fact, each town of any size had a plant for the gasification of coal (hence, the use of the term *town gas*). Most of the natural gas produced at the petroleum fields was vented to the air or burned in a flare stack. Only a small amount of the natural gas from the petroleum fields was pipelined to industrial areas for commercial use. It was only in the years after World War II that natural gas became a popular fuel commodity leading to its present recognition.

Currently, natural gas production is predicted to increase by almost 1% per year up to the year 2010, and gas-fired generators are predicted to overtake nuclear power as the secondary electricity source due to relatively low capital costs, high efficiency, and low emissions (EIA, 1995; Engelhard et al., 1993). One reason for the increase in natural gas usage as a power source is the predicted increase in electricity demand as well as growth in the industrial sector for cogeneration and other uses.

3.0 Other Energy Resources

It is predicted that fossil fuels will be the primary source of energy for the next several decades. Therefore, until other energy sources supplant coal, natural gas, and petroleum, the challenge is to develop technological concepts that will provide the maximum recovery of energy from these fossil fuel resources (Fulkerson et al., 1990). It is absolutely essential that energy from such resources be obtained not only cheaply but also efficiently (Dryden, 1975; Yeager and Baruch, 1987) and with minimal detriment to the environment.

To complete the energy resource scenario and to put the fossil energy resources into perspective, the following focus will be on the non-fossil fuel energy resources (Figure 2.1). These resources are derived from the sun and are generically referred to as nonfossil resources, renewable resources, or/and geophysical energy resources (Kahn, 1979; Golob and Brus, 1993; Mills and Diesendorf, 1994; Pickering and Owen, 1994). Since renewable energy resources are assumed to be "natural" and therefore in harmony with nature, there are environmental issues that need to be addressed when renewable energy resources are to be employed (Holdren et al., 1980).

Finally, there is the possibility of the generation of energy from unconventional sources. These resources include direct solar energy, hydroelectric energy, wind energy, geothermal energy, energy from biomass, and elements of minerals from which nuclear energy may be derived (Kruger, 1976; Morse and Simmons, 1976; Jones and Radding, 1980; Robinson, 1980; Dickson and Fanelli, 1990; Hafele, 1990; Weinberg and Williams, 1990; Himmel et al., 1994; Mills and Diesendorf, 1994). The timescale for the generation of electrical energy from such sources is generally unknown or, at best, open to much speculation and criticism.

Energy from nuclear sources has shown potential to be a significant energy source in the future. The technology is known (Knief, 1992) but has suffered some setbacks. Accidents and dubious claims of the ease with which energy may be derived from nuclear sources have reduced the credibility of the nuclear industry. Nevertheless, the potential still exists for the extraction of energy from nuclear sources.

Energy from nuclear sources is the result of nuclear fission, which uses uranium ore or refined uranium-235 as its basic energy source. The ore must be concentrated into fissionable isotopes through nuclear processing. Eventually, the spent material must be replaced, and the resources of uranium ore may be limited as determined by a specific test (American Society for Testing and Materials, 1995c, ASTM E 901). Hence there is a need to explore the potential for generating more of the fissionable isotopes. However, the use of such materials has become the core of a major controversy, which threatens to be the death-knell of the nuclear industry.

Unlike nuclear fission, which involves the use of fissionable material, nuclear fusion involves the fusion of hydrogen nuclei (Post, 1976). The concept has been suggested as having the potential to provide a clean and virtually unlimited supply of energy. However, it is generally felt (Halpern, 1980) that the technology is at the stage where several decades appear to be needed for the performance of believable laboratory scale experiments which are critical to the development of controlled energy from fusion.

Wood is one of the major natural resources of the earth and it has the potential to be a major source of energy in addition to being a source of chemicals (Tillman, 1978; Hewett et al., 1981). Wood is a renewable resource, but there are caveats that must be applied to the unlimited use of wood. One of these is the deforestation of large tracts of land and the ensuing imbalance in the carbon dioxide cycle. Wood ranks first in the world as a raw material for the manufacture of other products, which include the conventional wood products, cellophane, rayon, paper, methanol, plastics, and turpentine. Wood is composed of polysaccharides such as cellulose, part of which can be extracted for chemical use. A wide variety of chemical uses includes organic compounds and tannin, pigments, sugars, starch, cyclitols, gums, mucilages, pectins, galactans, terpenes, hydrocarbons, acids, esters, fats, fatty acids, aldehydes, resins, sterols, and waxes. Substantial amounts of methanol (often referred to as wood alcohol) are extracted from wood. Methanol, once a major source of liquid fuel, is again being considered for that use (Marsden, 1983; Cybulski, 1994; Skrzypek et al., 1994).

4.0 Mineral Resources

Mineral resources have been known and developed throughout historical time, especially ores containing copper (James and Thorpe, 1994). Mineral deposits, including ore bodies and potential ore, can be developed to provide economically valuable commodities (Bain 1987; Park, 1987; Agarwal, 1991). Therefore, extraction and maximum use must be planned for the benefit of the largest number of people. Such a policy is known as conservation, and it is influenced by many economic and political factors. As costs increase, people tend to use less; they retain materials for more essential uses, thus practicing conservation.

For convenience, raw materials are classified as fuels (see above), metals (including ferrous, nonferrous, light, as well as precious metals), and industrial minerals or nonmetallic elements and have a variety of uses. It is because of these various uses that metals find their way into the environment (Bowen, 1982; Ure and Berrow, 1982).

The increased demand for a particular metal, coupled with the necessity to utilize lower grade ores, has a significant effect upon the amount of ore that must be mined and processed to produce a unit of a product. Thus for every unit of the product, a leaner ore will produce more units of by-products, some of which can be used and some of which cannot. In keeping with Murphy's law, there is the tendency for the by-products to be unusable, thereby requiring discharge to the land and/or water systems and the accompanying environmental consequences.

A number of minerals other than those used to produce metals are also important sources of industrial and domestic products. For example, clays, which occur as suspended and sedimentary matter in water and as secondary minerals in soil, are also used for clarifying oils, as catalysts in petroleum processing, as fillers and coatings for paper, and in the manufacture of firebrick, pottery, sewer pipe, and floor tile.

Fluorine compounds are widely used in industry; for example, fluorspar (CaF_2) is used as a flux in steel manufacture. Freon-l2 (difluorodichloromethane) is widely used as a refrigerant and is considered to be a major stratospheric pollutant. Sodium fluoride is used for water fluoridation.

Mica is a complex aluminum silicate mineral that is transparent, tough, flexible, and elastic. Better grades of mica are cut into sheets and used in electronic apparatus, capacitors, generators, transformers, and motors. Finely divided mica is widely used in roofing, paint, welding rods, and many other applications. Phosphorus, along with nitrogen and potassium, is one of the major fertilizer elements. Many soils are deficient in phosphate. Its nonfertilizer applications include animal feed supplements, synthesis of detergent builders, and preparation of chemicals such as pesticides and medicines.

Sulfur can exist in many forms which undergo a complicated cycle (Figure 2.5). Sulfur may be expelled from volcanoes as sulfur dioxide or as hydrogen sulfide, which is oxidized to sulfur dioxide and sulfates in the atmosphere. Highly insoluble metal sulfides are oxidized upon exposure to atmospheric oxygen to relatively soluble metal sulfates. In the ground and in water, sulfate is converted to organic sulfur by plants and bacteria. Bacteria mediate transitions among sulfates, elemental sulfur, organic sulfur, and hydrogen sulfide. Sulfur may either escape to the atmosphere (usually as the oxides) and reappear in the form of sulfurous and sulfuric acids in rainfall (acid rain) or be retained as metal sulfides. The four most important sources of sulfur are (in decreasing order) deposits of elemental sulfur, hydrogen sulfide recovered from sour natural gas, organic sulfur recovered from petroleum, and pyrite (FeS_2).

The development of large machines for handling huge tonnages of rock and the use of refined explosives have improved mining methods, as illustrated in many coal mining operations (Speight, 1994) and in the mining of oil sand (Speight, 1990). Open-pit and strip mines are larger and more efficient than ever, and lower grades of ore are being recovered. It is common for a mine to produce 50,000 tons (45,000 metric tons) of ore a day, and some yield more than 100,000 tons (90,000 metric tons). Improved methods of transportation, concentration, and smelting permit less costly handling of large tonnages and increase recoveries. Newer smelting methods are cleaner, have as good or better recovery records, and may prove to be less expensive to operate.

The environmental aspects of mining many of these minerals need attention. Even the simplest mining procedure such as the recovery of gravel for asphalt highways, can have important environmental effects.

5.0 Resource Utilization

The increased use and popularity of mineral resources are due, no doubt, to the relative ease of accessibility, which has in many cases, remained relatively unchanged over the centuries. There are exceptions to the ready availability (petroleum is an example) because of various physical and political reasons, although the prognosis for a continuing petroleum industry remains optimistic through cooperation and planning (Al-Sowayegh, 1984; Kubursi and Naylor, 1985; Niblock and Lawless, 1985).

The relatively simple means by which many mineral resources can be utilized has also been a major factor in determining their popularity. And many of the commodities produced from mineral resources:

$$\text{Ore} \rightarrow \text{Mineral} \rightarrow \text{Commodity}$$

may be interchangeable on a purely physical basis insofar as one commodity can be

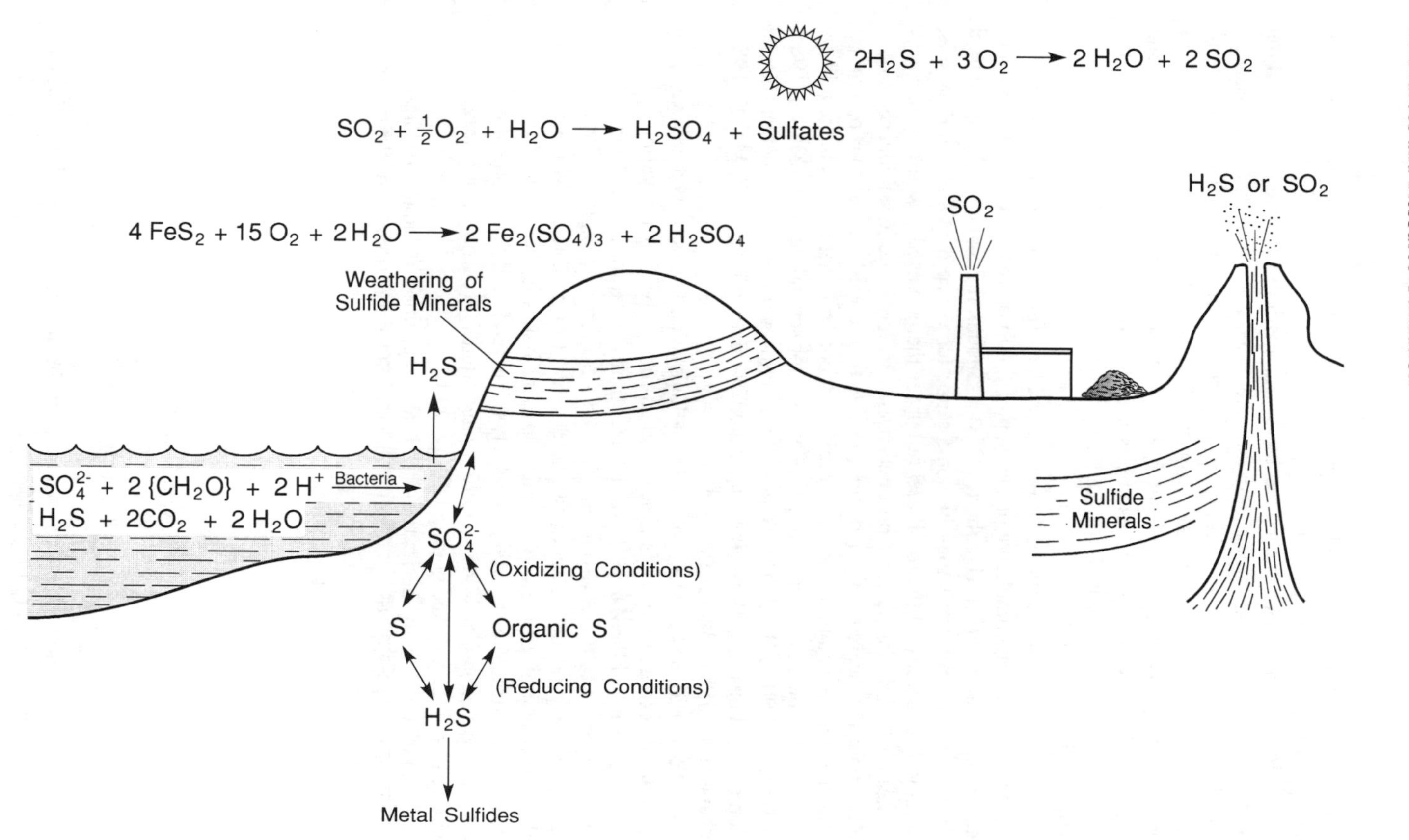

Figure 2.5: The sulfur cycle

converted to another, thereby finding supplemental uses that were not available in the early days of resource development.

The ease with which mineral resources (e.g., fossil fuels) were available to the world did, indeed, play a major role in their increased use (Cassedy and Grossman, 1990; Blunden and Reddish, 1991). The conversion of these natural products to usable products and, in some cases, to valuable chemicals served to increase their popularity, if coal, petroleum, and natural gas can be used as examples (Stobaugh and Yergin, 1979). More specific examples are the coal chemicals industry of the nineteenth century (Speight, 1994) and the petrochemical industry of the twentieth century (Speight, 1991).

There are projections that the era of fossil fuels (gas, petroleum, and coal) will be almost over when cumulative production of the fossil resources reaches 85% of their initial total reserves (Hubbert, 1973). Should the same apply to mineral resources in general, there may be some cause for concern. One major issue is, naturally, the scarcity of the fossil fuel resources that are used as sources of energy as a means of producing valuable products from a variety of nonfossil fuel resources. For example, the scarcity of petroleum (relative to a few decades ago) is real, but it seems that the remaining reserves of petroleum, coal, and natural gas make it likely that there will be an adequate supply of energy for several decades (Martin, 1985; MacDonald, 1990; OPEC, 1991; Banks 1992). However, this can only be achieved at some (as yet unknown) cost (Bending et al., 1987; Hertzmark, 1987; Meyers 1987; Sathaye et al., 1987; Banks, 1994).

To ameliorate the effects of the technologies that are employed in developing the mineral resources, there is a need for on-line analysis of the various product streams (Breen and DeMarco, 1992) to determine the nature of the effluents before they are released to the environment. The effects of, for example, acid rain on soil and water leave no doubt about the need to control its causes (Mohnen, 1988). Indeed, recognition of the need to address these issues is the driving force behind recent energy strategies as well as a variety of research and development programs (Katz, 1984; Stigliani and Shaw, 1990; United States Department of Energy, 1990; United States General Accounting Office, 1990).

The solution to the issue of acid rain deposition (Chapter 10) lies in the control of sulfur oxide (usually sulfur dioxide, SO_2) emissions as well as nitrogen oxide (NO_x) (Chapter 11) emissions. These gases react with the water in the atmosphere and the result is an acid:

$$SO_2 + H_2O = H_2SO_3$$
$$2SO_2 + O_2 = 2SO_3$$
$$SO_3 + H_2O = H_2SO_4$$
$$2NO + H_2O = 2HNO_2$$
$$2NO + O_2 = 2NO_2$$
$$NO_2 + H_2O = HNO_3$$

In fact, the combustion of coal can account for the majority of the sulfur oxides and nitrogen oxides released to the atmosphere.

Whichever technologies succeed in reducing the amounts of these gases in the atmosphere should also succeed in reducing the amounts of urban smog, those notorious brown and grey clouds that are easily recognizable at some considerable distances from urban areas, not only by their appearance but also by their odor.

The oxides of carbon (carbon monoxide, CO, and carbon dioxide, CO_2) are also of importance insofar as all carbon-based materials produce either or both of these gases during use. Both gases have the potential for harm to the environment. A reduction in the emissions of these gases, particularly carbon dioxide, which is the final combustion product of carbon-based mineral resources, has very little chance without a traumatic switch to noncarbon resources. However, such emissions can be moderated by trapping and recovering the carbon dioxide at the time of resource processing. It is also necessary to note here, on the positive side for natural gas, that methane produces less carbon dioxide per unit of energy than coal (Michaels, 1991).

Natural gas is far less abundant than coal and the inadvertent release of natural gas into the atmosphere (methane, CH_4, is a greenhouse gas) (Wuebbles and Edmonds, 1988; Graedel and Crutzen, 1989; Hileman, 1992) may tend to offset any of the advantages of its use. Indeed, the release into the atmosphere of other gases resulting from fossil fuel usage has been projected to cause environmental perturbations (Table 2.5) (Graedel and Crutzen, 1989) that may be difficult to ignore at the present rates of release.

Table 2.5: Gases which cause environmental changes

Gas	Greenhouse effect	Stratospheric ozone depletion	Acid deposition	Smog
Carbon monoxide (CO)				
Carbon dioxide (CO_2)	+	+/-		
Methane (CH_4)	+	+/-		
Nitric oxide (NO) and nitrogen dioxide (NO_2)		+/-	+	+
Nitrous oxide (N_2O)	+	+/-		
Sulfur dioxide (SO_2)	-	+		
Chlorofluorocarbons	+	+		
Ozone	+			+

\+ Positive contribution
\- Varies with conditions and chemistry, may not be a general contributor

Recognition of the production of these atmospheric pollutants in considerable quantities every year has led to the institution of national emission standards for all such pollutants (Walker and Wirl, 1994). Using sulfur dioxide as the example (Table 2.6) (Kyte, 1991), the various standards are not only very specific but will become more stringent with the passage of time (IEA Coal Research, 1991). Atmospheric pollution is being taken very seriously since the projections of uncontrolled pollutants into the atmosphere indicate serious problems in the twenty-first century (Graedel and Crutzen, 1991) and there is also the threat, or promise, of heavy fines and/or jail terms for any pollution-minded miscreants who seek to flout the laws!

There is a trend to the increased use of mineral resources that will require the most stringent approach yet to issues related to environmental protection. The necessity for the cleanup of process effluents is real. Indeed, the emissions resulting from the use of the various mineral resources have had deleterious effects on the environment and promise further detriment unless adequate curbs are taken to control not only the nature but also the amount of effluents being released into the environment.

Effluent products and by-products are produced in various industries (Austin, 1984; Probstein and Hicks, 1990; Speight, 1990, 1991, 1994). These products have the potential to contain quantities of noxious materials that are a severe detriment to the environment. All such noxious products need to be removed before use or discharge to the environment.

6.0 Environmental Aspects

The capacity of the environment to absorb the effluents and other impacts of energy technologies is not unlimited, as some would have us believe. The environment should be considered to be an extremely limited resource, and discharge of chemicals into it should be subject to severe constraints. Indeed, the declining quality of many raw materials dictates that more material must be processed to provide the needed products. The growing magnitude of effluents from industrial processes has moved above the line where the environment has the capability to absorb them without disruption. The general prognosis for environmental protection through an increased awareness of environmental issues is not pessimistic, however. In terms of gaseous emissions, the control of nitrogen oxides is achieved, to date, mainly through modification of the combustion process, and development of various flue gas treating processes. Indeed, it is anticipated that flue gas control technologies may achieve significantly higher control of gaseous emissions than currently being recorded. In fact, the current flue gas desulfurization techniques are capable of removing about 90% of emissions (Speight, 1991).

It is important that there be an understanding of the various types of emissions that escape to the atmosphere and to the surrounding environment. A number of inorganic emissions enter the atmosphere as the result of human activities. The most common of these are sulfur dioxide (SO_2), nitric oxide (NO), and nitrogen dioxide (NO_2). Other inorganic gaseous emissions include ammonia (NH_3), hydrogen sulfide

Table 2.6: National emissions standards for sulfur ($mg/SO_2/m^3$).

Country	New plants	Existing plants
Austria	200-400	200-400
Belgium[1]	400-2000(250)[2]	
Canada[3]	740	
Denmark[1]	400-2000	
Finland	400-660	660
France[1]	400-2000	
FRG[1]	400-2000	400-2500
Greece[1]	400-2000	
Ireland[1]	400-2000	
Italy[1]	400-2000	400-2000
Japan[4]		
Luxembourg[1]	400-2000	
Netherlands[1]	200-700	400-700
Poland	540	2700-4000
Portugal[1]	400-2000	
Spain[1]	2400-9000	2400-9000
Sweden	290	290-570
Switzerland	400-2000	400-2000
Taiwan	2145-4000	2145-4000
Turkey	400-2000	
United Kingdom[1]	400-2000	
United States	740-1480[5]	

1: European economic community countries
2: From 1995
3: Guidelines
4: Set on a plant-by-plant basis according to nationally-defined formulae
5: Clean Air Act Amendments, 1990

(H_2S), chlorine (Cl_2), hydrogen chloride (HCl), and hydrogen fluoride (HF), as well as other oxides of nitrogen.

The highest levels of carbon monoxide emissions from internal combustion engines tend to occur in congested urban areas during heavy road traffic periods (rush hours). Carbon monoxide emissions from automobiles may be lowered by employing a leaner air-fuel mixture, i.e., an air-fuel in which the weight ratio of air to fuel is relatively high. Newer model automobiles have catalytic exhaust reactors installed that cut down on carbon monoxide and nitrogen oxide emissions (Taylor, 1993; Noda, 1995). In more general terms, there is the need to increase the efficiency of automobile transportation (Nivola and Crandall, 1995) and, hopefully, would reduce the noxious emissions into the atmosphere.

Carbon monoxide is generally removed from the atmosphere by reaction with a hydroxyl radical:

$$CO + HO^{\cdot} = CO_2 + H^{\cdot}$$

and is accompanied by other chemical reactions, which lead to the regeneration of more hydroxyl radicals:

$$O_2 + H^{\cdot} = HOO^{\cdot}$$
$$HOO^{\cdot} + NO = HO^{\cdot} + NO_2$$
$$HOO^{\cdot} + HOO^{\cdot} = H_2O_2$$
$$H_2O_2 + h\nu = 2HO^{\cdot}$$

The sulfur cycle (Figure 2.5) involves primarily hydrogen sulfide, sulfur dioxide, sulfur trioxide, and sulfates. Sulfur compounds enter the atmosphere to a very large extent through human activities, primarily as sulfur dioxide from the combustion, or processing, of sulfur-containing fuels. Nonanthropogenic sulfur enters the atmosphere largely as hydrogen sulfide from volcanoes and from the biological decay of organic matter and reduction of sulfate.

Organic emissions may have a strong effect upon atmospheric quality. The effects of such emissions in the atmosphere may be divided into two major categories: (1) direct effects, such as effects caused by exposure to the emissions, and (2) the formation of secondary pollutants, especially smog (Chapter 10). Because of their widespread use in fuels, hydrocarbons predominate among organic atmospheric pollutants. Petroleum products, primarily gasoline, are the source of most of the anthropogenic hydrocarbon emissions found in the atmosphere.

The Clean Air Act of 1990 has already impacted the gasoline markets of the United States (Kortum and Miller, 1994; Piel, 1994). Additives and reformulation have already been implemented in some cases to meet regulations and emission standards required for compliance, and gasoline will continue to be reformulated. The regulations for reformulated gasoline have been divided into Phase I and II, starting in 1995 and 2000, respectively, with each phase having more stringent requirements. Phase I is divided into two compliance models which are designed to facilitate the determination of fuel formulation compliance without the expense of extensive motor performance testing. The simple model can be used from 1995 to 1997, and the complex model from 1997 on.

Alkanes are among the more stable hydrocarbons in the atmosphere. Because of their high vapor pressures, alkanes with six or fewer carbon atoms are normally present as gases, alkanes with twenty or more carbon atoms are present as aerosols or are adsorbed onto atmospheric particles. Alkanes with six to twenty-one carbon atoms per molecule may be present either as vapor or particles, depending upon conditions.

Alkenes enter the atmosphere from a variety of processes, including emissions from internal combustion engines and turbines, foundry operations, and petroleum refining. In addition to the direct release of alkenes, these hydrocarbons are commonly produced by the partial combustion and cracking (thermal decomposition) at high temperatures of alkanes, particularly in the internal combustion engine.

Unlike alkanes, alkenes are highly reactive in the atmosphere, especially in the presence of nitrogen oxides, hydroxyl radicals, and sunlight. In addition, the reaction of molecular oxygen to the resulting radical results in the formation of the extremely reactive hydroperoxyl radical. These radicals then participate in chain reactions such as those involved in the formation of smog (Chapter 10).

Single-ring aromatic compounds are important constituents of lead-free gasoline, which has largely replaced leaded gasoline. In fact, derivatives of naphthalene (which has two fused conjugated aromatic rings) and several compounds containing two or more unconjugated rings have been detected as atmospheric pollutants. One such compound, biphenyl, has been detected in diesel smoke.

Polynuclear aromatic hydrocarbons are present as aerosols in the atmosphere because of their extremely low vapor pressure. The partial combustion of coal, which has a hydrogen-to-carbon ratio less than 1, is a source of polynuclear aromatic compounds (White, 1983), which have considerable toxicity (Lee et al., 1981).

Carbonyl compounds, consisting of aldehydes and ketones, enter the atmosphere from internal combustion engine exhausts, incinerator emissions, spray painting, polymer manufacture, printing, petrochemicals manufacture, and lacquer manufacture. Formaldehyde and acetaldehyde are produced by microorganisms and acetaldehyde is emitted by some kinds of vegetation.

As expected from their low vapor pressures, the lighter organic halogen compounds are the most likely to be found in the atmosphere. The most abundant organic chlorine compounds in the atmosphere are methyl chloride (CH_3Cl) and carbon tetrachloride (CCl_4). Also found are methylene chloride (CH_2Cl_2), methyl bromide (CH_3Br), bromoform (tribromomethane), assorted chlorofluorocarbons and halogen-substituted ethylene compounds, such as trichloroethylene, vinyl chloride, perchloroethylene ($CCl_2{=}CCl_2$), and ethylene dibromide ($CH_2Br.CH_2Br$).

Many chlorofluorocarbons have been implicated in the halogen-atom-catalyzed destruction of atmospheric ozone which filters out cancer-causing ultraviolet radiation from the sun. Although inert in the lower atmosphere, chlorofluorocarbons undergo photodecomposition by the action of high-energy ultraviolet radiation in the stratosphere to release chlorine atoms. These atoms react with ozone to produce chlorine monoxide.

Although not highly significant as atmospheric contaminants on a large scale, organic sulfur compounds can cause local pollution problems because of their bad odors. Major sources of organic sulfur compounds in the atmosphere include microbial degradation, wood pulping, volatile matter evolved from plants, animal wastes, packing house and rendering plant wastes, starch manufacture, sewage treatment, and petroleum refining.

Aromatic amines are widely used as chemical intermediates, antioxidants, and curing agents in the manufacture of polymers (rubber and plastics), drugs, pesticides, dyes, pigments, and inks.

A large number of heterocyclic nitrogen compounds have been reported in tobacco smoke, and it is inferred that many of these compounds can enter the atmosphere from burning vegetation. Coke ovens are another major source of these compounds. In addition to the derivatives of pyridine, some of the heterocyclic nitrogen compounds are derivatives of pyrrole (Speight, 1994).

Particulate matter is a term that has come to stand for particles in the atmosphere. Particulate matter may be organic or inorganic and originates from a wide variety of sources and processes (Cawse, 1982). As one example, particulate matter is produced from mineral matter in the coal that is converted during combustion into finely divided inorganic material referred to as fly ash. Fly ash can be carried out of the stack with the hot exhaust gases and also has the ability to adsorb polynuclear aromatic hydrocarbons (Lee et al., 1993). Furthermore, the practice of burning finely divided coal can contribute to fly ash emissions. Fly ash contains several chemical species (such as arsenic, cadmium, mercury, thallium, and zinc compounds) that are of environmental significance in the soil, water, and atmosphere (Kothny, 1973; Gay and Davis, 1987; Sager, 1993; Sloss, 1995). Indeed, the United States Clean Air Act of 1990 (Chapter 12) requires the evaluation of several trace elements with a view to the introduction of relevant legislation in the future.

Particulate matter, which ranges in size from about 0.5 mm down to molecular dimensions, may consist of either solids or liquid droplets. Atmospheric aerosols are solid or liquid particles smaller than 100 microns (1 micron = 1 meter x 10^{-6}) in diameter. Pollutant particles in the 0.001 to 10 micron range are commonly suspended in the air near sources of pollution, such as the urban atmosphere, industrial plants, highways. and power plants. Very small solid particles include carbon black, silver iodide, combustion nuclei, and sea-salt nuclei. Larger particles include cement dust, windblown soil dust, foundry dust, and pulverized coal. The need for dust control is strong (Mody and Jakhete, 1988). Liquid particulate matter (mist) includes raindrops, fog, and sulfuric acid mist. Some particles are of biological origin, such as viruses, bacteria, bacterial spores, fungal spores, and pollen.

Atmospheric particles undergo a number of processes in the atmosphere. Sedimentation and scavenging by raindrops and other forms of precipitation are the major mechanisms for particle removal from the atmosphere; particulate matter also reacts with atmospheric gases.

Dispersion aerosols, such as dusts, formed from the disintegration of larger particles are usually above 1 micron in size. Typical processes for forming dispersion aerosols include evolution of dust from coal grinding, formation of spray in cooling towers, and blowing of dirt from dry soil.

Many dispersion aerosols originate from natural sources, such as sea spray, windblown dust, and volcanic dust. However, a variety of human activities break up material and disperse it to the atmosphere (Charlson and Wigley, 1994).

Metal oxides constitute a major class of inorganic particles in the atmosphere. These are formed whenever fuels containing metals are burned. For example, particulate iron oxide is formed during the combustion of pyrite-containing coal:

$$3FeS_2 + 8O_2 = Fe_3O_4 + 6SO_2$$

Organic vanadium in residual fuel oil is converted to particulate vanadium oxide. Part of the calcium carbonate in the ash fraction of coal is converted by heat to calcium oxide and is emitted to the atmosphere through the stack:

$$CaCO_3 = CaO + CO_2$$

A common process for the formation of aerosol mists involves the oxidation of atmospheric sulfur dioxide to sulfuric acid, a hygroscopic substance that accumulates atmospheric water to form small liquid droplets:

$$2SO_2 + O_2 + 2H_2O = 2H_2SO_4$$

In the presence of basic air pollutants, such as ammonia or calcium oxide, the sulfuric acid reacts to form ammonium or calcium salts. Under low-humidity conditions water is lost from these droplets, and a solid aerosol is formed.

A significant portion of organic particulate matter is produced by internal combustion engines. These products may include nitrogen-containing compounds and oxidized hydrocarbon polymers. Lubricating oil and its additives may also contribute to organic particulate matter. Particulate carbon as soot, carbon black, coke, and graphite originates from auto and truck exhausts, heating furnaces, incinerators, power plants, and steel and foundry operations and composes one of the more visible and troublesome particulate air pollutants. Because of its adsorbent properties, carbon can be a carrier of gaseous and other particulate pollutants. Particulate carbon surfaces may catalyze some heterogeneous atmospheric reactions, including the important conversion of sulfur dioxide to sulfate.

Another mineral worthy of note as a pollutant is asbestos. Asbestos is of concern as an air pollutant because when inhaled, the fibrils can lodge in human tissue and cause asbestosis (a pneumonia condition), mesothelioma (tumor of the mesothelial tissue lining the chest cavity adjacent to the lungs), and bronchogenic carcinomas (cancer originating with the air passages in the lungs). Because asbestos has long been indicated in asbestosis (similar to silicosis) and, in more recent years, is considered a carcinogen, several countries have issued regulations that restrict its use. For many years, it was used in fireproof fabrics, brake lining, gaskets, roofing, insulation, paint fillers, reinforcing agents in rubber and plastics, and in electrolytic diaphragm cells.

By way of definition, *asbestos* is the name given to a group of fibrous silicate minerals, typically those of the serpentine group. Chemically, asbestos is a group of impure magnesium silicate minerals and colors may be white, gray, green, or brown, sp. gr. 2.5, noncombustible. Serpentine asbestos is the mineral chrysotile, a magnesium silicate. The fibers are strong and flexible. Spinning is possible with the longer fibers. Amphibole asbestos includes various silicates of magnesium, iron, calcium, and sodium. The fibers are generally brittle and cannot be spun, but are more resistant to chemicals and to heat than serpentine asbestos.

Some of the metals found predominantly as particulate matter in polluted atmospheres are known to be hazardous to human health. All of these except beryllium are so-called heavy metals. Lead is the toxic metal of greatest concern in the urban atmosphere because it comes closest to being present at a toxic level; mercury ranks second. With the reduction in the use of leaded fuels, atmospheric lead is of less concern than it used to be. Other heavy metals include beryllium, cadmium, chromium, vanadium, nickel, and arsenic (a metalloid). Arsenic can have a significant effect on the environment, whether the metalloid is released to the land, the water, or the air (Nriagu, 1994a, 1994b).

Atmospheric mercury is of concern because of its toxicity, volatility, and mobility. Some atmospheric mercury is associated with particulate matter. One means by which mercury can enter the atmosphere is as volatile elemental mercury from coal combustion (American Society for Testing and Materials, 1995b, ASTM D 3684). Volatile organic mercury compounds such as dimethyl mercury $[(CH_3)_2Hg]$ and methyl mercury salts such as methyl mercury bromide (CH_3HgBr) are also encountered in the atmosphere.

A significant natural source of radionucleides in the atmosphere is radon, a noble gas product of radium decay. Radon may enter the atmosphere as either of two isotopes, ^{222}Rn (half-life 3.8 days) and ^{220}Rn (half-life 54.5 seconds). Both are alpha emitters in decay chains that produce a series of daughter products and terminate with stable isotopes of lead. The initial decay products, isotopes of polonium (^{218}Po and ^{216}Po) are nongaseous and adhere readily to atmospheric particulate matter. Therefore some of the radioactivity detected in these particles is of natural origin.

One of the more serious problems in connection with radon is that of radioactivity originating from uranium mine tailings that have been used in some areas as backfill, soil conditioner, and a base for building foundations. Radon produced by the decay of radium exudes from foundations and walls constructed on tailings. Some medical authorities have suggested that in areas where uranium mill tailings have been used in residential construction the rate of birth defects and the occurrence of infant cancer are significantly higher than normal.

The combustion of fossil fuels introduces radioactivity into the atmosphere in the form of radionucleides contained in fly ash. Large coal-fired power plants lacking ash-control equipment may introduce up to several hundred millicuries of radionucleides into the atmosphere each year, far more than either an equivalent nuclear or an equivalent oil-fired power plant.

The radioactive noble gas krypton (^{85}Kr, half-life 10.3 years) is emitted into the atmosphere by the operation of nuclear reactors and the processing of spent reactor fuels. In general, other radionucleides produced by reactor operation are either chemically reactive and can be removed from the reactor effluent or have such short half-lives that a short time delay prior to emission prevents their leaving the reactor.

The above-ground detonation of nuclear weapons can add large amounts of radioactive particulate matter to the atmosphere. Although there are general agreements between several nations to cease testing nuclear devices in the atmosphere, not all nations agree to this proposition.

It is considered likely that most of the environmental impact of mineral resource development (including the hazards of coal mining, gaseous emissions, acid precipitation) could be substantially abated. A considerable investment in retrofitting or replacing existing facilities and equipment might be needed. However, replacing coal with natural gas, which releases less carbon dioxide per unit of energy, is at best a short-term solution.

Minimizing the carbon dioxide emissions from many mineral resources (which must also include emissions of hydrogen sulfide) would require revamping of the current technology. However, it is possible, and a conscious goal must be to improve the efficiency with which the resources are transformed and consumed and by shifting to alternative products.

It is known that the production of transportation fuels from biomass sources (Sheehan 1994) can also produce hydrocarbon emissions during biomass fuel combustion (Olsson and Persson, 1991). Ethanol (ethyl alcohol; grain alcohol) is a major biomass fuel (Geller, 1985), but in the current market conditions it remains significantly more expensive than gasoline. The widespread use of biomass (alcohol) fuels is constrained by the size of the resource base from which they are produced. There is also the argument that methyl alcohol (methanol) produced from natural gas releases as much carbon dioxide as does gasoline. The counter argument is that methanol-fueled engines have a greater potential for improvements than gasoline engines (for example, they can function at higher levels of compression).

There is a need for systematic identification and evaluation of the potential impacts of proposed projects, plans, programs, policies, or legislative actions upon the physical-chemical, biological, cultural, and socioeconomic components of the environment. Also known as environmental impact assessment (EIA), this procedure includes the consideration of measures to mitigate undesirable impacts. The primary purpose of environmental impact assessment is to encourage consideration of the environment in planning and decision making and ultimately to arrive at actions that are environmentally compatible.

Impact assessment refers to the interpretation of the significance of anticipated changes related to the proposed project. Impact interpretation can be based upon the systematic application of the definition of significance, as described earlier. For some types of anticipated impacts, there are specific numerical standards or criteria that can be used as a basis for impact interpretation. Examples include air-quality standards, environmental noise criteria, surface-water and groundwater quality standards, and wastewater discharge standards for particular facilities. The latter is especially important since the quality of wastewater can vary with the nature of the facility (Table 2.7).

In this context, it is necessary to consider the impacts related to the biological environment and the potential significance of the loss of particular habitats, including wetland areas (Dennison and Berry, 1993). Another basis for assessment is public input, which could be received through the scoping process or through public participation programs.

Identifying and evaluating potential impact mitigation measures should also be an activity in the process of environmental impact assessment. Mitigation can be defined as the sequential consideration of the following measures: (1) avoiding the impact by not taking a certain action or partial action; (2) minimizing impacts by

Table 2.7: General character of industrial wastewater

Industry	Processes or waste	Effect
Brewery and distillery	Malt and fermented liquors	Organic load
Chemical	General	Stable organics, phenols, inks
Dairy	Milk processing, bottling, butter and cheese making	Acid
Dyeing	Spent dye, sizing, bleach	Color, acid or alkaline
Food processing	Canning and freezing	Organic load
Laundry	Washing	Alkaline
Leather tanning	Leather cleaning and tanning	Organic load, acid and alkaline
Meat packing	Slaughter, preparation	Organic load
Paper	Pulp and paper manufacturing	Organic load, wood
Steel	Pickling, plating, and others	Acids
Textile manufacture	Wool scouring, dyeing	Organic load, alkaline

limiting the degree or magnitude of the action and its implementation; (3) rectifying the impact by repairing, rehabilitating, or restoring the affected environment; (4) reducing or eliminating the impact over time by preservation and maintenance operations during the life of the action; and (5) compensating for the impact by replacing or providing substitute resources or environments. Examples of mitigation measures include pipeline routing to avoid archeological resources, inclusion of pollution-control equipment on airborne and liquid discharges, reductions in project size, revegetation programs, wildlife protection plans, erosion control measures, remediation activities, and creation of artificial wetlands.

A key activity in environmental impact assessment is associated with selecting the proposed action from alternatives that have been evaluated (Inhaber, 1982). In public projects, there is considerable emphasis on the evaluation of alternatives; in fact, various regulations (Majumdar, 1993) indicate that the analysis of alternatives represents the heart of the impact assessment process. Conversely, for many private developments, the range of alternatives may be limited. However, there are potential alternative measures that could be evaluated, including those relating to project size and design features even if location alternatives are not available.

Environmental impact studies need to address a minimum of two alternatives, and can include more than fifty alternatives. Typical studies address three to five alternatives. The minimum number usually represents a choice between construction and operation of a project versus project rejection. The alternatives may encompass a wide range of considerations such as site location; design of a site; construction, operation, and decommissioning options; project size; phasing in of operational units; timing of activities; and the "what if there is no project" option.

All of the aforementioned arguments for and against the use of resources do not belie the obvious. Mineral resources, including fossil fuel resources, will be sources of valuable products for some time to come. They must be used judiciously insofar as they are sources of pollutants, many of which are gases and which must not be released to the atmosphere.

Obviously, much work is needed to accommodate the change to a different fuel source. In the meantime, we use the available resources, all the while working to improve efficiency and to ensure that there is no damage to the environment. Such is the nature of resource development and use and the expectancy of environmental protection.

7.0 References

Abraham, H. 1945. *Asphalts and Allied Substances*. Van Nostrand and Co., New York.

Afonso, R., Dusatko, G., and Pohl, J. 1993. *Combustion Science and Technology*. 93: 41.

Agarwal, J.C. 1991. In *Energy and the Environment in the 21st Century*. p. 389. Edited by J.W. Tester, D.O. Wood, and N.A. Ferrari. MIT Press, Cambridge, Massachusetts.

Agricola, Georgius (Bauer, Georg). 1556. *De Re Metallica*. Froben, Basel, Switzerland.

Al-Sowayegh, A. 1984. *Arab Petro-politics*. Croom Helm Ltd., London, England.

American Society for Testing and Materials. 1995a. ASTM D 388. Classification of Coals by Rank. *Annual Book of ASTM Standards*. Vol. 05.05. American Society for Testing and Materials, Philadelphia, Pennsylvania.

American Society for Testing and Materials. 1995b. ASTM D 3684. Test Method for Total Mercury in Coal by the Oxygen Bomb Combustion/Atomic Absorption Method. *Annual Book of ASTM Standards*. Vol. 05.05. American Society for Testing and Materials, Philadelphia, Pennsylvania.

American Society for Testing and Materials. 1995c. ASTM E 901. Classification System for Uranium Resources. *Annual Book of ASTM Standards*. Vol. 12.02. American Society for Testing and Materials, Philadelphia, Pennsylvania.

Austin, G.T. 1984. *Shreve's Chemical Process Industries*. McGraw-Hill, New York.

Bain, G.W. 1987. In *Encyclopedia of Science and Technology*. 6th ed. Vol. 11, p. 224. Edited by S.P. Parker. McGraw-Hill, New York.

Banks, F.E. 1992. *Opec Bulletin* XXIII(2): 20.

Banks, F.E. 1994. *Energy Sources* 16: 241.

Bending, R.C., Cattell, R.J., and Eden, R.J.. 1987. *Annual Review of Energy*. 12: 185.

Berkowitz, N. 1979. *An Introduction to Coal technology*. Academic Press, New York.

Bisio, A., and Boots, S.R. (eds.). 1995. *Energy Technology and the Environment*. John Wiley & Sons Inc., New York.

Blunden, J., and Reddish, A. (eds.). 1991. *Energy, Resources and Environment*. Hodder and Stoughton, London, England.

Bowen, H.M.J. 1982. Chap. 2 in *Environmental Chemistry*. Vol. 2. Specialist Periodical Reports. The Chemical Society, London, England.

Breen, J.J., and DeMarco, M.J. (eds.). 1992. *Pollution Prevention in Industrial Processes: The Role of Process Analytical Chemistry*. Symposium Series No. 508. American Chemical Society, Washington, D.C.

Burlington Northern Railroad. 1995. *Energy, Coal and Low Cost Electricity Facts*. Burlington Northern Railroad, Fort Worth, Texas.

Cassedy, E.S., and Grossman, P.Z.. 1990. *Introduction to Energy Resources, Technology and Society*. Cambridge University Press, Cambridge, England.

Cawse, P.A. 1982. Chap. 1 in *Environmental Chemistry*. Vol. 2. Specialist Periodical Reports. The Chemical Society, London, England.
Chagger H.K., Goddard, P.R. Murdoch, P.L., and Williams, A. 1993. *Fuel*. 72: 1451.
Charlson, R.J., and Wigley, T.M.L. 1994. *Scientific American* 270(2): 48.
Chenier, P.J. 1992.Chap. 25. *Survey of Industrial Chemistry*. 2nd ed. VCH Publishers, New York.
Cooper, R.N. 1994. *Environment and Resource Policies for the World Economy*. Brookings Institution, Washington, D.C.
Cruden, A. 1930. *Complete Concordance to the Bible*. Butterworth Press, London, England.
Cybulski, A. 1994. *Catalysis Reviews. Science and Engineering* 36: 557.
Davidson, R. 1993. *Organic Sulphur in Coal*. Report No. IEACR/60. International Energy Agency Coal Research, London, England.
Davidson, R. 1994. *Nitrogen in Coal*. Report No. IEAPER/08. International Energy Agency Coal Research, London, England.
Dennison, M.S. and Berry, J.F. (eds.). 1993. *Wetlands: Guide to Science, Law, and Technology*. Noyes Data Corp., Park Ridge, New Jersey.
Dickson, M.H., and Fanel, M. (eds.). 1990. *Small Geothermal Resources: A Guide to Development and Utilization*. UNITAR/UNDP Centre on Small Energy Resources, Rome, Italy.
Dovers, S. 1994. In *Sustainable Energy Systems*. p. 13. Edited by S. Dovers. Cambridge University Press, Cambridge, England.
Dryden, I.G.C. 1975. *The Efficient Use of Energy*. IPC Business Press, Guildford, Surrey, England.
Easterbrook, G. 1995. *A Moment on the Earth: The Coming Age of Environmental Optimism*. Viking Press, New York.
EIA. 1988. *International Energy Outlook: Projections to 2000*. Report No. DOE/EIA-0484(87). United States Department of Energy, Energy Information Administration, Washington, D.C.
EIA. 1989. *Annual Energy Outlook: Long Term Projections*. Report No. DOE/EIA-0383(89). United States Department of Energy, Energy Information Administration, Washington, D.C.
EIA. 1991a. *Annual Energy Review 1990*. Report No. DOE/EIA-0384(90). United States Department of Energy, Energy Information Administration, Washington, D.C.
EIA. 1991b. *Annual Outlook for Oil and Gas 1991*. Report No. DOE/EIA-0517(91). United States Department of Energy, Energy Information Administration, Washington, D.C.
EIA. 1992. *Annual Energy Outlook: With Projections to 2010*. Report No. DOE/EIA-0383(92). United States Department of Energy, Energy Information Administration, Washington, D.C.
EIA. 1994. *Energy Use and Carbon Emissions: Non-OECD Countries*. Report No. DOE/EIA-0587(94). United States Department of Energy, Energy Information Administration, Washington, D.C.
EIA. 1995. *Annual Energy Outlook*. Report No. DOE/EIA-0383(95). United States

Department of Energy, Energy Information Administration, Washington, D.C.
Engelhard, J., Schiffer, H.W., and Schloesser, N. 1993. *Energie* 45(10): 18.
Forbes, R.J. 1958. *Studies in Early Petroleum Chemistry*. E.J. Brill, Leyden, The Netherlands.
Forbes, R.J. 1959. *More Studies in Early Petroleum Chemistry*. E.J. Brill, Leyden, The Netherlands.
Forbes, R.J. 1964. *Studies in Ancient Technology. Vol. I.* E.J. Brill, Leyden, The Netherlands.
Fulkerson, W., Judkins, R.R., and Sanghvi, M.K. 1990. Chap. 8 in *Energy for Planet Earth*. W.H. Freeman, New York.
Galloway, R.L. 1882. *A History of Coal Mining in Great Britain*. Macmillan & Co., London, England.
Gay, A.J., and Davis, P.B. 1987. In *Coal Science and Chemistry*. p. 221. Edited by A Volborth. Elsevier, Amsterdam.
Geller, H.S. 1985. *Annual Review of Energy* 10: 135.
Gibbons, J.H., Blair, P.D. and Gwin, H.L. 1989. *Scientific American* 261(3): 136.
Golob, R., and Brus, E. 1993. *The Almanac of Renewable Energy*. Henry Holt, New York.
Graedel, T.E., and Crutzen, P.J. . 1989. *Scientific American* 261(3): 58.
Graedel, T.E., and Crutzen, P.J. 1991. In *Energy and the Environment in the 21st Century*. p. 13. Edited by J.W. Tester, D.O. Wood, and N.A. Ferrari. MIT Press, Cambridge, Massachusetts.
Gray, P.E., Tester, J.W., and Wood, D.O. 1991. In *Energy and the Environment in the 21st Century*. p. 120. Edited by J.W. Tester, D.O. Wood, and N.A. Ferrari. MIT Press, Cambridge, Massachusetts.
Grenon, M. 1976. *Annual Review of Energy* 2: 67.
Hafele, W. 1990. Chap. 9 in *Energy for Planet Earth*. W.H. Freeman, New York.
Halpern, G.M. 1980. Fusion Energy. In *Encyclopedia of Chemical Technology*. Volume 11, p. 590. John Wiley & Sons Inc., New York. .
Herodotus, 447 B.C. *Historia. The History of Herodotus*. (Translated by George Rawlinson. 1956. Edited by M. Kamroff Editor). Tudor Publishing, New York.
Hertzmark, D.I. 1987. *Annual Review of Energy* 12: 23.
Hessley, R.K., J.W. Reasoner, and J.T. Riley. 1986. *Coal Science*. John Wiley & Sons Inc., New York.
Hessley, R.K. 1990. In *Fuel Science and Technology Handbook*. Edited by J.G. Speight. Marcel Dekker Inc., New York.
Hileman, B. 1992. *Chemical and Engineering News* 70(6): 26.
Hewett, C.E., High, C.J., Marshall, N., and Wildermuth, R. 1981. *Annual Review of Energy* 6: 139.
Himmel, M.E., Baker, J.O., and Overend, R.P. 1994. *Enzymatic Conversion of Biomass for Fuels Production*. Symposium Series No. 566. American Chemical Society, Washington, D.C.
Holdren, J.P., Morris, G., and Mintzer, I. 1980. *Annual Review of Energy* 5: 241.
Hubbert, M.K. 1973. *American Association of Petroleum Geologists Bulletin* 57(9):

1843.
IEA Coal Research. 1991. *Emissions Standards Data Base*. International Energy Agency Coal Research, London England.
Inhaber, H. 1982. *Energy Risk Assessment*. Gordon and Breach, New York.
Jahn, E.C., and Strauss, R.W. 1983. Chap. 15 in *Regal's Handbook of Industrial Chemistry*. Edited by J.A. Kent. Van Nostrand Reinhold, New York.
James, P., and Thorpe, N. 1994. *Ancient Inventions*. Ballantine Books, New York.
Johansson, T.D., H. Kelly, A.K.N. Reddy, and R.H. Williams. (eds.). 1993. *Renewable Energy for Fuels and Electricity*. EarthScan Publications, United Nations, New York.
Jones, J.L., and Radding, S.B. (eds.). 1980. *Thermal Conversion of Solid Wastes and Biomass*. Symposium Series No. 130. American Chemical Society, Washington, D.C.
Kahn, E. 1979. *Annual Review of Energy* 1: 313.
Kasibhatla, P.S., Levy, H., and Moxin, W.J. 1993. *Journal of Geophysical Research. (Atmosphere)* 98(D4): 7165.
Katz, J.E. 1984. *Congress and National Energy Policy*. Transaction Books, New Brunswick, New Jersey.
Knief, R.A. 1992. *Nuclear Engineering: Theory and Technology of Commercial Nuclear Power*. Taylor & Francis, Washington, D.C.
Kortum, D.J., and Miller, M.G. 1994. Preprints. *American Chemical Society Division of Fuel Chemistry* 39(2): 267.
Kothny, E.L. (ed.). 1973. *Trace Elements in the Environment*. Advances in Chemistry Series No. 123. American Chemical Society, Washington, D.C.
Kruger, P. 1976. *Annual Review of Energy* 1: 159.
Kubursi, A.A., and Naylor, T. (eds.). 1985. *Cooperation and Development in the Energy Sector*. Croom Helm, London, England.
Kyte, W.S. 1991. *Desulphurisation 2: Technologies and Strategies for Reducing Sulphur Emissions*. Institute of Chemical Engineers. Rugby, Warwickshire, England.
Lee, M.L., Novotny, M.V., and Bartle, K.D. 1981.Chap. 3 in *Analytical Chemistry of Polycyclic Aromatic Compounds*. Academic Press, New York.
Lee, W.M.G., Yuan, Y.S., and Chen, J.C. 1993. *Journal of Environmental Science and Health, Part A* A28: 1017.
Lincoln, B. 1785. *Memoirs of the American Academy of Arts and Sciences*. Vol. I, p. 372.
Linden, H.R., Bodle, W.W., Lee, B.S., and Vyas, K.C. 1976. *Annual Review of Energy* 1: 65.
Lockhart, L. 1939. *Journal of the Institute of Petroleum* 25: 1.
Lowinger, T.C., and Hinman, G.W. (eds.). 1994. *Nuclear Power at the Crossroads: Challenges and Prospects for the Twenty-First Century*. International Research Center for Energy and Economic Development, Boulder, Colorado.
Lyman, W.J., Reehl, W.F., and Rosenblatt, D.H. 1990. *Handbook of Chemical Property Estimation Methods*. American Chemical Society, Washington, D.C.
MacDonald, G.J. 1990. *Annual Review of Energy* 15: 53.

Majumdar, S.B. 1993. *Regulatory Requirements for Hazardous Materials*. McGraw-Hill, New York.

Manowitz, B., and Lipfert, F.W. 1990. Chap. 3 in *Geochemistry of Sulfur in Fossil Fuels*. Edited by W.L. Orr and C.M. White. American Chemical Society, Washington, D.C.

Marsden, S.S., Jr. 1983. *Annual Review of Energy* 8: 333.

Martin, A.J. 1985. Chap. 1 in *Prospects for the World Oil Industry*. Edited by T. Niblock and Lawless, R. Croom Helm, London, England.

Meyerhoff, A.A., and Meyerhoff, H.A. 1987. In *Encyclopedia of Science and Technology*. Vol. 11. p. 217. Edited by S.P. Parker. McGraw-Hill, New York.

Meyers, S. 1987. *Annual Review of Energy* 12: 81.

Michaels, P.J. 1991. *Journal of Coal Quality* 10(1): 1.

Mills, D., and M. Diesendorf. 1994. In *Sustainable Energy Systems*. Edited by S. Dovers. p. 112. Cambridge University Press, Cambridge, England.

Mody, V., and R. Jakhete. 1988. *Dust Control Handbook*. Noyes Data Corp., Park Ridge, New Jersey.

Mohnen, V.A. 1988. *Scientific American* 259(2): 30.

Mooney, H. 1988. *Towards an Understanding of Global Change*. National Academy Press, Washington, D.C.

Morse, F.H., and Simmons, M.K.. 1976. *Annual Review of Energy* 1: 131.

Niblock, T., and Lawless, R. (eds.). 1985. *Prospects for the World Oil Industry*. Croom Helm, London, England.

Nivola, P.S., and Crandall, R.W. 1995. *The Extra Mile: Rethinking Energy Policy for Automotive Transportation*. Brookings Institution, Washington, D.C.

Noda, F. 1995. Technical Review. *Toyota Motor Corp*. 44(2): 2.

Nriagu, J.O. (ed.). 1994a. *Arsenic in the Environment. Part I: Cycling and Characterization*. New York: John Wiley.

Nriagu, J.O. (ed.). 1994b. *Arsenic in the Environment. Part II: Human Health and Ecosystem effects*. John Wiley & Sons Inc., New York.

Olsson, M., and Persson, E.M. 1991. Report No. SVF 410. Stiftelsen Vaermetelnisk Forsk., Stockholm, Sweden. Also published in *Energy Abstracts*. 1992. 17(12): 332595.

OPEC. 1991. *Facts and Figures: A Graphical Analysis of World Energy up to 1990*. Organization of Petroleum Exporting Countries, Vienna, Austria. .

Palmer, C.A., Finkelman, R.B., Krasnow, M.R., and Eble, C.F. 1995. Preprints. *American Chemical Society Division of Fuel Chemistry* 40(4): 803.

Park, C.F. Jr. 1987. In *Encyclopedia of Science and Technology*. Vol. 11, p. 231. Edited by S.P. Parker. McGraw-Hill, New York.

Piel, W.J. 1994. Preprints. *American Chemical Society Division of Fuel Chemistry* 39(2): 273.

Pickering, K.T., and Owen, L.A. 1994. *Global Environmental Issues*. Routledge Publishers, New York.

Post, R.F. 1976. *Annual Review of Energy* 1: 213.

Probstein, R.F., and Hicks, R.E. 1990. *Synthetic Fuels*. pH Press, Cambridge, Massachusetts.

Rahn, F.J. 1987. In *Encyclopedia of Science and Technology*. Vol. 12. p. 171. Edited by S.P. Parker. McGraw-Hill, New York.

Robinson, J.S. (ed.). 1980. *Fuels from Biomass: Technology and Feasibility*. Noyes Data Corp., Park Ridge, New Jersey.

Sager, M. 1993. *Fuel* 72: 1227.

Sathaye J., Ghirardi, A., and Schipper, L. 1987. *Annual Review of Energy* 12: 253.

Scheil, V., and A. Gauthier. 1909. *Annales de Tukulti Ninip II*. Paris.

Schmidt, R.A., and Hill, G.R. 1976. *Annual Review of Energy* 1: 37.

Schroder, O. 1920. *Keilschriftetexte aus Assur verscheidenen XIV*. Leipzig.

Sheehan, J.J. 1994. Chap. 1 in *Enzymatic Conversion of Biomass for Fuels Production*. Edited by M.E. Himmel, J.O. Baker, and R.P. Overend. Symposium Series No. 566. American Chemical Society, Washington, D.C.

Singer, C., Holmyard, E.J., Hall, A.R., and Williams, T.I. 1958. *A History of Technology*. Vols. I-V. Clarendon Press, Oxford, England.

Skrzypek, J., Sloczynski, A.J., and Ledakowicz, S. 1994. *Methanol Synthesis*. Polish Academy of Sciences, Gliwice, Poland.

Sloss, L. 1995. Preprints. *American Chemical Society Division of Fuel Chemistry* 40(4): 793.

Speight, J.G. (ed.). 1990. *Fuel Science and Technology Handbook*. Marcel Dekker Inc., New York.

Speight, J.G. 1991. *The Chemistry and Technology of Petroleum*. 2nd ed. Marcel Dekker Inc., New York.

Speight, J.G. 1993. *Gas Processing: Environmental Aspects and Methods*. Butterworth Heinemann, Oxford England.

Speight, J.G. 1994. *The Chemistry and Technology of Coal*. 2nd ed. Marcel Dekker Inc., New York.

Stigliani, W.M., and Shaw, R.W. 1990. *Annual Review of Energy* 15: 201.

Stobagh, R., and Yergin, D. (eds.). 1979. *Energy Future*. Random House, New York.

Taylor, K.C. 1993. *Catalysis Reviews. Science and Engineering* 35: 457.

Tester, J.W., Wood, D.O., and Ferrari, N.A. (eds.). 1991. *Energy and the Environment in the 21st Century*. MIT Press, Cambridge, Massachusetts.

Tillman, D.A. 1978. *Wood as an Energy Resource*. Academic Press, New York.

United States Department of Energy. 1990. *Gas Research Program: Implementation Plan*. DOE/FE-0187P. United States Department of Energy, Washington, D.C.

United States General Accounting Office. 1990. *Energy Policy: Developing Strategies for Energy Policies in the 1990s*. Report to Congressional Committees. GAO/RCED-90-85. United States General Accounting Office, Washington, D.C. June.

Ure, A.M., and Berrow, M.L. 1982. Chap. 3 in *Environmental Chemistry*. Vol. 2. Specialist Periodical Reports. The Chemical Society, London, England.

Walker, I.O., and Wirl, F. 1994. *OPEC Bulletin* XXV(10): 16.

Walker, S. 1993. *Major Coalfields of the World*. Report No. IEACR/51. Internaitonal Energy Agency Coal Research, London, England.

Weinberg, C.J., and Williams, R.H. 1990. Chap. 10 in *Energy for Planet Earth*. W.H. Freeman, New York.
White, C.M. 1983. Chap. 13 in *Handbook of Polycyclic Hydrocarbons*. Edited by A. Bjorseth. Marcel Dekker Inc., New York.
Williamson, J., and Wigley, F. (eds.). 1994. *The Impact of Ash Deposition on Coal Fired Plants*. Taylor and Francis, Washington, D.C.
Wood, T.S., and Baldwin, S. 1985. *Annual Review of Energy* 10: 407.
Wuebbles, D.J., and Edmonds, J. 1988. *A Primer on Greenhouse Gases*. Report No. DOE/NBB-0083. Office of Energy Research, United States Department of Energy, Washington, D.C.
Yeager, K.E., and Baruch, S.B. 1987. *Annual Review of Energy* 12: 471.
Yergin, D. 1992. *The Prize: The Epic Quest for Oil, Money, and Power*. Simon & Schuster, New York.
Zebroski, E., and Levenson, M. 1976. *Annual Review of Energy* 1: 101.

Part II Ecosystems

Chapter 3

Land Systems

1.0 Introduction

Land systems (also referred to as the *geosphere*, and sometimes as the *lithosphere*) are those parts of the earth upon which the human species, and many other animal species, live and from which they extract most of their food and energy. In this respect, the land systems may be expected to be the ecosystems that receive the highest degree of anthropogenic stress. This is true, in part, but the water systems and the atmosphere are also pollutant receptors, and to single out one system as the major recipient of pollutants would be misleading, even erroneous.

Land systems were at one time believed to contain unlimited resources and would be able to sustain human activities and survive development beyond imaginable time. As resource development continued, it became evident that this was not the case. Land systems are now recognized as being extremely fragile and have suffered irreparable damage as a result of human activities.

Two atmospheric phenomena, viz., the *greenhouse effect* and the tendency for *acid deposition* (*acid rain*) have the potential to cause major changes in the life forms that inhabit the land, as well as affect the life forms that are dependent upon the water in acidified lakes. This example illustrates many of the complex relationships that exist between the land systems, the water systems, and the atmosphere (Chapter 1). A very general rule is *pollute the one and you pollute the other(s)*.

One of the greater impacts of humans upon land systems has been the creation of desert areas by misuse/abuse of land where the yearly water deposition (rainfall or snow) borders on the arid/semiarid allocation. The result is a noticeable decrease in groundwater availability, a reduction in the availability of surface water, soil erosion, and a decrease in the types, and overall amount, of vegetation.

The problem is severe in many parts of the world where large arid and semiarid areas are experiencing the need for water to prevent desert formation. The southeastern part of the state of Wyoming, for example, generally receives approximately 13 inches (0.3 m) of water per year, an amount sufficient to designate that area as semiarid. Careless use of this scarce water resource could result in the formation of a desert in the Laramie valley. Thus, with increasing population and industrialization, one of the more important aspects of the use of the land systems is concerned with the protection of water resources. Mining, agricultural, chemical, and radioactive wastes all have the potential for contaminating surface water and groundwater. There are limitations to the earth's tolerance for continual increases in the human population (Choucri, 1991).

Even though environmental issues related to the land systems may be discussed in generalities, there are some specific items that need to be understood.

Land systems are composed predominantly of four subsystems: minerals, sedimentary strata, clays, and soil. Each subsystem is, in effect, a different physical and chemical system, and each subsystem can be expected to react differently to pollutants and/or contaminants. Some knowledge of each of these subsystems is essential if the different modes of reaction of chemical waste are to be understood.

2.0 Minerals

The land systems are divided into *strata* (layers) consisting of the solid iron-rich inner core, molten outer core, mantle, and crust. The latter is the outer skin (ranging in depth from 5 to 40 km) of the land systems that is, to some degree, accessible to humans. The crust consists of *rocks*, a rock being a mass of a pure mineral or an aggregate of two or more *minerals*.

Minerals (Table 3.1) are naturally occurring inorganic solids with well-defined crystalline structures resulting from definite internal structures and compositions (Ramsdell, 1987). Most minerals are silicates, such as quartz (SiO_2) or potassium aluminum silicate (orthoclase, $K.Al.Si_3O_8$). Oxygen and silicon make up about 47% and 28% by weight (w/w), respectively, of the land systems. Other prominent elements are aluminum (8%), iron (5%), calcium (4%), sodium (3%), potassium (3%), magnesium (2%) and titanium (0.4%) (Table 3.2). These nine elements make up approximately 99% of the earth's crust.

The color of a mineral in its powdered form as observed when the mineral is rubbed across an unglazed porcelain plate is known as streak. Color can vary widely due to the presence of impurities. Hardness is expressed on a scale (the Moh scale), which is a criterion for comparing the resistance of materials to crushing (Fahrenwald, 1934; Perry and Chilton, 1973). The scale ranges from 1 to 10 and is based upon minerals that vary from talc (hardness 1) to diamond (hardness 10). Cleavage, in the mineral sense, denotes the manner in which minerals break along planes and the angles in which these planes intersect. Most minerals fracture irregularly, although some fracture along smooth curved surfaces or into fibers or splinters. Specific gravity (density relative to that of water) is another important physical characteristic of minerals.

Evaporites are soluble salts that precipitate from solution, for example, as the result of the evaporation of seawater. The most common evaporite is halite (sodium chloride, NaCl) and other simple evaporite minerals are sylvite (potassium chloride, KCl), thenardite (sodium sulfate, Na_2SO_4), and anhydrite (calcium sulfate, $CaSO_4$). Many evaporites are hydrates, including bischofite (magnesium chloride hexahydrate, $MgCl_2.6H_2O$), gypsum (calcium sulfate dihydrate), keiserite (magnesium sulfate monohydrate, $MgSO_4.H_2O$), and epsomite (magnesium sulfate heptahydrate, $MgSO_4.7H_2O$). Double salts, such as carnallite (potassium magnesium chloride hexahydrate, $KCl.MgCl_2.6H_2O$), polyhalite (potassium magnesium calcium sulfate dihydrate, $K_2SO_4.MgSO_4.2CaSO_4.2H_2O$), and loeweite (sodium magnesium sulfate pentadecahydrate, $6Na_2SO_4.7MgSO_4.15H_2O$) are very common in evaporites.

The solidification of molten rock (*magma*) produces igneous rock. Common

Table 3.1: Major mineral groups

Mineral group	Examples	Formula
Silicates	Quartz	SiO_2
	Olivine	$(Mg,Fe)_2Si_4$
	Potassium feldspar	$KAlSi_3O_8$
Oxides	Corundum	Al_2O_3
	Magnetite	Fe_3O_4
Carbonates	Calcite	$CaCO_3$
	Dolomite	$CaCO_3 \cdot MgCO_3$
Sulfides	Pyrite	FeS_2
	Galena	PbS
	Gypsum	$CaSO_4 \cdot 2H_2O$
Halides	Halite	NaCl
	Fluorite	CaF_2
Native elements	Copper	Cu
	Sulfur	S

Table 3.2: Major elements in the earth's crust

Element	Symbol	%
Oxygen	O	46.6
Silicon	Si	27.7
Aluminum	Al	8.1
Iron	Fe	5.0
Calcium	Ca	3.6
Sodium	Na	2.8
Potassium	K	2.6
Magnesium	Mg	2.1
Titanium	Ti	0.4

igneous rocks are granite, basalt, quartz (SiO_2), feldspar ($CaNaKAlSi_3O_8$), and magnetite (Fe_3O_4). Igneous rocks are formed under chemically-reducing conditions of high temperature and high pressure. Exposed igneous rocks are not in chemical equilibrium with their surroundings and disintegrate by *weathering*. The stage of weathering to which a mineral is exposed depends upon time, chemical conditions, including exposure to air, carbon dioxide, and water, as well as physical conditions, such as temperature and mixing with water and air. Weathering is generally a slow process because igneous rocks are often hard, nonporous, and of low reactivity. Erosion from wind, water, or glaciers picks up materials from weathering rocks and converts it to sedimentary rock and soil, which in contrast to the parent igneous rocks are porous, soft, and chemically reactive. Heat and pressure convert sedimentary rock to metamorphic rock.

Sedimentary rocks may be detrital rocks consisting of solid particles eroded from igneous rocks as a consequence of weathering; quartz is the most likely to remain chemically intact through weathering and transport from its original location. A second kind of sedimentary rock consists of chemical sedimentary rocks produced by the precipitation or coagulation of dissolved weathering products. When sedimentary rocks contain organic material (such as the residues of plant and animal remains/decay) they are usually designated *organic sedimentary rocks*. Petroleum may be found in such sediments.

Carbonate minerals of calcium and magnesium (limestone, $CaCO_3$, or dolomite, $CaCO_3.MgCO_3$) are especially abundant in sedimentary rocks. Reactive and soluble minerals such as carbonates, gypsum, olivine, feldspars, and iron-rich substances can survive only early weathering. This stage is characterized by dry conditions, low leaching, absence of organic matter, reducing conditions, and limited time of exposure. Quartz, vermiculite, and smectites can survive the intermediate stage of weathering manifested by retention of silica, sodium, potassium, magnesium, calcium, and iron (Fe^{2+}). These substances are mobilized in advanced-stage weathering, other characteristics being: intense leaching by fresh water, low pH, aluminum hydroxy oxides, and silica.

3.0 Sedimentary Strata

Sedimentary strata (*sediments*) typically consist of mixtures of clay, silt, sand, organic matter, and various minerals. They may vary in composition from pure mineral matter to predominantly organic matter (Siever, 1987).

Vast areas of land, as well as lake and stream sediments, are formed from sedimentary rocks. The properties of these masses of material depend strongly upon their origins and transport. Water is the main vehicle of sediment transport, although wind can also be significant.

Physical, chemical, and biological processes may all result in the deposition of sediments in the bottom regions of bodies of water. Sedimentary material may be washed into a body of water by erosion or through sloughing (caving in) of the shore. Thus clay, sand, organic matter, and other materials may be washed into a lake and settle out as layers of sediment.

Many minerals, organic pollutants, proteinaceous materials, some algae, and bacteria exist in the form of very small particles suspended in water. These are classified as colloidal particles if they have characteristics of both species in solution and larger particles in suspension, range in diameter from about 0.001 micron (1 micron = 1 meter x 10^{-6}) to about 1 micron, and scatter white light as a light blue hue observed at right angles to the incident light.

The action of flowing water in streams cuts away stream banks and carries sedimentary materials for great distances. These sedimentary materials may be (1) dissolved material from sediment-forming minerals in solution; (2) suspended material from solid sedimentary materials carried along in suspension; and (3) bed material dragged along the bottom of the stream channel.

Streams mobilize sedimentary materials through erosion, then transport materials along with stream flow, and release them in a solid form during deposition. Deposits of stream-borne sediments are called *alluvium*. As conditions such as lowered stream velocity begin to favor deposition, larger particles are released first. This results in sorting such that particles of a similar size and type tend to occur together in alluvium deposits. Much sediment is deposited in flood plains where streams overflow their banks. The Nile is an example of a river that is capable of mobilizing material every year during the spring flood. The Pharaohs of Biblical times looked to the Nile to replenish the growing areas of Egypt with the organic rich silt that was deposited every spring.

Bottom sediments are important sources of inorganic and organic matter in streams, freshwater impoundments, estuaries, and oceans. It is incorrect to consider bottom sediments simply as wet soil. Unlike soil, which is in contact with the atmosphere and is aerobic, the environment around a bottom sediment is anaerobic, and the sediment is subject to reducing conditions. Bottom sediments undergo continuous leaching, whereas soil does not.

Sediments and suspended particles are important repositories for trace amounts of metals such as chromium, cadmium, copper, molybdenum, nickel, cobalt, and manganese. These metals may exist as discrete compounds: ions held by cation-exchanging clays, bound to hydrated oxides of iron or manganese, or chelated by insoluble humic substances.

4.0 Clays

Clays are silicate minerals, usually containing aluminum, that constitute one of the most significant classes of secondary minerals (Ebert, 1987). Clays are defined by size, as are other inorganic constituents of the earth (Table 3.3). Secondary minerals are formed by the alteration of parent mineral matter. Olivine, augite, hornblende, and feldspars all form clays.

Clay materials constitute the most important class of common minerals occurring as colloidal matter in water. They consist largely of hydrated aluminum and silicon oxides (Figure 3.1) and are secondary minerals, which are formed by weathering and other processes acting on primary rocks. The most abundant clay minerals are illite, montmorillonite, chlorite, and kaolinite. These clay minerals are distinguished from each other by fairly complex chemical formulae, structure, and chemical and physical properties. The three major groups of clay minerals are montmorillonite [$AlMgNa(OH)_2Si_4O_{10}$], illite [$Al_3KSi_8O_{10}(OH)_4$], and kaolinite [$Al_2Si_2O_5(OH)_4$]. Many clays contain large amounts of sodium, potassium, magnesium, calcium, and iron, as well as trace quantities of other metals.

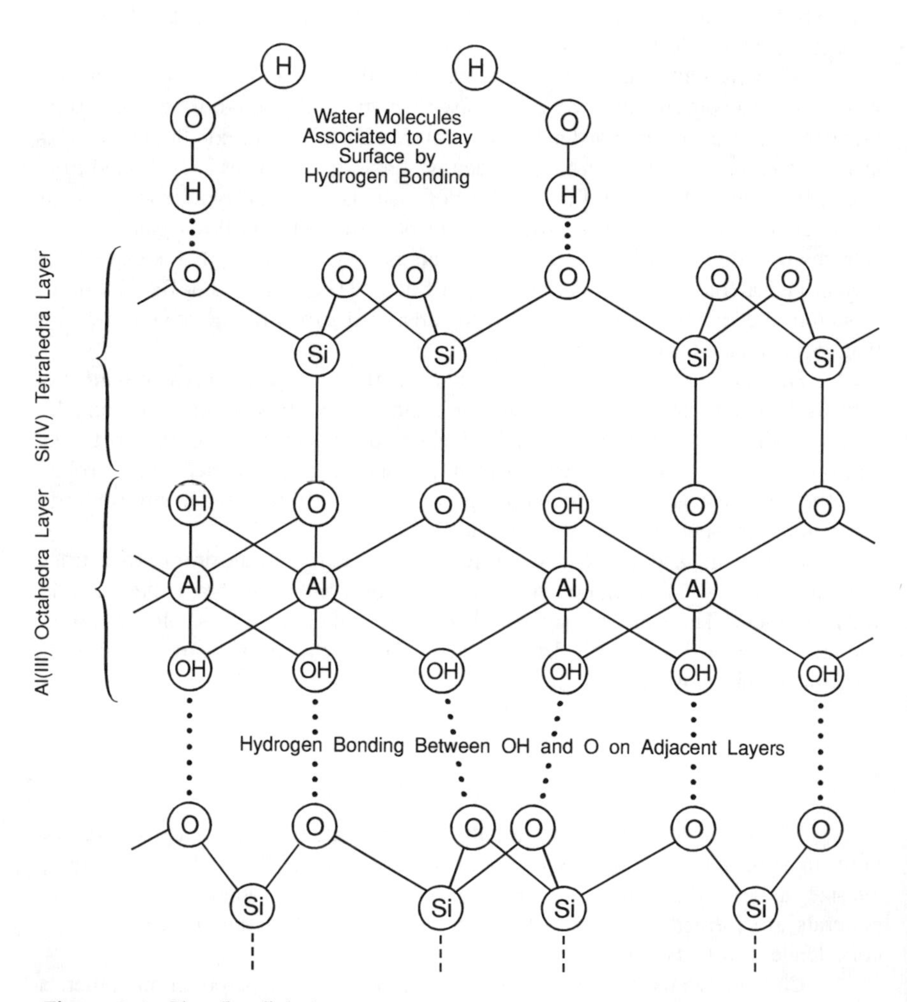

Figure 3.1: Clay (kaolinite) structure

Table 3.3: Approximate sizes of various rock types

Name	Diameter, mm
Boulder	Larger than 256
Cobble	64 to 256
Pebble (gravel)	2 to 64
Sand	1/16 to 2
Silt	1/256 to 1/16
Clay	Size less than 1/256

Physically, clays consist of very fine grains having layered structures of sheets of silicon oxide alternating with sheets of aluminum oxide. Two or three sheets make up unit layers. Some clays, particularly montmorillonite, may absorb large quantities of water between the unit layers and cause swelling of the clay. The silicon oxide sheets are made up of tetrahedra in which each silicon atom is surrounded by four oxygen atoms. Of the four oxygen atoms in each tetrahedron, three are shared with other silicon atoms that are components of other tetrahedra. This sheet is called the tetrahedral sheet. The aluminum oxide is contained in an octahedral sheet, so named because each aluminum atom is surrounded by six oxygen atoms in an octahedral configuration. The structure is such that some of the oxygen atoms are shared between aluminum atoms and some are shared with the tetrahedral sheet (Figure 3.1). Structurally, clays may be classified as either two-layer clays in which oxygen atoms are shared between a tetrahedral sheet and an adjacent octahedral sheet, or three-layer clays, in which an octahedral sheet shares oxygen atoms with tetrahedral sheets on either side.

Because of their structure and high surface area per unit weight, clays have a strong tendency to adsorb chemical species from water. As a result of their physical and chemical reactivity with organic and inorganic substrates (Theng, 1974; Anderson and Rubin, 1981), clays play a role in the transport and reactions of biological wastes, organic chemicals, gases, and other pollutant species in water. However, clay minerals also may effectively immobilize dissolved chemicals in water and so exert a purifying action. Some microbial processes occur at clay particle surfaces, and in some cases, adsorption of organic compounds by clay inhibits biodegradation. Thus clay may play a role in the microbial degradation, or non-degradation, of organic wastes.

Clays bind cations such as calcium, magnesium, and ammonium (NH_4^+), which protects the cations from leaching by water but keeps them available in soil as plant nutrients. Since many clays are readily suspended in water as colloidal particles, they may be leached from soil or carried to lower soil layers.

5.0 Soil

The most important part of any land system is the soil insofar as it is the medium upon which virtually all terrestrial organisms depend for their existence. The productivity of soil is strongly affected by environmental conditions and pollutants (Smith, 1987).

Soil is the receptor of chemical waste from landfill leachate, lagoons, and other sources (Borchardt, 1995). In some cases, land farming of degradable organic wastes is practiced as a means of disposal and degradation. The degradable material is worked into the soil, and soil microbial processes bring about its degradation, as for example by the application of sewage and fertilizer-rich sewage sludge to soil.

Soil covers most of the land surfaces as a continuum and each soil grades into the rock materials below and into other soils at its margins, where changes occur in relief, groundwater, vegetation, rock types, or other factors that influence the development of soils.

Soil is a finely divided rock-derived material containing organic matter and capable of supporting vegetation. Soils are independent natural bodies, each with a unique morphology resulting from a particular combination of climate, living plants and animals, parent rock materials, relief, the groundwater, and age.

Soil is a variable mixture of minerals, organic matter, and water (Figure 3.2) and is the final product of the weathering action of physical, chemical, and biological processes on rocks. The organic portion of soil consists of plant biomass in various stages of decay. High populations of bacteria, fungi, and animals such as earthworms may be found in soil.

Approximately 35% of the volume of a typical soil is composed of air-filled pores. As a point of reference, the atmosphere at sea level contains 21% volume per volume (v/v) oxygen and <0.1% v/v carbon dioxide. These amounts may be quite different in soil air because of the decay of organic matter, which produces carbon dioxide by consumption of oxygen. As a result, the oxygen content of air in soil may be as low as 15% v/v and the carbon dioxide content may be several percent.

Soils are structured in terms of *horizons* (Figure 3.3) or layers, more or less parallel to the surface and differing from those above and below in one or more properties, such as color, texture, structure, consistency, porosity, and reaction. The horizons may be thick or thin. They may be prominent or so weak that they can be detected only in the laboratory. The succession of horizons is called the *soil profile*. In general, the boundary of soils with the underlying rock or rock material occurs at depths ranging from 1 to 6 feet (0.3 to 1.8 meters), though the extremes lie outside this range. Rainwater percolating through soil carries dissolved and colloidal solids to lower horizons where they are deposited.

The top layer of soil, typically several inches in thickness, is known as the *A horizon*, or *topsoil*. This is the layer of maximum biological activity in the soil and contains most of the soil organic matter. Metal ions, which can originate from biological systems (Sigel and Sigel, 1995) as well as from waste materials, and clay particles in the A horizon are subject to considerable leaching.

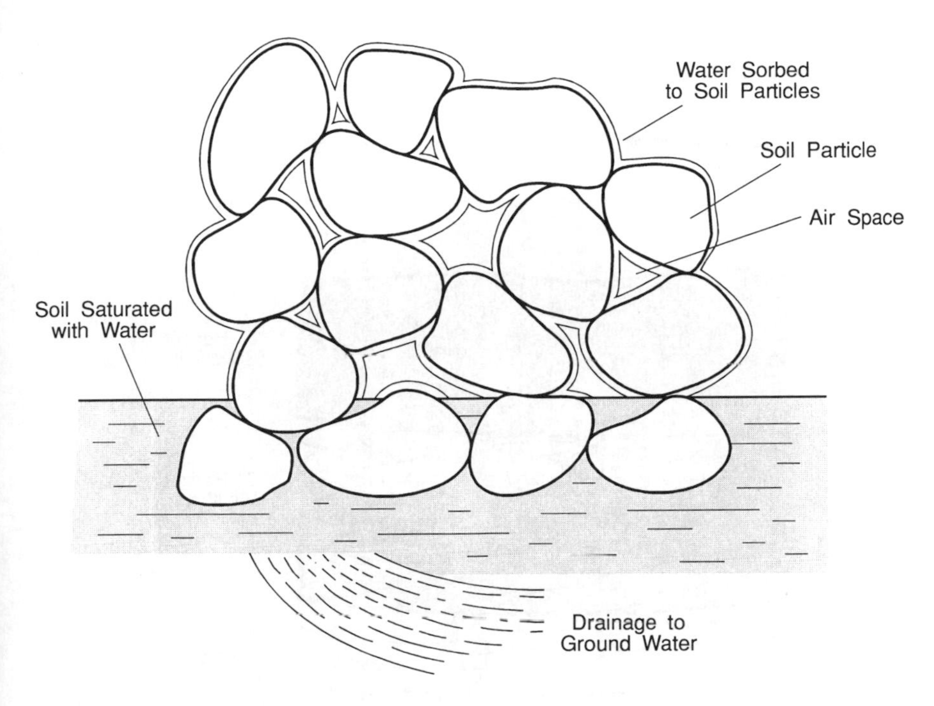

Figure 3.2: Generalized structure of soil

The next layer is the *B horizon*, or *subsoil*. It receives material such as organic matter, salts, and clay particles leached from the topsoil. The *C horizon* is composed of weathered parent rocks from which the soil originated.

One of the more important classes of productive soils is the *podzol* type, which are soils formed in temperate-to-cold climates under relatively high rainfall conditions under coniferous or mixed forest or heath vegetation. These generally rich soils tend to be acidic (pH 3.5-4.5) such that alkali and alkaline earth metals and, to a lesser extent, aluminum and iron, are leached from the A horizon, leaving kaolinite as the predominant clay mineral. At somewhat higher pH in the B horizon, hydrated iron oxides and clays are redeposited.

The mechanical properties of soil are largely determined by particle size (Smith, 1987), the four major categories of which are (1) gravel (2-64 mm), (2) sand (0.06-2 mm), (3) silt (0.06-0.006 mm), and (4) clay (<0.002 mm). Clays represent a size fraction rather than a specific class of mineral matter.

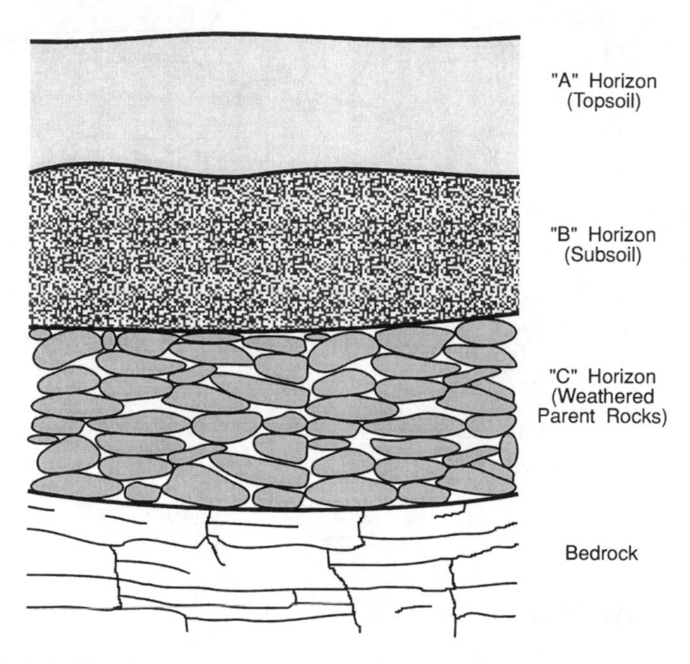

Figure 3.3: Soil horizons

Cation exchange in soil is the mechanism by which potassium, calcium, magnesium, and essential trace-level metals are made available to plants. When nutrient metal ions are taken up by plant roots, a hydrogen ion is exchanged for the metal ions. This process, and the leaching of calcium, magnesium, and other metal ions from the soil by water containing carbonic acid, tends to make the soil acidic.

In addition to being the site of most food production, soil is the receptor of large quantities of pollutants (Borchardt, 1995), such as particulate matter from power plant smokestacks. Fertilizers and some other materials applied to soil (Chapter 6) often contribute to water and air pollution (Bourke et al., 1992). Therefore, soil is a key component of environmental chemical cycles; soils are open systems that undergo continual exchange of matter and energy with the atmosphere, hydrosphere, and biosphere.

5.1 Inorganic Matter

The weathering of parent rocks and minerals to form the inorganic soil components results ultimately in the formation of inorganic colloids. These colloids are repositories of water and plant nutrients, which may be made available to plants as needed. Inorganic soil colloids often absorb toxic substances in soil, thus playing a role in detoxification of substances that otherwise would harm plants. The abundance and nature of inorganic colloidal material in soil are obviously important factors in determining soil productivity.

The uptake of plant nutrients by roots involves interactions with the water and inorganic phases. For example, a nutrient held by an inorganic colloid material has to traverse the mineral/water interface and then the water/root interface. This process is often strongly influenced by the ionic structure of soil inorganic matter.

The most common elements in the earth's crust are oxygen, silicon, aluminum, iron, calcium, sodium, potassium, and magnesium. Therefore, minerals composed of these elements (particularly silicon and oxygen) constitute most of the mineral fraction of the soil. Common soil mineral constituents are finely divided quartz (SiO_2), orthoclase ($KAlSi_3O_8$), albite ($NaAl.Si_3O_8$), epidote [$4CaO3(AlFe)_2O_36SiO_2H_2O$], goethite [$FeO(OH)$], magnetite ($Fe_3O_4$), calcium and magnesium carbonates ($CaCO_3$; $CaCO_3.MgCO_3$), and oxides of manganese (MnO) and titanium (TiO).

5.2 Organic Matter

The solid fraction of typical productive soil is ~5% w/w organic matter and ~95% w/w inorganic matter. Some soils, such as peat, may contain as much as 95 % organic material, whereas other soils contain as little as 1%.

Though typically comprising less than 5% w/w, organic matter in soil (Table 3.4) largely determines soil productivity. It serves as a source of food for microorganisms, undergoes chemical reactions such as ion exchange, and influences the physical properties of soil. Some organic compounds even contribute to the weathering of mineral matter, the process by which soil is formed.

The accumulation of organic matter in soil is strongly influenced by temperature and by the availability of oxygen. Since the rate of biodegradation decreases with decreasing temperature, organic matter does not degrade rapidly in colder climates and tends to build up in soil. In water and in waterlogged soils, decaying vegetation does not have easy access to oxygen, and organic matter accumulates. The organic content may reach 90% w/w in areas where plants grow and decay in soil saturated with water. Of the organic components, humus is by far the most significant.

Humus is a generic term for the water-insoluble material that makes up the bulk of soil organic matter (Stevenson, 1994). Humus is composed of a base-soluble fraction (*humic and fulvic acids*) and an insoluble fraction (*humin*) and is the residue from the biodegradation (by bacteria and fungi) of plant material. The bulk of plant biomass consists of relatively degradable cellulose and degradation-resistant *lignin*,

Table 3.4: Major types of organic compounds found in soil

Compound type	Composition	Significance
Humus	Degradation-resistant residue from plant decay, largely C/H/O	Improves physical properties and is reservoir of fixed nitrogen
Fats, resins, and waxes	Extractable by organic solvents	May adversely affect soil physical properties by repelling water, perhaps phytotoxic
Saccharides	Cellulose, starches, gums	Major food source for soil microorganisms, soil stabilizers
N-containing organics	Nitrogen bound to humus, amino acids, amino sugars,	Provides nitrogen for soil fertility
Phosphorous compounds	Phosphate esters, inositol phosphates (phytic acid) phospholipids	Sources of plant phosphate

a complex polymeric substance that is second only to carbohydrates in natural abundance (Sarkanen and Ludwig, 1971). Lignin is a precursor of soil humus.

An increase in the nitrogen/carbon ratio is a significant feature of the transformation of plant biomass to humus. During the humification process, microorganisms convert organic carbon to carbon dioxide to obtain energy. Simultaneously, the bacterial action incorporates bound nitrogen with the compounds produced by the decay processes. The result is a nitrogen/carbon ratio of ~ 1/10 upon completion of humification. As a general rule, therefore, humus is relatively rich in organically bound nitrogen.

Humic materials in soil strongly adsorb many solutes in soil water and have a particular affinity for heavy polyvalent cations. Soil humic substances may contain levels of uranium more than 10,000 times that of the water with which they are in equilibrium. Thus, water becomes depleted of its cations (or purified) in passing through humic-rich soils. Humic substances in soils also have a strong affinity for organic compounds with low water solubility such as the types of herbicides widely used to kill weeds.

In some cases, there is a strong interaction between the organic and inorganic portions of soil. This is especially true of the strong complexes formed between clays and humic (fulvic) acid compounds. In many soils, 50-100% of the soil carbon is in the form of complexes with clay. These complexes play a role in determining the physical properties of soil, soil fertility, and stabilization of soil organic matter.

One of the mechanisms for the chemical binding between clay colloidal particles and humic organic particles is probably of the flocculation type, in which anionic organic molecules with carboxylic amid functional groups serve as bridges in combination with cations to bind clay colloidal particles together as a floc.

The synthesis, chemical reactions, and biodegradation of humic materials are affected by interaction with clays. The lower-molecular-weight fulvic acids may be bound in the spaces in layers in clay particles.

5.3 Nutrients

One of the most important functions of soil in supporting plant growth is to provide essential plant *nutrients* (Table 3.5). *Macronutrients* are those elements that occur in standard levels in plant materials or in fluids in the plant. *Micronutrients* are elements that are essential only at very low levels and generally are required for the functioning of essential enzymes.

The elements generally recognized as essential macronutrients for plants are carbon, hydrogen, oxygen, nitrogen, phosphorus, potassium, calcium, magnesium, and sulfur. Carbon, hydrogen, and oxygen are obtained from the atmosphere. The other essential macronutrients must be obtained from soil. Of these, nitrogen, phosphorus, and potassium are the most likely to be lacking and are commonly added to soil as fertilizers.

Table 3.5: Plant nutrients

Nutrient	Source
Macronutrients	
Carbon (CO_2)	Atmosphere, decay
Hydrogen	Water
Oxygen	Water
Nitrogen (NO_3^-)	Decay, atmosphere (from nitrogen-fixing organisms), pollutants
Phosphorus (phosphate)	Decay, minerals, pollutants
Potassium	Minerals, pollutants
Sulfur (sulfate)	Minerals
Magnesium	Minerals
Calcium	Minerals
Micronutrients	
B, Cl, Co, Cu, Fe	Minerals, pollutants

Calcium-deficient soils are relatively uncommon. Liming, a process used to treat acid soils by the addition of lime (CaO), provides a more than adequate calcium supply for plants. However, calcium uptake by plants and leaching by carbonic acid may produce a calcium deficiency in soil. Acid soils may still contain an appreciable level of calcium that, because of competition by hydrogen ion, is not available to plants. Treatment of acid soil to restore the pH to near neutrality generally remedies the calcium deficiency. In alkaline soils, the presence of high levels of sodium, magnesium, and potassium sometimes produces calcium deficiency because these ions compete with calcium for availability to plants.

Although magnesium makes up ~2 % of the land system, most of it is in the form of mineral chemicals. Generally, exchangeable magnesium is considered available to plants and is held by ion-exchanging organic matter or clays. The availability of magnesium to plants depends upon the calcium/magnesium ratio. If this ratio is too high, magnesium may not be available to plants, and magnesium deficiency results. Similarly, excessive levels of potassium or sodium may cause magnesium deficiency.

Sulfur is assimilated by plants as the sulfate ion (SO_4^{2-}). In addition, in areas where the atmosphere is contaminated with sulfur dioxide, sulfur may be absorbed as sulfur dioxide by plant leaves. Atmospheric sulfur dioxide levels have been high enough to kill vegetation in some areas. However, in soils with an unexpected sulfur deficiency some experiments designed to show sulfur dioxide toxicity to plants have resulted in increased plant growth.

Soils deficient in sulfur do not support plant growth very well, largely because sulfur is a component of some essential amino acids and of thiamin and biotin. Sulfate ions are generally present in the soil as immobilized insoluble sulfate minerals or as soluble salts, which are readily leached from the soil and lost as soil water runoff. Unlike the case of nutrient cations such as potassium ions, little sulfate is adsorbed to the soil (that is, bound by ion exchange binding) where it is resistant to leaching while still available for assimilation by plant roots. Soil sulfur deficiencies have been found in a number of regions of the world. Whereas most fertilizers at one time contained sulfur, its use in commercial fertilizers is declining. If this trend continues, it is possible that sulfur will become a limiting nutrient in more cases.

The reaction of pyrite (FeS_2) with acid in acid-sulfate soils may release hydrogen sulfide, which is very toxic to plants and which kills many beneficial microorganisms. Toxic hydrogen sulfide can also be produced by reduction of sulfate ion through microorganism-mediated reactions with organic matter. Production of hydrogen sulfide in flooded soils may be inhibited by treatment with oxidizing compounds, one of the most effective of which is potassium nitrate (KNO_3).

Nitrogen, phosphorus, and potassium are plant nutrients obtained from soil that are so important for crop productivity that they are commonly added to soil as fertilizers. In most soils, more than 90% of the nitrogen is organic and participates in the nitrogen cycle (Chapter 1). This organic nitrogen is primarily the product of the biodegradation of dead plants and animals. It is eventually hydrolyzed to ammonia, which can be oxidized to nitrate by the action of bacteria in the soil.

Nitrogen bound to soil humus is especially important in maintaining soil fertility. Unlike potassium or phosphate, nitrogen is not a significant product of mineral weathering. Nitrogen-fixing organisms ordinarily cannot supply sufficient nitrogen to meet peak demand. Inorganic nitrogen from fertilizers and rainwater is often largely lost by leaching. Soil humus, however, serves as a reservoir of nitrogen required by plants.

Nitrogen is an essential component of proteins and other constituents of living matter. Plants and cereals grown on nitrogen-rich soils not only provide higher yields, but are often substantially richer in protein and therefore more nutritious. Nitrogen is most generally available to plants as nitrate ion. Some plants such as rice may utilize ammonium nitrogen; however, other plants are poisoned by this form of nitrogen. When nitrogen is applied to soils in the ammonium forms, nitrifying bacteria perform an essential function in converting it to available nitrate ions.

Nitrogen fixation is the process by which atmospheric nitrogen is converted to nitrogen compounds available to plants, i.e.,

$$N_2 \rightarrow NO_3^-$$

Human activities are resulting in the fixation of a great deal more nitrogen than would otherwise be the case. Artificial sources now account for 30-40% of all nitrogen fixed and these include (1) chemical fertilizer manufacture, (2) combustion of nitrogen-containing fuels, and (3) the increased cultivation of nitrogen-fixing legumes. An issue that can arise as a result of the increased fixation of nitrogen is the possible effect upon the atmospheric ozone layer by nitrogen dioxide released during denitrification of fixed nitrogen.

Although the percentage of phosphorus in plant materials is relatively low, it is essential to plants. Phosphorus, like nitrogen, must be present in a simple inorganic form, namely orthophosphate, before it can be taken up by plants. Orthophosphate is most available to plants at pH values near neutrality. It is believed that in relatively acidic soils, orthophosphate ions are precipitated or adsorbed by species of aluminum (Al^{3+}) and iron (Fe^{3+}). In general, because of these reactions, little phosphorus applied as fertilizer leaches from the soil. This is important from the standpoint of both water pollution and utilization of phosphate fertilizers.

Potassium is one of the most abundant elements in the earth's crust, of which it makes up ~3%. However, much of this potassium is not easily available to plants. For example, some silicate minerals contain strongly bound potassium. Exchangeable potassium held by clay minerals is relatively more available to plants.

Relatively high levels of potassium are utilized by growing plants. Potassium activates some enzymes and plays a role in the water balance in plants. It is also essential for some carbohydrate transformations. Crop yields are generally greatly reduced in potassium-deficient soils. The higher the productivity of the crop, the more potassium is removed from soil. When nitrogen fertilizers are added to soils to increase productivity, removal of potassium is enhanced. Therefore, potassium may become a limiting nutrient in soils heavily fertilized with other nutrients.

Boron, chlorine, copper, iron, manganese, molybdenum (for N-fixation), sodium, vanadium, and zinc are considered essential plant micronutrients. These elements are needed by plants only at very low levels and frequently are toxic at higher levels. Most of these elements function as components of essential enzymes. Manganese, iron, chlorine, zinc, and vanadium may be involved in photosynthesis.

Iron and manganese occur in a number of soil minerals. Sodium and chlorine (as chloride) occur naturally in soil and are transported as atmospheric particulate matter from marine sprays. Some of the other micronutrients and trace elements are found in primary (unweathered) minerals that occur in soil. Boron is substituted for silicon in mica and is present in tourmaline, a mineral. Copper is present in feldspar, amphibole, olivine, pyroxene, and mica; it also occurs as trace levels of copper sulfides in silicate minerals. Molybdenum occurs as molybdenite (MoS_2). Vanadium is substituted for iron or aluminum in oxides, pyroxenes, amphiboles, and mica. Zinc is present as the result of isomorphic substitution for magnesium, iron, and manganese in oxides, amphiboles, olivines, and pyroxenes and as traces of zinc sulfide in silicates. Other trace-level elements that occur as specific minerals, sulfide inclusions, or by isomorphic substitution for other elements in minerals are chromium, cobalt, arsenic, selenium, nickel, lead, and cadmium.

Crop fertilizers contain nitrogen, phosphorus, and potassium as major components. Magnesium, sulfate, and micronutrients may also be added. Fertilizers designate numbers, such as 6-12-8, to show the respective percentages of the nutrients expressed as nitrogen (6%), phosphorus (as P_2O_5, 12%), and potassium (as K_2O, 8%).

5.4 Water

Although not usually recognized as a nutrient, water (Chapter 4) is very important for the condition of the soil and the maturation of the flora and fauna (Kramer and Boyer, 1995). Therefore, in the present context, water can also be classed as a nutrient.

Large quantities of water are required for the production of most plant materials. Water is part of the three-phase, solid-liquid-gas system that makes up soil. It is the basic transport medium for carrying essential plant nutrients from solid soil particles into plant roots and to the farthest reaches of the plant's leaf structure. Water enters the atmosphere from the plant's leaves, a process called transpiration.

Normally, because of the small size of soil particles and the presence of small capillaries and pores in the soil, the water phase is not totally independent of soil solid matter. The availability of water to plants is governed by gradients arising from capillary and gravitational forces. The availability of nutrient solutes in water depends upon concentration gradients and electrical potential gradients. Water present in larger spaces in soil is relatively more available to plants and readily drains away. Water held in smaller pores, or between the unit layers of clay particles is held much more strongly. Soils high in organic matter may hold appreciably more water than other soils, but it is relatively less available to plants because of physical and chemical sorption of the water by the organic matter.

There is a very strong interaction between clays and water in soil. Water is absorbed on the surfaces of clay particles. Because of the high surface/volume ratio of colloidal clay particles, a great deal of water may be bound in this manner. Water is also held between the unit layers of expanding clays, such the montmorillonite clays.

As soil becomes waterlogged (water-saturated), it undergoes drastic changes in physical, chemical, and biological properties. Oxygen in such soil is rapidly used up by the respiration of microorganisms that degrade soil organic matter. In such soils, the bonds holding soil colloidal particles together are broken, which causes disruption of soil structure. Thus the excess water in such soils is detrimental to plant growth, and the soil does not contain the air required by most plant roots. Most useful crops, with the notable exception of rice, cannot grow on waterlogged soils.

The decay of organic matter in soil increases the equilibrium level of dissolved carbon dioxide in groundwater. This lowers the pH and contributes to weathering of carbonate minerals, particularly calcium carbonate. Carbon dioxide also shifts the equilibrium of the process by which roots absorb metal ions from soil.

5.0 References

Anderson, M.A., and A.J. Rubin (eds.). 1981. *Adsorption of Inorganics at Solid-Liquid Interfaces*. Ann Arbor Science Publishers, Ann Arbor, Michigan.

Borchardt, J.K. 1995. *Today's Chemist at Work* 4(3): 47.

Bourke, J.B., A.S. Felsot, A.S., Gilding, T.J., Jensen, J.K., and Seiber, J.N. (eds.). 1992. *Pesticide Waste Management*. Symposium Series No. 510. American Chemical Society, Washington, D.C.

Choucri, N. 1991. In *Energy and the Environment in the 21st Century*. p. 91. Edited by J.W. Tester, D.O. Wood, and N.A. Ferrari. MIT Press, Cambridge, Massachusetts.

Ebert, D.D. 1987. In *Encyclopedia of Science and Technology*. 6th ed. Vol. 16, p. 658. Edited by S.P. Parker. McGraw-Hill, New York.

Fahrenwald, J. 1934. *Transactions. American Institute of Mechanical Engineers* 112: 88.

Kramer, P.J., and Boyer, J.S. 1995. *Water Relations of Plants and Soils*. Academic Press, San Diego, California.

Perry, R.H., and Chilton, C.H. (eds.). 1973. *Chemical Engineers Handbook*. Pp. 8-9. McGraw-Hill, New York.

Ramsdell, L.S. 1987. In *Encyclopedia of Science and Technology*. 6th ed. Vol. 11, p. 214. Edited by S.P. Parker. McGraw-Hill, New York.

Sarkanen, K.V., and Ludwig, C.H. (eds.). 1971. *Lignins: Occurrence, Formation, Structure and Reactions*. John Wiley & Sons Inc., New York.

Siever, R. 1987. In *Encyclopedia of Science and Technology*. 6th ed. Vol. 16, p. 198. Edited by S.P. Parker. McGraw-Hill, New York.

Sigel, H., and Sigel, A. (eds.). 1995. *Metal Ions in Biological Systems*. Marcel Dekker Inc., New York.

Smith, G.D. 1987. In *Encyclopedia of Science and Technology*. 6th ed. Vol. 16, p. 507. Edited by S.P. Parker. McGraw-Hill, New York.

Stevenson, F.J. 1994. *Humus Chemistry*. John Wiley & Sons Inc., New York.

Theng, B.K.G. 1974. *The Chemistry of Clay-Organic Reactions*. John Wiley & Sons Inc., New York.

Chapter 4

Water Systems

1.0 Introduction

Water is a complex system of chemical species that has been studied considerably (Eglinton, 1975; Jenne, 1979; Melchior and Bassett, 1990). Indeed, the public has been introduced to the issues of water pollution for the past several years, not the least of which is the pollution of Boston Harbor, which played a significant role in discussions and debates during presidential elections within recent memory (Easterbrook, 1995).

Water is essential to all living organisms. Precipitation and evaporation rates, humidity, and available soil moisture are factors governing water availability for various life forms. Precipitation varies in relation to the position and movement of air masses and weather systems, location relative to mountain ranges (rain shadow effect), and altitude. Seasonal distribution of rainfall is as important as the total amount; rainfall evenly distributed throughout the year usually results in greater availability.

The water that humans use is primarily fresh surface water and groundwater. In arid regions a small fraction of the water supply comes from the ocean, a source that is likely to become more important as the world's supply of fresh water dwindles relative to demand. Saline or brackish groundwater may also be utilized in some areas.

Water pollution refers to any change in natural waters that may impair further use of the waters, caused by the introduction of organic or inorganic substances (Table 4.1) or a change in temperature of the water (see American Society for Testing and Materials, 1995). Obviously, humans have also been polluting water since the early days of civilization. The development of towns and cities in close proximity to rivers caused the rivers to become polluted by human waste and effluents. Indeed, whole civilizations have disappeared not only because of water shortages resulting from changes in the climate but also because of waterborne diseases, such as cholera and typhoid (Cartwright and Biddiss, 1972). In fact, one can wonder whether or not the castle moat actually protected the inmates from the attackers or whether the disease-ridden moat was fatal to both sides!

Serious epidemics of waterborne diseases such as cholera, dysentery, and typhoid fever were caused by underground seepage from privy vaults into town wells (James and Thorpe, 1994). Such direct bacterial infections through water systems can be traced back for several centuries, even though the germ or bacterium as the cause of disease was not proved for nearly another century.

The indirect reuse of wastewater that has been discharged into rivers is common around the world, and it is acceptable as long as the discharges are treated

Table 4.1: Sources and types of chemical waste

Industry	Category of pollutant	Characteristics	Type of Treatment
Metal finishing, plating, rayon processing, steel mills, tanneries	Dissolved metals	Generally cations of Al, Cr, Cu, Fe, or Zn in low-pH solution	Precipitate with lime, followed by sedimentation
Tanneries, plating, metal finishing		Chromates	Reduce with ferrous iron sulfate or sulfur dioxide; then precipitate with lime, followed by sedimentation
Plating, foundry		Cyanide (generally with metal complexes)	Oxidize with chlorine or hypochlorite; lime treatment for precipitating metals
Mining, phosphate, steel milling, power plants (fly ash), beet-sugar processing pulp, foundry	Toxic materials	Dense, rapid settling	Plain sedimentation
Pulp and paper, textile, petroleum, food plants, steel mills, mining, chemical plants	Suspended solids	Colloidal	Chemical coagulation followed by sedimentation or flotation
Petroleum and petro-chemical, laundry, meat packing, machining (cutting oil), car washing, dairies, food plants		Oily material or light weight solids	Flotation (with chemical treatment if necessary)
Beet- and cane-sugar plants dairies, meat packing, pulp and paper, canning, chemical plants, brewing, petroleum and petrochemical, tanneries	Organic matter	Vary with industry; some are easily oxidized biologically; others require special techniques	Trickling filter; activated sludge; conventional, high-rate, contact stabiliza-tion, aerobic digestion
Chemical and Petrochemical		Very strong organic wastes	Anaerobic treatment followed by aerobic treatment

and contamination from chemical waste is avoided (Lacy, 1983). Less acceptable is the direct reuse of wastewater for potable use, even after a high level of treatment. Of most concern here is the possibility of an outbreak of disease.

Water pollutants can be divided among some general classifications (Table 4.2), and water pollution control is closely allied with the water supplies of communities and industries because both generally share the same water resources (Noyes, 1991). There is great similarity in the pipe systems that bring water to each home or business property and in the systems of sewers or drains that subsequently collect the wastewater and conduct it to a treatment facility. Treatment should prepare the flow for return to the environment, so that the receiving watercourse will be suitable for beneficial uses such as general recreation, and safe for subsequent use by downstream communities or industries. In considering water pollution, it is useful to keep in mind an overall picture of possible pollutant cycles (Chapter 1), which involve the major routes of pollutant interchange among the biotic, terrestrial, atmospheric, and aquatic environments.

Problems with water quantity and quality continue and in some respects are becoming more serious. These problems include increased water use due to population growth, contamination of drinking water by improperly discarded waste

Table 4.2: Water pollutants and their effects

Class of pollutant	Significance
Acidity, alkalinity, salinity (in excess)	Water quality, aquatic live
Algal nutrients	Eutrophication
Asbestos	Human health
Biochemical oxygen demand	Water quality, oxygen levels
Chemical carcinogens	Incidence of cancer
Detergents	Eutrophication,wildlife,esthetics
Inorganic pollutants	Toxicity, aquatic biota
Metal-organic combinations	Metal transport
Pathogens	Health effects
Pesticides	Toxicity, aquatic biota, wildlife
Petroleum wastes	Effect of wildlife, esthetics
Polychlorinated biphenyls	Possible biological effects
Radionuclides	Toxicity
Sediments	Water quality, aquatic biota,
Sewage	Water quality, oxygen levels
Taste, odor, and color	Esthetics
Trace elements	Health, aquatic biota
Trace organic pollutants	Toxicity
	wildlife

materials, and destruction of wildlife by water pollution. Water supply can be considered to be one part of a hydrologic cycle (Figure 4.1) in which a major portion of the water occurs in the oceans (Pickering and Owen, 1994). Water (as water vapor) is also present in the atmosphere and is contained as ice and snow in snowpack, glaciers, and the polar ice caps. Surface water is found in lakes, streams, and reservoirs. Groundwater is located in aquifers underground. However, for the present purposes, water supply is generally considered to occur in four accessible locations: groundwater, rivers, lakes, and oceans.

Groundwater is a vital resource that plays a crucial role in geochemical processes, such as the formation of secondary minerals. The nature, quality and mobility of groundwater are all strongly dependent upon the rock formations in which the water is held. Groundwater is the part of the hydrosphere most vulnerable to damage from chemical wastes. Although surface-water supplies are subject to contamination, groundwater can become almost irreversibly contaminated by the improper disposal of chemicals on land. Once there is penetration into aquatic systems, chemical species are subject to a number of chemical and biochemical processes. These include acid-base, oxidation-reduction, precipitation-dissolution, and hydrolysis reactions, as well as biodegradation.

Under many circumstances, biochemical processes largely determine the fates of chemical species in water. The most important such processes are those mediated by microorganisms. In particular, the oxidation of biodegradable organic wastes in water generally occurs by means of microorganism-mediated biochemical reactions. Bacteria produce organic acids and chelating agents, such as citrate, which have the effect of solubilizing heavy metal ions. Some mobile methylated forms, such as compounds of methylated arsenic and mercury, are produced by bacterial action.

Physically important characteristics of such formations are *porosity*, which determines the percentage of rock volume available to contain water, and *permeability*, which describes the ease of flow of the water through the rock. High permeability is usually associated with high porosity, however, clays tend to have low permeability even when a large percentage of the volume is filled with water.

Most groundwater originates from precipitation in the form of rain or snow. Water from precipitation that is not lost by evaporation, transpiration, or to steam runoff may infiltrate into the ground. Initial amounts of water from precipitation onto dry soil are held very tightly as a film on the surfaces and in the micropores of soil particles. At intermediate levels the soil particles are covered with films of water, but air is still present in larger voids in the soil. The region in which such water is held is called the unsaturated zone or zone of aeration, and the water present in it is *vadose water*.

At lower depths, in the presence of adequate amounts of water, all voids are filled to produce a zone of saturation, the upper level of which is the *water table*. The water table is crucial in explaining and predicting the flow of wells and springs and the levels of streams and lakes. It is also an important factor in determining the extent to which pollutant and hazardous chemicals are likely to be transported by water. The water table tends to follow the general contours of the surface topography and varies with differences in permeability and water infiltration. The water table is at surface

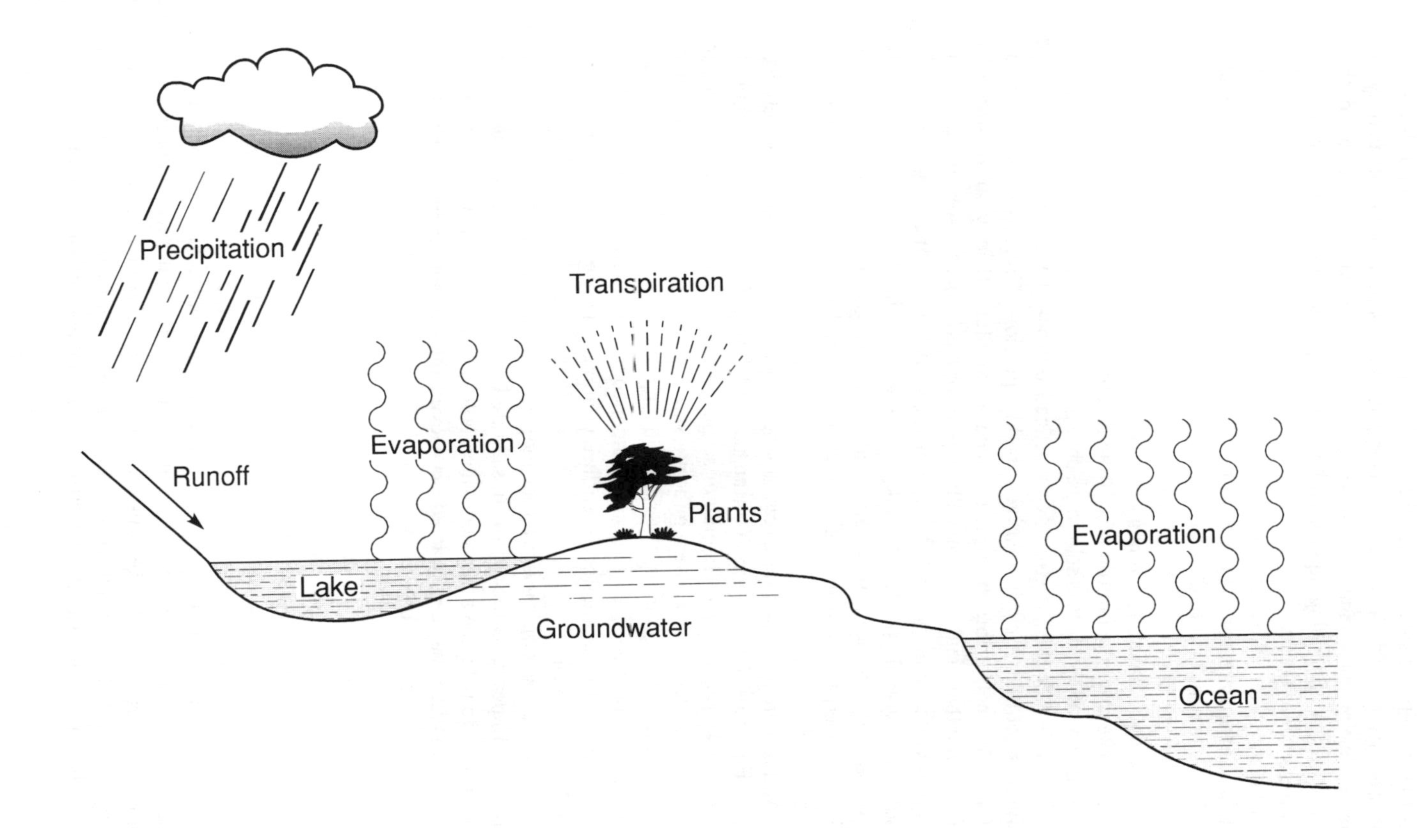

Figure 4.1: The water (hydrological) cycle

level in the vicinity of swamps and frequently above the surface where lakes and streams are encountered. The water level in such bodies may be maintained by the water table. In-flowing streams or reservoirs are located above the water table; they lose water to the underlying aquifer and cause an upward bulge in the water table beneath the surface water.

Groundwater flow is an important consideration in determining the accessibility of the water for use and transport of pollutants from underground waste sites. Various parts of a body of groundwater are in hydraulic contact so that a change in pressure at one point will tend to affect the pressure and level at another point. For example, infiltration from a heavy, localized rainfall may affect water level at a point remote from the infiltration. Groundwater flow occurs as the result of the natural tendency of the water table to assume even levels by the action of gravity.

Groundwater flow is strongly influenced by rock permeability. Porous or extensively fractured rock is relatively highly pervious. Because water can be extracted from such a formation, it is called an aquifer. Briefly, an *aquifer* is a subsurface zone that yields economically important amounts of water to wells. The term is synonymous with water-bearing formation. An aquifer may be porous rock, unconsolidated gravel, fractured rock, or cavernous limestone. Economically important amounts of water may vary from less than a gallon per minute for cattle water in the desert to thousands of gallons per minute for industrial, irrigation, or municipal use.

Aquifers are important reservoirs storing large amounts of water relatively free from evaporation loss or pollution. If the annual withdrawal from an aquifer regularly exceeds the replenishment from rainfall or seepage from streams, the water stored in the aquifer will be depleted. Lowering the pressure in an aquifer by overpumping may cause the aquifer and confining layers of silt or clay to be compressed under the weight of the overburden. The resulting subsidence of the ground surface may cause structural damage to the aquifer, to surface buildings, damage to wells, and cause other problems as well.

By contrast, an *aquiclude* is a rock formation that is too impermeable to yield groundwater. Impervious rock in the unsaturated zone may retain water infiltrating from the surface to produce a perched water table that is above the main water table and from which water may be extracted. However, the amounts of water that can be extracted from such a formation are limited and the water is vulnerable to contamination.

A *river* is a natural, freshwater surface stream that has considerable volume compared with its smaller tributaries, known as brooks, creeks, branches, or forks. Rivers are usually the main stems and larger tributaries of the drainage systems that convey surface runoff from the land. Rivers flow from headwater areas of small tributaries to their mouths, where they may discharge into the ocean, a major lake, or a desert basin. Rivers flowing to the ocean drain about two-thirds of the land systems. The remainder of the land either is covered by ice or drains to closed basins (common in desert regions). Regions draining to the sea are termed *exoreic*, while those regions which drain into interior closed basins are *endoreic*. *Areic* regions are those which lack surface streams because of low rainfall or lithologic conditions.

There is a strong connection between the *hydrosphere*, where water is found, and the *lithosphere*, or land; human activities affect both. For example, disturbance of land by conversion of grasslands or forests to agricultural land or intensification of agricultural production may reduce vegetation cover, decreasing transpiration (loss of water vapor by plants) and affecting the microclimate. The result is increased rain runoff, erosion, and accumulation of silt in bodies of water. The nutrient cycles may be accelerated, leading to nutrient enrichment of surface waters. This, in turn, can profoundly affect the chemical and biological characteristics of bodies of water.

Groundwater and surface water have different characteristics: substances either dissolve in surface water or become suspended in it on its way to the ocean. Surface water in a lake or reservoir that contains mineral nutrients may support a heavy growth of algae. Surface water with a high level of biodegradable organic material, used as food by bacteria, normally contains a large population of bacteria. All these factors have a profound effect upon the quality of surface water.

Groundwater may dissolve minerals from the formations through which it passes. Most microorganisms originally present in groundwater are gradually filtered out as it seeps through mineral formations. Occasionally, the content of undesirable salts may become excessively high in groundwater, although it is generally superior to surface water as a domestic water source.

The movement of water from waste landfills to aquifers is also an important process whereby pollutants in the landfill leachate may be adsorbed by solid material through which the water passes.

2.0 Properties

The study of water (*hydrology*) is divided into a number of subcategories. *Limnology* is the branch of science dealing with the characteristics of fresh water, including its biological, chemical, and physical properties. *Oceanography* is the science of the ocean and its physical and chemical characteristics.

Water has a number of unique properties (Table 4.3) that make it suitable for living organisms and a solvent for many materials. Thus it is the basic transport medium for nutrients and waste products in life processes (Stollenwerk and Kipp, 1990). A point that must not be missed, however, is the very real potential for water to transport toxins (Hall and Strichartz, 1990).

The high dielectric constant of water affects its solvent properties, in that most ionic materials are dissociated in water. In addition, water has a high heat capacity; thus, a relatively large amount of heat is required to appreciably change the temperature of a mass of water. Hence, a body of water can have a stabilizing effect upon the temperature of nearby geographic regions. The high heat capacity of water also prevents sudden changes of temperature in large bodies of water and thereby protects aquatic organisms from the shock of abrupt temperature variations. The high heat of vaporization of water (585 cal/g at 20°C, 68°F) also stabilizes the temperature of bodies of water and influences the transfer of heat and water vapor between bodies of water and the atmosphere.

Table 4.3: Properties of water

Property	Effects and Significance
Solvent	Transport of nutrients and waste products
High dielectric constant	High solubility of ionic substances
High surface tension	Controlling factor in physiology and surface surface phenomena
Transparent to visible and longer wavelength uv light	Colorless, allowing light required for photosynthesis to reach considerable depths
Maximum at 4°C	Ice floats; vertical circulation restricted in stratified bodies of water
High heat of evaporation	Determines transfer of heat and water to the atmosphere
High latent heat of fusion	Temperature stabilized at the freezing point of water
High heat capacity	Stabilization of temperatures

Water has its maximum density at 4°C (39°F), a temperature above its freezing point (0°C, 32°F). The fortunate consequence of this fact is that ice floats, so that few large bodies of water ever freeze solid. Furthermore, the pattern of vertical circulation of water in lakes, a determining factor in their chemistry and biology, is governed largely by the unique temperature-density relationship of water (Baker, 1993).

The physical condition of a body of water strongly influences the chemical and biological processes that occur in there. Surface water occurs primarily in streams, lakes, and reservoirs. Lakes may be classified as *oligotrophic*, *eutrophic*, or *dystrophic*, an order that often parallels the life of the lake. Oligotrophic lakes are deep, generally clear, deficient in nutrients, and without much biological activity. Eutrophic lakes have more nutrients, support more life, and are more turbid. Dystrophic lakes are shallow, clogged with plant life, and normally contain colored water with a low pH.

Wetlands are flooded areas in which the water is shallow enough to enable growth of bottom-rooted plants (Dennison and Berry, 1993; Easterbrook, 1995). *Wet flatlands* are areas where mesophytic vegetation is more important than open water and which are commonly developed in filled lakes, glacial pits, and potholes, or in poorly drained coastal plains or floodplains. The term *swamp* is usually applied to a wetland where trees and shrubs are an important part of the vegetative association, and the term bog implies lack of solid foundation. Some bogs consist of a thick zone of vegetation floating on water.

Unique plant associations characterize wetlands in various climates and exhibit marked zonation characteristics around the edge in response to different thicknesses of the saturated zone above the firm base of soil material. Coastal marshes covered with vegetation adapted to saline water are common on all continents.

Some constructed reservoirs are very similar to lakes, while others differ a great deal from them. Reservoirs with a large volume relative to their inflow and outflow are called *storage reservoirs*. Reservoirs with a large rate of flow-through compared to their volume are called *run-of-the-river reservoirs*. The physical, chemical, and biological properties of water in the two types of reservoirs may vary appreciably. Water in storage reservoirs more closely resembles lake water, whereas water in run-of-the-river reservoirs is much like river water.

Impounding water in reservoirs may have some profound effects upon water quality. These changes result from factors such as different velocities, varied detention time, and altered surface-to-volume ratios relative to the streams that were impounded. Some resulting beneficial changes due to impoundment are a decrease in the level of organic matter, a reduction in turbidity, and a decrease in hardness (calcium and magnesium content). Some detrimental changes are lower oxygen levels due to decreased re-aeration, decreased mixing, accumulation of pollutants, lack of a bottom scour produced by flowing water scrubbing a stream bottom, and increased growth of algae. Algal growth may be enhanced when suspended solids settle from impounded water, causing increased exposure of the algae to sunlight. Stagnant water in the bottom of a reservoir may be of low quality. Oxygen levels frequently go to almost zero near the bottom, and hydrogen sulfide is produced by the reduction of sulfur compounds in the low-oxygen environment. Insoluble iron (Fe^{3+}) and manganese (Mn^{4+}) species are reduced to soluble iron (Fe^{2+}) and manganese (Mn^{2+}) ions, which must be removed prior to using the water.

Estuaries constitute another type of body of water, consisting of arms of the ocean into which streams flow. The mixing of fresh and salt water gives estuaries unique chemical and biological properties. Estuaries are the breeding grounds of much marine life, which makes their preservation very important.

Water's unique temperature-density relationship results in the formation of distinct layers within nonflowing bodies of water (Figure 4.2). During the summer a surface layer (*epilimnion*) is heated by solar radiation and, because of its lower density, floats upon the bottom layer (*hypolimnion*). This phenomenon is called thermal stratification.

When an appreciable temperature difference exists between the two layers, they do not mix but behave independently and have very different chemical and biological properties. The epilimnion, which is exposed to light, may have a heavy growth of algae. As a result of exposure to the atmosphere and (during daylight hours) because of the photosynthetic activity of algae, the epilimnion contains relatively higher levels of dissolved oxygen and generally is aerobic. In the hypolimnion, bacterial action on biodegradable organic material may cause the water to become anaerobic. As a consequence, chemical species in a relatively reduced form tend to predominate in the hypolimnion.

The shear plane, or layer between epilimnion and hypolimnion, is called the *thermocline*. During the autumn, when the epilimnion cools, a point is reached at which the temperatures of the epilimnion and hypolimnion are equal. This disappearance of thermal stratification causes the entire body of water to behave as

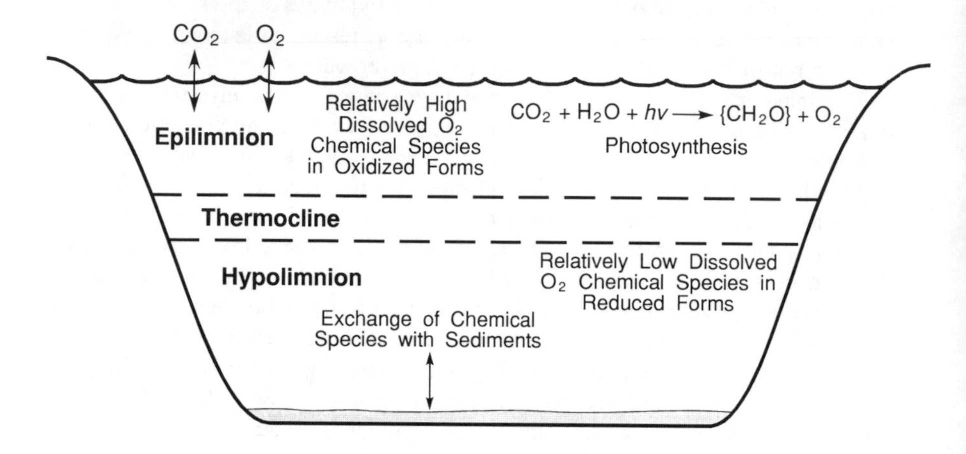

Figure 4.2: The stratification of a lake

a hydrological unit, and the resultant mixing is known as *overturn*, which generally occurs in the spring. During the overturn, the chemical and physical characteristics of the body of water become much more uniform, and a number of chemical, physical, and biological changes may result. Biological activity may increase from the mixing of nutrients. Changes in water composition during overturn may cause disruption in water treatment processes (Cheremisinoff, 1995).

An increasing amount of attention has been given to thermal pollution, the raising of the temperature of a waterway by heat discharged from the cooling system or effluent wastes of an industrial installation. This rise in temperature may upset the ecological balance of the waterway sufficiently to pose a threat to the native life-forms. This problem has been especially noted in the vicinity of nuclear power plants. Thermal pollution may be combated by allowing wastewater to cool in large cooling towers before emptying it into the waterway.

The chemistry and biology of the oceans are unique because of the high salt content, depth, and other factors. The environmental problems of the oceans have increased greatly in recent years because of ocean dumping of pollutants, oil spills, and increased utilization of natural resources from the oceans. Many of the pollutants have considerable lifetimes in the ocean (Figure 4.3).

In many cases, rivers, lakes, and oceans have become polluted from discharges of liquid wastes from residential, commercial, and industrial sources (Baker 1993; Seidl et al., 1995). Many of these bodies of water have been reclaimed because of the construction of new wastewater treatment facilities. Both physical-chemical and biological processes are used to remove organic matter from the liquid wastewater stream. It is this organic matter that causes a depletion of oxygen in rivers and lakes, with consequent anaerobic conditions leading to fish kills and noxious odors.

The physical-chemical and biological processes remove an abundance of the organic matter, along with floatable scum and grease from the waste stream. Chemical disinfection inactivates bacteria, viruses, and protozoa in the waste stream. The physical and chemical processes used for removing solids from the waste stream include screening, sand and grit separation, chemical coagulation, and plain sedimentation. Biological processes include activated sludge, contact towers, and biological discs.

The application of these treatment processes to residential, commercial, and industrial wastewaters has reduced the pollution load on many rivers, lakes, and harbors. There do remain, however, cases of large metropolitan areas that have not completed their cleanup campaigns, and in many other regions, discharges of untreated mixtures of sanitary wastes and storm waters occur during heavy rains. These discharges emanate from large conduits that carry both sanitary and storm wastes. The impurities that run off the land during a storm mix with the sanitary

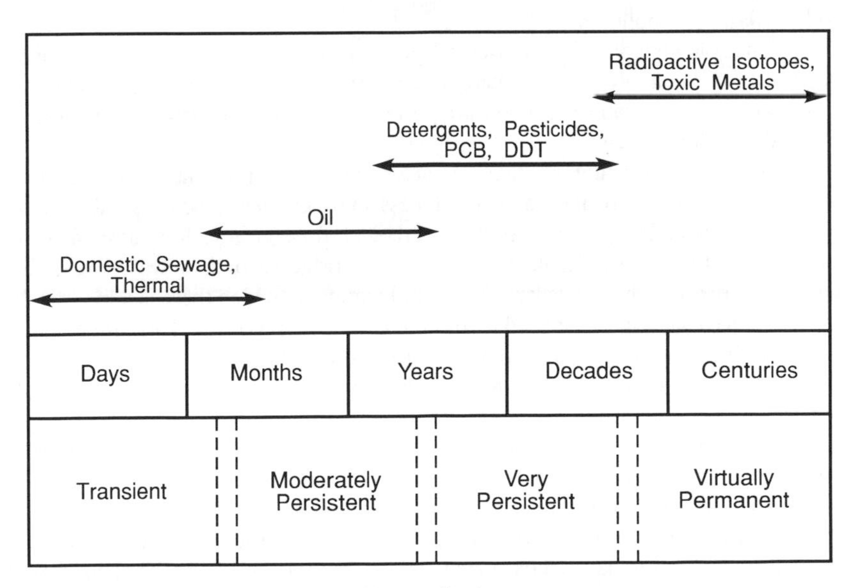

Figure 4.3: Relative lifetimes of pollutants in the oceans

wastes in the pipes, and the quantity of wastewater overtaxes the carrying capacity of the conduits. A portion of this mixture flows into the nearest body of water before it can reach the wastewater treatment facility. Where wastewater must be discharged into lakes and dry streams, a higher level of treatment must be employed, including removal of nutrients and colloidal matter.

2.1 Gases

Dissolved gases (oxygen for fish and carbon dioxide for photosynthetic algae) are crucial to the welfare of living species in water. Carbon dioxide is produced by respiratory processes in waters and sediments and can also enter water from the atmosphere. Carbon dioxide is required for the photosynthetic production of biomass by algae and in some cases is a limiting factor. High levels of carbon dioxide produced by the degradation of organic matter in water can cause excessive algal growth and productivity.

Dissolved oxygen frequently is the key substance in determining the extent and kinds of life in a body of water. Oxygen deficiency is fatal to many aquatic animals such as fish. The presence of oxygen can be equally fatal to many kinds of anaerobic bacteria. Without an appreciable level of dissolved oxygen, many kinds of aquatic organisms cannot exist in water. Dissolved oxygen is consumed by the degradation of organic matter in water. Many fish kills are caused not from the direct toxicity of pollutants but from a deficiency of oxygen because of its consumption in the biodegradation of pollutants.

The solubility of oxygen in water depends upon water temperature, the partial pressure of oxygen in the atmosphere, and the salt content of the water. Water in equilibrium with air cannot contain a high level of dissolved oxygen compared to many other solute species.

If oxygen-consuming processes are occurring in the water, the dissolved oxygen level may rapidly approach zero unless some efficient mechanism for the re-aeration of water is operative, such as turbulent flow in a shallow stream or air pumped into the aeration tank of an activated sludge secondary waste treatment facility. The problem becomes largely one of kinetics, in which there is a limit to the rate at which oxygen is transferred across the air-water interface. This rate depends upon turbulence, air bubble size, temperature, and other factors.

The temperature effect on the solubility of gases in water is especially important in the case of oxygen. At higher temperatures, the decreased solubility of oxygen, combined with the increased respiration rate of aquatic organisms, frequently causes a condition in which a higher demand for oxygen accompanied by lower solubility of the gas in water results in severe oxygen depletion.

In addition to a wide array of other chemicals, carbon dioxide, bicarbonate ion, and carbonate ion have an extremely important influence upon the chemistry of water and wastewater (Benefield et al., 1982; Schwarzenbach et al., 1993; Huang et al., 1995). Many minerals are deposited as salts of the carbonate ion. Algae in water utilize dissolved carbon dioxide in the synthesis of biomass.

Carbon dioxide is only about 0.035% v/v of normal dry air. As a consequence of the low level of atmospheric carbon dioxide, water totally lacking in alkalinity in equilibrium with the atmosphere contains only a very low level of carbon dioxide. However, the formation of bicarbonate and carbonate greatly increases the solubility of carbon dioxide. High concentrations of free carbon dioxide in water may adversely affect respiration and gas exchange of aquatic animals.

Carbon dioxide is a weak acid in water. Because of the presence of carbon dioxide in air and its production from microbial decay of organic matter, dissolved carbon dioxide is present in virtually all natural waters and wastewaters. Carbon dioxide is a weak acid, so that rainfall from even an absolutely unpolluted atmosphere is slightly acidic due to the presence of dissolved carbon dioxide.

A large share of the carbon dioxide found in water is a product of the breakdown of organic matter by bacteria. Even algae, which utilize carbon dioxide in photosynthesis, produce it through their metabolic processes in the absence of light. As water seeps through layers of decaying organic matter while infiltrating the ground, it may dissolve a great deal of carbon dioxide produced by the respiration of organisms in the soil. Later, as water goes through limestone formations, it dissolves calcium carbonate because of the presence of the dissolved carbon dioxide. This process is the one by which limestone caves are formed.

2.2 Acidity and Alkalinity

Acidity as applied to natural water and wastewater is the capacity of the water to neutralize hydroxyl ions (OH^-). It is analogous to *alkalinity*, the capacity to neutralize the hydrogen ion (H^+).

Although virtually all water has some alkalinity, acidic water is not frequently encountered, except in cases of severe pollution. Acidity generally results from the presence of weak acids such as phosphoric acid, carbon dioxide, hydrogen sulfide, proteins, fatty acids, and acidic metal ions, particularly Fe^{3+}. Acidity is more difficult to determine than alkalinity. One reason for this difficulty is that two of the major contributors to acidity are carbon dioxide and hydrogen sulfide, both volatile solutes that are readily lost from solution.

From the pollution standpoint, acids are the most important contributors to acidity. The term *free mineral acid* is applied to strong acids such as sulfuric acid (H_2SO_4) and hydrochloric acid (hydrogen chloride, HCl) in water. Acidic water from mines is a common water pollutant that contains an appreciable concentration of free mineral acid.

The capacity of water to accept hydrogen ions (H^+) is called alkalinity. Alkalinity is important in water treatment and in the chemistry and biology of natural waters. Frequently, the alkalinity of water must be known to calculate the quantities of chemicals to be added in treating the water. Highly alkaline water often has a high pH and generally contains elevated levels of dissolved solids. These characteristics may be detrimental for water to be used in boilers, food processing, and municipal water systems.

Alkalinity serves as a pH buffer and reservoir for inorganic carbon, thus helping to determine the ability of a water body to support algal growth and other aquatic life. It is used by biologists as a measure of water fertility. Generally, the basic species responsible for alkalinity in water are bicarbonate ions, carbonate ions, and hydroxide ions. Other, usually minor, contributors to alkalinity are ammonia and the conjugate bases of phosphoric, silicic, boric, and organic acids.

2.3 Chemical Species

Metal ions in aqueous solution seek to reach a state of maximum stability through chemical reactions. Acid-base, precipitation, complex formation, and oxidation-reduction reactions all provide the means through which metal ions in water are transformed to more stable forms.

Hydrated metal ions, particularly those with a high positive charge tend to lose protons in aqueous solution, and fit the definition of Bronsted acids. Hydrated trivalent metal ions, such as iron (Fe^{3+}), generally lack at least one hydrogen ion at neutral pH values or above. For tetravalent metal ions, the completely protonated forms are rare even at very low pH values. The tendency of hydrated metal ions to behave as acids may have a profound effect upon the aquatic environment. A good example is acidic mine water which derives part of its acidity from the character of hydrated iron (Fe^{3+}).

Of the cations found in most freshwater systems, calcium (Ca^{2+}) generally has the highest concentration. Calcium is a key element in many geochemical processes, and minerals constitute the primary sources of calcium ions in water. Among the primary contributing minerals are: gypsum ($CaSO_4.2H_2O$), anhydrite ($CaSO_4$), dolomite ($CaCO_3MgCO_3$), calcite ($CaCO_3$) and aragonite ($CaCO_3$), the latter two being different mineral forms (*polymorphs*) of calcium carbonate. Water containing a high level of carbon dioxide readily dissolves calcium from its carbonate minerals. Calcium (Ca^{2+}) ions, along with magnesium (Mg^{2+}) ions and sometimes iron (Fe^{2+}) ions, account for water hardness. The most common manifestation of water hardness is the precipitate formed by soap in hard water.

A number of other chemical species are present naturally, and are significant, in water (Table 4.4); some of these can also be pollutants. In particular, chelating agents are common potential water pollutants and occur in sewage effluent and industrial wastewater such as metal plating wastewater. Chelating agents are complex-forming agents and have the ability to solubilize heavy metals. The formation of soluble complexes increases the leaching of heavy metals from waste disposal sites and reduces the efficiency with which heavy metals are removed with sludge in conventional biological waste treatment. Although chelating agents are never entirely specific for a particular metal ion, some complicated chelating agents of biological origin approach almost complete specificity for certain metal ions. One example of such a chelating agent is ferrichrome, synthesized by and extracted from fungi, which forms extremely stable chelates with iron (Fe^{3+}).

Many organic compounds interact with suspended material and sediments in

Table 4.4: Chemical species commonly occurring in water

Substance	Sources
Aluminum	Aluminum containing minerals
Chloride	Minerals, pollution
Fluoride	Minerals, water additive
Iron	Minerals, mine water
Magnesium	Minerals, such as dolomite ($CaCO_3.MgCO_3$)
Manganese	Minerals, decay of nitrogenous organic matter, pollution
Potassium	Mineral matter, fertilizer runoff, forest fire runoff
Phosphorus	Minerals, fertilizer runoff, domestic wastes (detergents)
Silicon	Minerals, such as sodium feldspar albite, $NaAlSi_3O_8$,
Sulfur	Minerals, pollutants, acid mine water, acid rain
Sodium	Minerals, pollution

bodies of water. Settling suspended material containing adsorbed organic matter carries organic compounds into the sediment of a stream or lake. This phenomenon is largely responsible for the presence of herbicides in sediments containing contaminated soil particles eroded from cropland. Some organic compounds are carried into sediments by the remains of organisms or by fecal pellets from the zooplankton that have accumulated organic contaminants. Suspended particulate matter affects the mobility of organic compounds adsorbed onto particles. Furthermore, adsorbed organic matter undergoes chemical degradation and biodegradation at different rates and by different pathways compared to organic matter in solution.

The most common types of sediments considered for their organic binding abilities are clays (Chapter 3), organic humic substances, and clay-humic complexes. Both clays and humic substances act as cation exchangers. Therefore these materials adsorb cationic organic compounds through ion exchange. This is a relatively strong sorption mechanism, greatly reducing the mobility and biological activity of the organic compound. When adsorbed by clays, cationic organic compounds are generally held between the layers of the clay mineral structure, where their biological activity is essentially zero.

3.0 Aquatic Organisms

Microorganisms (*algae*, *bacteria*, and *fungi*) are living catalysts that enable a vast number of chemical processes to occur in water and soil. The living organisms (*biota*) in an aquatic ecosystem may be classified as either *autotrophic* or *heterotrophic*.

Autotrophic organisms utilize solar or chemical energy to fix elements from simple, nonliving inorganic material into complex life molecules that compose living organisms. Algae are typical autotrophic aquatic organisms. Generally, carbon dioxide (CO_2), nitrate (NO_3^-), and phosphates ($H_2PO_4^-$, HPO_4^{2-}) are sources of carbon, nitrogen, and phosphorus, respectively, for autotrophic organisms. Organisms that utilize solar energy to synthesize organic matter from inorganic materials are called producers.

Heterotrophic organisms utilize the organic substances produced by autotrophic organisms as energy sources and as the raw materials for the synthesis of their own biomass. Decomposers (or reducers) are a subclass of the heterotrophic organisms and consist chiefly of bacteria and fungi, which ultimately break down material of biological origin to the simple compounds originally fixed by the autotrophic organisms.

3.1 Algae

Algae may be considered as generally microscopic organisms that subsist on inorganic nutrients and produce organic matter from carbon dioxide by photosynthesis. The general nutrient requirements of algae are carbon (from carbon dioxide or bicarbonate, nitrogen (generally as nitrate), phosphorus (as some form of orthophosphate), sulfur (as sulfate), and trace elements including sodium, potassium, calcium, magnesium, iron, cobalt, and molybdenum.

In the absence of light, algae metabolize organic matter in the same manner as do nonphotosynthetic organisms. Thus algae may satisfy their metabolic demands by utilizing chemical energy from the degradation of stored starches or oils, or from the consumption of algal protoplasm itself. In the absence of photosynthesis, the metabolic process consumes oxygen, so during the hours of darkness, an aquatic system with a heavy growth of algae may become depleted in oxygen.

3.2 Bacteria

Bacteria occur individually or grow as groups ranging from two to millions of individual cells. Individual bacteria cells are very small and may be observed only through a microscope. Most bacteria fall into the size range of 0.5-3.0 microns (1 micron = 1 meter x 10^{-6}). However, considering all species, a size range of 0.3-50 microns is observed. In general, it is assumed that a filter with 0.45-micron pores will remove all bacteria from water passing through it.

The metabolic activity of bacteria is greatly influenced by their small size. Their surface-to-volume ratio is extremely large, so that the inside of a bacterial cell is highly accessible to a chemical substance in the surrounding medium. Thus for the same reason that a finely divided catalyst is more efficient than a more coarsely divided one, bacteria may bring about very rapid chemical reactions compared to those mediated by larger organisms. Bacteria excrete enzymes that can act outside the cell (exoenzymes) and break down solid food material to soluble components that can penetrate bacterial cell walls, where the digestion process is completed.

Bacteria obtain the energy needed for their metabolic processes and reproduction by mediating redox reactions. Nature provides a large number of such reactions, and bacterial species have evolved that utilize many of these. As a consequence of their participation in such reactions, bacteria are involved in many biochemical processes in water and soil.

Bacteria are essential participants in many important elemental cycles in nature, including those of nitrogen, carbon, and sulfur. They are responsible for the formation of many mineral deposits, including some of iron and manganese. On a smaller scale, some of these deposits form through bacterial action in natural water systems and even in pipes used to transport water.

3.3 Fungi

Fungi are non-photosynthetic organisms and frequently possess a filamentous structure. Some fungi are as simple as the microscopic unicellular yeasts, whereas other fungi form large, intricate toadstools. The microscopic filamentous structures of fungi generally are much larger than bacteria and usually are 5-10 microns in width. Fungi are aerobic (oxygen-requiring) organisms and generally can thrive in more acidic media than can bacteria. They are also more tolerant of higher concentrations of heavy metal ions than are bacteria.

Perhaps the most important function of fungi in the environment is the breakdown of cellulose in wood and other plant materials. To accomplish this, fungal cells secrete a biological catalyst (enzyme) called cellulase. This enzyme breaks insoluble cellulose down to soluble carbohydrates that can be absorbed by the fungal cell. Because it acts outside the organism, it is called an extracellular enzyme or exoenzyme.

Although fungi do not grow well in water, they play an important role in determining the composition of natural waters and wastewaters because of the large amount of their decomposition products that enter water. An example of such a product is humic material, which interacts with hydrogen ions and metals.

3.4 General Effects

The ability of a body of water to produce living material is known as its productivity. Productivity results from a combination of physical and chemical factors. Water of low productivity generally is desirable for water supplies or for swimming. Relatively high productivity is required for the support of fish. Excessive productivity can result in choking by weeds and odor problems. The growth of algae may become quite high in very productive waters, with the result that the concurrent decomposition of dead algae reduces oxygen levels in the water to very low values (*eutrophication*).

In inland lakes the issue of eutrophication is due mainly to excessive, but inadvertent, introduction of domestic and industrial wastes, runoff from fertilized agricultural and urban areas, precipitation, and groundwater. The interaction of the natural processes within the lake with the artificial disturbance caused by human activities complicates the overall problem and leads to an accelerated rate of deterioration in lakes. Since a population increase necessitates an expanded utilization of lakes and streams, cultural eutrophication has become one of the major water resource problems throughout the world. *Cultural eutrophication* is reflected in changes in species composition, population sizes, and productivity in groups of organisms throughout the aquatic ecosystem. Thus the biological changes caused by excessive fertilization are of considerable interest from both practical and academic viewpoints.

Life forms higher than algae and bacteria (e.g., fish) comprise a comparatively small fraction of the biomass in most aquatic systems. The influence of these higher life forms upon aquatic chemistry is minimal. However, aquatic life is strongly influenced by the physical and chemical properties of the body of water in which it lives. Temperature, transparency, and turbulence are the three main physical properties affecting aquatic life. Very low water temperatures result in very slow biological processes, whereas very high temperatures are fatal to most organisms. A difference of only a few degrees can produce large differences in the kinds of organisms present. Thermal discharges of hot water from power plants (cooling water) frequently kill temperature-sensitive fish while increasing the growth of fish and other species that are adapted to higher temperatures.

The transparency of water is particularly important in determining the growth of algae. Turbid water may not be very productive of biomass, even though it has the nutrients and optimum temperature. Turbulence is an important factor in mixing and transport processes in water. Some small organisms (plankton) depend upon water currents for their own mobility.

Water turbulence is largely responsible for the transport of nutrients to living organisms and of waste products away from them. It plays a role in the transport of oxygen, carbon dioxide, and other gases through a body of water and in the exchange of these gases at the water-atmosphere interface. Moderate turbulence is generally beneficial to aquatic life.

Biochemical oxygen demand, another important water-quality parameter, refers to the amount of oxygen utilized when the organic matter in a given volume of water is degraded biologically. A body of water with a high biochemical oxygen demand, and no means of rapidly replenishing the oxygen, obviously cannot sustain organisms that require oxygen.

The levels of nutrients in water frequently determine its productivity. Aquatic plant life requires an adequate supply of carbon (carbon dioxide, CO_2), nitrogen (nitrate, NO_3^-), phosphorus (orthophosphate, HPO_4^{2-}), and trace elements such as iron. In many cases, phosphorus is the limiting nutrient and is generally controlled in attempts to limit excess productivity.

The salinity of water also determines the kinds of life forms present. Irrigation waters may pick up harmful levels of salt. Marine life obviously requires or tolerates salt water, whereas many freshwater organisms are intolerant of salt.

A majority of the important chemical reactions occurring in water, particularly those involving organic matter and oxidation-reduction processes, occur through bacterial intermediaries. Algae are the primary producers of biological organic matter (biomass) in water. Microorganisms are responsible for the formation of many sediment and mineral deposits; they also play the dominant role in secondary waste treatment.

Another example of waste control by oxidation involves oxidation of chemical waste materials in supercritical water. Supercritical water oxidation has been used to convert organic materials to carbon dioxide and is claimed to achieve destruction of more than 99.9% of chemicals such as polyols and amines that occur in liquid waste streams (Stadig, 1995).

4.0 Biodegradation Processes

The biodegradation of organic matter in the aquatic and terrestrial environments is a crucial life process. Some organic pollutants are *biocidal*. For example, in addition to killing harmful fungi, fungicides frequently harm beneficial saprophytic fungi (fungi that decompose dead organic matter) and bacteria. Herbicides are designed for plant control, and insecticides are used to control insects.

The biodegradation of petroleum is essential to the elimination of oil spills. This oil is degraded by both marine bacteria and filamentous fungi. The physical form of crude oil makes a large difference in its potential for degradation. Degradation in water occurs at the water-oil interface. Therefore thick layers of crude oil prevent contact with bacterial enzymes and oxygen. Apparently, bacteria synthesize an emulsifier that keeps the oil dispersed in the water as a fine colloid and therefore accessible to the bacterial cells.

Some of the most important microorganism-mediated chemical reactions in aquatic and soil environments are those involving nitrogen compounds and the cycle of such compounds throughout the biosphere (Chapter 1). Among the biochemical transformations in the nitrogen cycle are (1) nitrogen fixation, whereby molecular nitrogen is fixed as organic nitrogen; (2) nitrification, the process of oxidizing ammonia to nitrate; (3) nitrate reduction, in which nitrogen in nitrate ions is reduced to nitrogen in a lower oxidation state; and (4) denitrification, the reduction of nitrate and nitrite to ammonia.

Biodegradation of phosphorus compounds is important in the environment for two reasons. First, it provides a source of algal nutrient orthophosphate from the hydrolysis of polyphosphates. Second, biodegradation deactivates highly toxic organophosphate compounds, such as the organophosphate insecticides. The organophosphorus compounds of greatest environmental concern tend to be sulfur-containing phosphorothionate and phosphorodithioate ester insecticides. These are used because they exhibit higher ratios of insect:mammal toxicity than do their nonsulfur analogs. The biodegradation of these compounds is an important environmental chemical process. Fortunately, unlike the organic halogen insecticides that they largely displaced, the organophosphates readily undergo biodegradation and do not accumulate.

Sulfate ions (SO_4^{2-}) are found in varying concentration in practically all natural waters. Organic sulfur compounds, those of natural origin and pollutant species, are very common in natural aquatic systems, and the degradation of these compounds is an important microbial process. Sometimes the degradation products, such as the odorous and toxic hydrogen sulfide, cause serious problems with water quality.

One consequence of bacterial action on metal compounds is the occurrence of drainage of acidic aqueous solutions from mines. *Acid mine drainage* is one of the most common and damaging problems in the waters flowing from coal mines and draining from the spoil piles (mine tippage, gob piles) left over from coal processing and washing. These waters are highly acidic and have the ability to sterilize the surrounding land and water systems with ensuing serious (often fatal) effects on the flora and fauna. Acidic mine water results from the presence of sulfuric acid produced by the oxidation of pyrite (FeS_2). Microorganisms are closely involved in the overall process. The prevention and cure of acid mine water are major challenges facing the environmental chemist.

Selenium is also subject to bacterial oxidation and reduction. These transitions are important because selenium is a crucial element in nutrition, particularly of livestock. Microorganisms are closely involved with the selenium cycle, and microbial reduction of oxidized forms of selenium has been known for some time. A soil dwelling strain of *Bacillus megaterium* has been found to be capable of oxidizing elemental selenium to selenite (SeO).

5.0 References

American Society for Testing and Materials. 1995. *Annual Book of ASTM Standards*. Vol. 11.02. American Society for Testing and Materials, Philadelphia, Pennsylvania.

Baker, L.A. (ed.). 1993. *Environmental Chemistry of Lakes and Reservoirs*. Symposium Series No. 237. American Chemical Society, Washington, D.C.

Benefield, L.D., Judkins, J.F., and Weand, B.L. 1982. *Process Chemistry for Water and Wastewater Treatment*. Prentice-Hall, Englewood Cliffs, New Jersey.

Cartwright, F., and Biddiss, M. 1972. *Disease and History*. Dorset Press, New York.

Cheremisinoff, P. 1995. *Handbook of Water and Wastewater Treatment Technology*. Marcel Dekker Inc., New York.

Dennison, M.S. and Berry, J.F. (eds.). 1993. *Wetlands: Guide to Science, Law, and Technology*. Noyes Data Corp., Park Ridge, New Jersey.

Easterbrook, G. 1995. *A Moment on the Earth: The Coming Age of Environmental Optimism*. Viking Press, New York.

Eglinton, G. (ed.). 1975. *Environmental Chemistry. Vol. 1*. Specialist Periodical Reports. The Chemical Society, London, England.

Hall, S., and Strichartz G. (eds.). 1990. *Marine Toxins: Origin, Structure, and Molecular Pharmacology*. Symposium Series No. 418. American Chemical Society, Washington, D.C.

Huang, C.P., O'Melia, C.R., and Morgan, J.J. (eds.). 1995. *Aquatic Chemistry: Interfacial and Interspecies Processes*. Advances in Chemistry Series No. 244. Washington, D.C.: American Chemical Society.

James, P., and Thorpe, N. 1994. *Ancient Inventions*. Ballantine Books, New York.

Jenne, E.D. (ed.). 1979. *Chemical Modeling in Aqueous Systems*. Symposium Series No. 93. American Chemical Society, Washington, D.C.

Lacy, W.J. 1983. In Riegel's *Handbook of Industrial Chemistry*. p. 14. Edited J.A. Kent. Van Nostrand Reinhold, New York.

Melchior, D.C., and Bassett, R.L. (eds.). 1990. *Chemical Modeling of Aqueous Systems II*. Symposium Series No. 416. American Chemical Society, Washington, D.C.

Noyes, R. (ed.). 1991. *Handbook of Pollution Control Processes*. Noyes Data Corp., Park Ridge, New Jersey.

Pickering, K.T., and Owen, L.A. 1994. *Global Environmental Issues*. Routledge Publishers, New York.

Schwarzenbach, R.P., Gschwend, P.M., and Imboden, D.M. 1993. *Environmental Organic Chemistry*. John Wiley & Sons Inc., New York.

Seidl, P.R., Gottlieb, O.R., and Kaplan, M.A.C. (eds.). 1995. *Chemistry of the Amazon: Biodiversity, Natural Products, and Environmental Issues*. Symposium Series No. 588. Amercian Chemical Society, Washington, D.C.

Stollenwerk, K.G., and K.L. Kipp. 1990. Chap. 19 in *Chemical Modeling of Aqueous Systems II*. Edited by D.C. Melchior and R.L. Bassett. Symposium Series No. 416. American Chemical Society, Washington, D.C.

Stadig, W.P. 1995. *Chemical Processing* 58(8): 34.

Chapter 5

The Atmosphere

1.0 Introduction

The atmosphere is essential to life. It also serves a vital protective function to living species and one can wonder how life on earth would have evolved had the atmosphere been different in composition. Perhaps the absence of recognizable life forms on the planets adjacent to earth provide an answer to this thought.

The composition of the atmosphere is primarily nitrogen (N_2) and oxygen (O_2). The concentration of water vapor (H_2O) is highly variable, especially near the surface. There are many minor constituents or trace gases (Dasgupta, 1993), such as neon (Ne), helium (He), krypton (Kr), and xenon (Xe), that are inert, and others, such as carbon dioxide (CO_2), methane (CH_4), hydrogen (H_2), nitrous oxide (NO), carbon monoxide (CO), and sulfur dioxide (SO_2), that play an important role in radiative and biological processes. In addition, the atmosphere contains organic species other than methane (Westberg and Zimmerman, 1993) that can play an important role in nucleation and in the formation of smog as well as participate in the greenhouse effect (Chapter 10).

Dry air within several kilometers of ground level consists of two major components: nitrogen (78.1% v/v) and oxygen (21.0%v/v). Other components include argon (0.9% v/v), carbon dioxide (0.04% v/v) and the noble gases neon, helium, krypton, and xenon (< 0.01% v/v in toto). Atmospheric air may contain 0.1-5% v/v water, with a normal range of 1-3% v/v.

The atmosphere is life supporting by virtue of the relatively high proportion of molecular oxygen that exists in spite of the presence of gases such as nitrogen, methane, hydrogen and other gases that are capable of existing in various oxide forms. The chemical disequilibrium is maintained by the continuous production of gases from biological processes. Thus floral and faunal chemistry plays a major role in maintaining the composition of the atmosphere.

Below about 60 miles (100 km) in altitude, the proportion of the major constituents in the atmosphere is very uniform. The region is known as the *homosphere*, to distinguish it from the *heterosphere* above 60 miles, where the relative amounts of the major constituents change with height. In the homosphere there are sufficient atmospheric motions and a molecular free path short enough to maintain uniformity in composition. Above the boundary between the homosphere and the heterosphere, known as the *homopause* (or *turbopause*), the collision frequency (mean free path) of an individual molecule is sufficiently long to allow a partial separation (by gravity) of the (relatively) lighter molecules from the heavier molecules. Hence the average molecular weight of the heterosphere decreases with height, the lighter molecules dominating the composition at higher elevation.

In addition to the gaseous constituents, the atmosphere suspends many solid and liquid particles. An *aerosol* is a colloidal system in which a gas, frequently air (such as is evident in the formation of smog; Chapter 10), is the continuous medium and particles of solids or liquids are dispersed in it (Hidy et al., 1979; Flagan, 1993; Charlson and Wigley, 1994). The particle size distribution of atmospheric aerosols has a multimodal character, usually with a bimodal mass, volume, or surface area distribution, and frequently trimodal surface area distribution near sources of fresh combustion aerosols. Measurement of aerosol particle sizes requires a combination of techniques and can be difficult (Cahill and Wakabayashi, 1993).

The nuclei mode (<0.03 micron) (1 micron = 1 meter x 10^{-6}) is formed by condensation of vapors from high-temperature processes, or by gaseous reaction products. The intermediate or accumulation mode (0.1-1.0 micron) is formed by coagulation of nuclei. The larger particles (>1.0 micron) fall out, whereas the very fine particles (smaller than 0.1 micron) then agglomerate to form larger particles which remain suspended.

Aerosols are created by gas-to-particle reactions and are lifted from the surface by winds. A portion of these aerosols can become centers of condensation or deposition in the growth of water and ice clouds. Cloud droplets and ice crystals are made primarily of water and trace amounts of particles and dissolved gases. The diameters of the particles range from a few microns to 100 microns. Water or ice particles larger than about 100 microns begin to fall because of gravity and may result in precipitation at the surface.

2.0 Composition and Character

The atmosphere is the source of oxygen for respiration and carbon dioxide for photosynthesis. Nitrogen is often (erroneously) considered to be a nonrelevant gas in the atmosphere but is employed by nitrogen-fixing bacteria and ammonia-manufacturing plants to produce chemically bound nitrogen, necessary in many life-supporting molecules. As a basic part of the hydrologic cycle (Chapter 4), the atmosphere transports water from the oceans to land.

Unfortunately, the atmosphere also has been the recipient of many chemical pollutants (ranging from sulfur dioxide to fluorocarbon refrigerants) (Newman, 1993) and detection of the pollutants involves a series of standard tests, many of which are applicable to the emissions testing of stacks and flues (American Society for Testing and Materials, 1995). Such chemical intrusions into the atmosphere result in damage to life and materials (through corrosion) and alter the characteristics of the atmosphere as well.

In its protective role, the atmosphere absorbs most of the cosmic rays from outer space and most of the electromagnetic radiation from the sun. However, sufficient transmission of radiation is allowed in the regions of 300-2500 nanometers (1 nanometer (nm) = 1 meter x 10^{-9}) (near-ultraviolet, visible, and near-infrared radiation) and 0.01-40 nm (radio waves). Absorption of electromagnetic radiation below 300 nm removes the damaging ultraviolet radiation. The overall effect is that the atmosphere stabilizes the temperature of the land surfaces thereby preventing the potential for extremes of temperature.

Atmospheric pressure decreases as an approximately exponential function of altitude, which largely determines the characteristics of the atmosphere. Thus:

$$P_h = P_o^{e/mgh/RT}$$

where P_h is the pressure at any given height, P_o is the pressure at zero altitude (sea level), m is the average gram molecular mass of air (28.97 g/mole in the troposphere), g is the acceleration of gravity (981 cm x sec^{-1} at sea level), h is the altitude (in cm or meters or kilometers), R is the gas constant (8.314 x 10^7 erg x deg^{-1} x $mole^{-1}$), and T is the absolute temperature. Furthermore:

$$\log P_h = \log P_o - (mgh \times 10^5)/2.303RT$$

or at sea level where the pressure is 1 atm:

$$\log P_h = (mgh \times 10^5)/2.303RT$$

The characteristics of the atmosphere vary widely with altitude, time (season), location (latitude), and even solar activity. At very high altitudes normally reactive species, such as atomic oxygen, persist for long periods of time. At such altitudes, the pressure is very low, and a reactive species will travel long distances before it collides with a potential reactant (the mean free path).

2.1 Structure

The atmosphere is stratified on the basis of temperature-density relationships resulting from interrelationships between physical and photochemical (light-induced chemical phenomena) processes in air (Zamaraev et al., 1994). The lowest layer of the atmosphere (Figure 5.1), extending from sea level to an altitude of 635 miles (1016 km), is the *troposphere*, which is characterized by a generally homogeneous composition of major gases other than water and decreasing temperature with increasing altitude from the heat-radiating surface of the land. The upper limit of the troposphere, which has a temperature minimum of about -56°C (-70°F), varies in altitude by a half mile (a kilometer) or so with atmospheric temperature, underlying terrestrial surface, and time. The homogeneous composition of the troposphere results from constant mixing by circulating air masses. However, the water vapor content of the troposphere is extremely variable because of cloud formation, precipitation, and evaporation of water from terrestrial water bodies.

At altitudes of ~30 miles (~50 km) and more, ions are so prevalent that the region is called the *ionosphere*. Ultraviolet light is the primary producer of ions in the ionosphere. Thus during periods of darkness, the positive ions slowly recombine with free electrons. The process is especially rapid in the lower regions of the ionosphere,

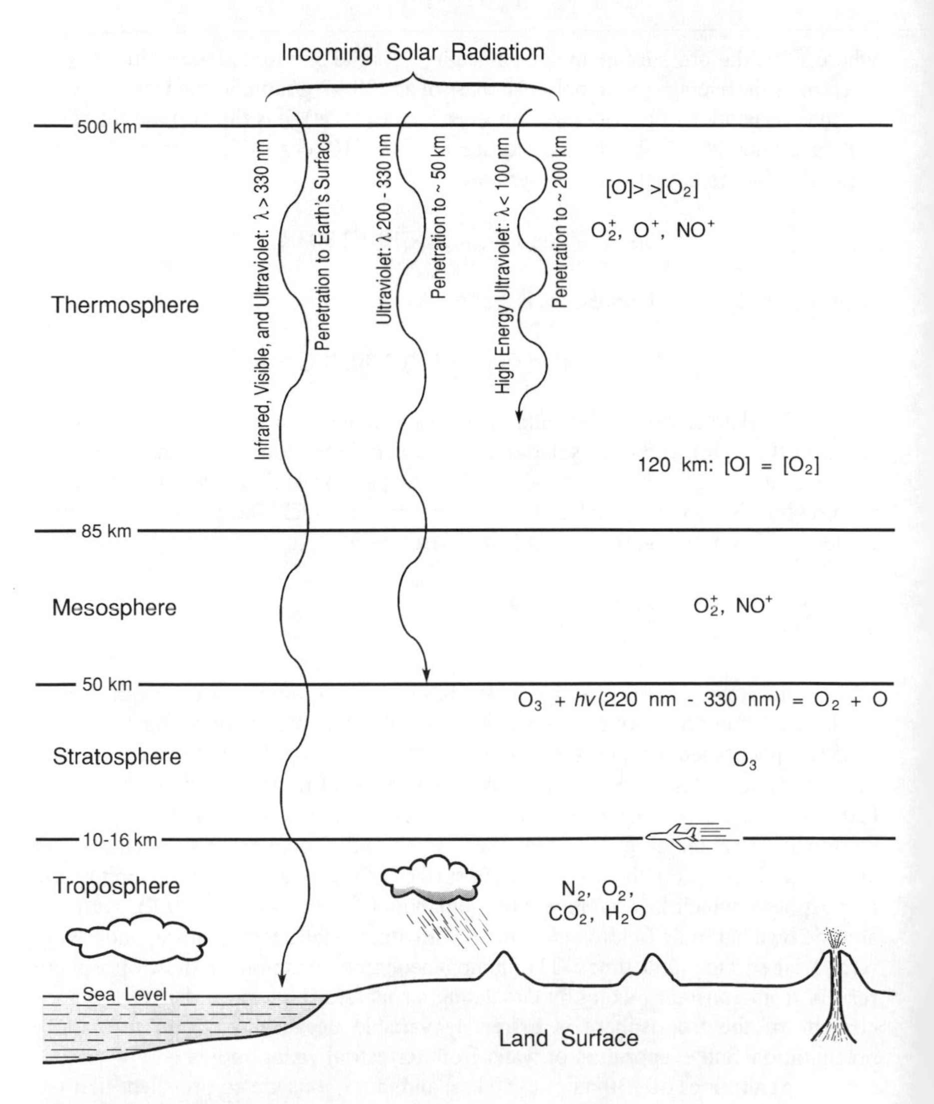

Figure 5.1: The structure of the atmosphere

where the concentration of species is relatively high. Thus the lower limit of the ionosphere lifts at night and makes possible the transmission of radio waves over much greater distances.

The very cold temperature of the *tropopause* layer at the top of the troposphere serves as a barrier that causes water vapor to condense to ice. Therefore the vapor cannot reach altitudes at which it would dissociate through the action of intense high-energy ultraviolet radiation.

The atmospheric layer directly above the troposphere is the *stratosphere*, in which the temperature rises to a maximum of -2°C (28°F) with increasing altitude. This phenomenon is due to the presence of ozone (O_3) which may reach a level of around 10 parts per million (ppm) v/v in the midrange of the stratosphere. The heating effect is caused by the absorption of ultraviolet radiation energy by ozone. The situation can become even more complicated due to the formation of additional ozone in urban plumes (Colbeck and MacKenzie, 1994).

2.2 Oxygen/Ozone

Ozone is found in trace quantities throughout the atmosphere, the largest concentrations being located in a layer in the lower stratosphere between the altitudes of 9 and 18 miles (15 and 30 km) (Figure 5.1). This ozone results from the dissociation by solar ultraviolet radiation of molecular oxygen in the upper atmosphere and nitrogen dioxide in the lower atmosphere.

Although present in only trace quantities, atmospheric ozone plays a critical role for the biosphere by absorbing the ultraviolet radiation with a wavelength (λ) in the range 240-320 nm which would otherwise be transmitted to the land surface. This radiation is lethal to simple unicellular organisms (algae, bacteria, protozoa) and to the surface cells of higher plants and animals. It also damages deoxyribonucleic acid (DNA), the genetic material of cells, and is responsible for sunburn in human skin with the ensuing potential for skin cancer.

Ozone also plays a role in heating the upper atmosphere by absorbing solar ultraviolet and visible radiation ($\lambda < 710$ nm and thermal infrared radiation ($\lambda = \sim 9.6$ microns). In consequence, the temperature increases steadily from about -50°C (-60°F) at the *tropopause* (5-10 miles or 8-16 km altitude) to about 7°C (45°F) at the *stratopause* (30 miles or 50 km altitude). This ozone heating provides the major energy source for driving the circulation of the upper stratosphere and mesosphere.

The absence of high levels of radiation-absorbing species in the *mesosphere* immediately above the stratosphere results in a further temperature decrease to about -92°C (-134°F) at an altitude ~53 miles (~85 km). The upper regions of the mesosphere define a region called the exosphere from which molecules and ions can completely escape the atmosphere. Extending to the far outer reaches of the atmosphere is the *thermosphere*, in which the highly rarified gas reaches temperatures as high as 1200°C (2200°F) by the absorption of very energetic radiation of wavelengths less than approximately 500 nm by gas species.

Approximately 50% of the solar radiation entering the atmosphere reaches the land surface. The remainder is either reflected away from the earth or absorbed into the atmosphere, after which the energy is radiated away from the earth. In either case, the effect may appear to be the same. Most of the solar energy that is transmitted through the atmosphere must, at some time, be returned to space in order to maintain a heat balance. In addition, <1% of the total energy interacting with the atmosphere reaches the land surface by convection and conduction processes, and this must be returned to space. Thus, energy transport, which is crucial to the disposition of energy within the atmosphere (as well as within the land and water systems) involves three major physical mechanisms: conduction, convection, and radiation.

Briefly, *conduction* of energy occurs through the interaction of adjacent atoms or molecules without the bulk movement of matter. On the other hand, *convection* involves the movement of whole masses of air, which may be either relatively warm or cold. For example, convection is the means by which temperature variations occur when large masses of air move across a land surface or ocean. *Radiation*, in the current context, is the mechanism by which heat is transported away from the earth, usually after conduction and convection effects have transported the heat to the atmosphere.

Horizontally moving air (wind) and vertically moving air (air current) are very much involved with air pollution phenomena. Wind carries and disperses air pollutants. Prevailing wind direction is an important factor in determining the areas most affected by a source of chemical pollutants.

Atmospheric water, present as vapor, liquid, or ice, provides the humidity and the relative humidity, which expressed as a percentage, describes the amount of water vapor in the air as a ratio of the maximum amount that the air can hold at that temperature. Clouds normally form when the air can no longer hold water in vapor form, and the water forms very small aerosol droplets.

Clouds are important absorbers and reflectors of radiation (heat) and their formation is affected by human activities, especially particulate matter pollution and emission of deliquescent gases such as sulfur dioxide (SO_2) and hydrogen chloride (HCl). In fact, concern over the emissions of potentially toxic substances from all sources and their possible effects in the environment has led to the introduction of legislative controls in several countries (Maier, 1990; Kyte, 1991; Nilsson, 1991). In the United States, passage of the Amendments to the Clean Air Act introduced stricter controls for present and future years (Chapter 12).

Condensation of water vapor, which forms clouds, must occur prior to the formation of precipitation in the form of rain or snow. For this condensation to happen, air must be cooled below the dew point, and nuclei of condensation must be present. These nuclei are hygroscopic substances such as salts, sulfuric acid droplets, and some organic materials, including bacterial cells.

Distinct air masses are a major feature of the troposphere and these air masses are uniform and horizontally homogeneous. The temperature and water vapor content are particularly uniform. These characteristics are determined by the nature of the surface over which a large air mass forms. Polar continental air masses form over cold land regions; polar maritime air masses form over polar oceans. Air masses originating in the tropics may be similarly classified as tropical continental air masses or tropical maritime air masses. The movement of air masses and the conditions that form them may have important effects upon pollutant reactions and dispersal.

The movement of air across the land surface is crucial insofar as, when air movement ceases, stagnation can occur with a resultant buildup of chemical pollutants in localized regions. Although the temperature of air relatively near the land surface normally decreases with increasing altitude, certain atmospheric conditions can result in the opposite condition, increasing temperature with increasing altitude. Such conditions are characterized by high atmospheric stability and are known as *temperature inversions*. Typically, an inversion occurs by the collision of a warm air mass (warm front) with a cold air mass (cold front). The warm air mass overrides the cold air mass in the frontal area, producing the inversion. Temperature inversions result in air stagnation and the trapping of pollutants in local areas.

At this time, it is appropriate to define and use the terms: reservoir, burden, and flux. A *reservoir* is a domain, such as the atmosphere, where a pollutant may reside for an indeterminate time. The amount of a specific pollutant in a reservoir is known as the *burden*, which is usually expressed in units of 10^6 metric tons (10^{12} g called teragrams, Tg). The rate of transfer of a pollutant from one sphere or domain to another is called the *flux*, which is frequently expressed in units of teragrams per year.

The chemical reactions that occur in the atmosphere (Batta, 1975; Prinn, 1991; Birks et al., 1993) primarily involve, on paper, a series of simple chemical reactions. Indeed, the chemistry of the solar system (Lunine, 1995) appears to involve simple chemical reactions also. However, the simple paper chemistry belies the fact that both the atmosphere and the solar system are extremely complex systems. In addition, atmospheric chemistry suffers from the influence of human activities to a greater extent than the solar system.

In terms of the nonhuman effects on atmospheric chemistry, one of the issues least understood is the role of the absorption of light by chemical species which brings about photochemical reactions. These reactions would not occur under atmospheric (particularly temperature) conditions in the absence of light.

Nitrogen dioxide (NO_2) is one of the most photochemically active species found in a polluted atmosphere and is an essential participant in the smog-formation process. A species such as nitrogen dioxide may absorb photo energy (hv) to produce an excited molecule:

$$NO_2 + hv = NO_2^*$$

These highly reactive species are strongly involved in chemical reactions in the atmosphere. The other two species are atoms or molecular fragments with unshared electrons (free radicals) and ions consisting of ionized atoms or molecular fragments.

The free radicals are generated when the absorption of light promotes an orbital electron (usually paired in the stable molecular state) to a vacant orbital of higher energy. This interaction gives rise to an excited singlet state, in which the promoted electron has a spin opposite to that of its former orbital partner. In other cases the spin of the promoted electron is reversed, such that it has the same spin as its former partner and gives rise to an excited triplet state.

Although ions are produced in the upper atmosphere primarily by the action of energetic electromagnetic radiation, they may also be produced in the troposphere by the shearing of water droplets during precipitation. The shearing may be caused by the compression of descending masses of cold air or by strong winds over hot, dry land masses.

Atmospheric oxygen takes part in energy-producing reactions, such as the burning of fossil fuels:

$$CH_{4(natural\ gas)} + 2O_2 = CO_2 + 2H_2O$$

Atmospheric oxygen is also utilized by aerobic organisms in the degradation of organic material. Some oxidative weathering processes consume oxygen, and oxygen is returned to the atmosphere through plant photosynthesis.

Because of the extremely rarefied atmosphere and the effects of ionizing radiation, elemental oxygen in the upper atmosphere exists to a large extent in forms other than diatomic oxygen, O_2. These forms in the upper atmosphere are oxygen atoms, $O^{\cdot}$, excited oxygen molecules, $O_2{}^*$, and ozone, O_3.

Above about 19 miles (30 km), oxygen is dissociated during the daytime by photo energy:

$$O_2 + hv = O + O$$

and these oxygen atoms then form ozone:

$$O + O_2 + M = O_3 + M$$

where M is an arbitrary third-body molecule required to conserve the energy of the reaction. Ozone has a short lifetime during the day because of photodissociation:

$$O_3 + hv = O_2 + O$$

Above 54 miles (90 km), where oxygen is a minor component of the atmosphere, there is no net destruction of ozone, and the oxygen is almost exclusively converted back to ozone.

Nitric oxide (NO) and nitrogen dioxide (NO_2) tend to promote ozone removal:

$$NO + O_3 = NO_2 + O_2$$
$$NO_2 + O = NO + O_2$$

Ozone destruction also involves chlorine atoms ($Cl^{\cdot}$) and hypochlorite species ($ClO^{\cdot}$) produced by decomposition in the stratosphere of chlorine species from various sources (e.g., $CFCl_3$, CF_2Cl_2, CH_3CCl_3, $CHCl_3$, and CCl_4):

$$Cl + O_3 = ClO + O_2$$
$$ClO + O = Cl + O_2$$
$$ClO + h\nu = Cl + O$$
$$Cl + CH_4 = HCl + CH_3$$
$$OH + HCl = Cl + H_2O$$
$$ClO + NO_2 + M = ClNO_3 + M$$
$$ClNO_3 + h\nu = ClO + NO_2$$

In addition to these chemical reactions, ~1% of the atomic oxygen is removed by downward transport of ozone into the troposphere, where it is destroyed at or near the ground.

One of the objections to the use of supersonic aircraft, designed to fly in the lower stratosphere (such as the Anglo-French Concorde and Russian Tupolev-144), was related to the production of nitric oxide and nitrogen dioxide by the thermal decomposition of air (N_2 and O_2) in their engines. It was projected that a fleet of such aircraft would inject nitrogen oxides into the stratosphere at a rate some three times greater than that from other sources. The ensuing perturbation to the atmospheric nitrogen cycle would cause depletion of the stratospheric ozone. There is the alternate view that the supersonic aircraft have a much smaller effect on stratospheric ozone than originally predicted. Future plans for such aircraft include serious attempts to mitigate the potential effects on the atmosphere (Zurer, 1995).

There has also been concern that industrial production of the chlorofluoromethanes Freon 11 ($CFCl_3$) and Freon 12 (CF_2Cl_2) may cause substantial changes to the natural chlorine cycle. The freons have been used principally as refrigerants and aerosol-can propellants. Once they are released into the atmosphere, their only presently recognized removal mechanism involves photodissociation in the stratosphere:

$$CFCl_3 + h\nu = CFCl_2 + Cl$$
$$CF_2Cl_2 + h\nu = CF_2Cl + Cl$$

A number of other potential ozone-altering processes with anthropogenic origins have also been identified. For example, increases in the levels of carbon dioxide in the atmosphere due to the combustion of carbon-based (fossil) fuel can lead to stratospheric cooling and the ensuing decrease in the ozone destruction rate. Atmospheric testing of nuclear weapons produces nitric oxide and nitrogen dioxide which passes into the stratosphere. In general, any process that produces chemicals that have the potential to react directly (or indirectly) to increase/decrease the level of ozone in the atmosphere has the potential to cause environmental effects.

Ozone is produced during the oxidation of unburned hydrocarbons (RH):

$$RH + OH = R + H_2O$$
$$R + O_2 = RO_2$$
$$RO_2 + NO = RO + NO_2$$
$$NO_2 + hv = NO + O$$
$$O + O_2 = O_3$$

Ozone is also generated by the oxidation of carbon monoxide:

$$CO + OH = CO_2 + H$$
$$H + O_2 = HO_2$$
$$HO_2 + NO = OH + NO_2$$
$$NO_2 + hv = NO + O$$
$$O + O_2 = O_3$$

Some of the ozone in the troposphere is derived by injection of ozone from the stratosphere. Tropospheric ozone is involved in the formation of photochemical smog, in the production of acid rain (Chapter 10) from sulfur oxides and from nitrogen oxides, and in reactions with chemical pollutants.

Acid rain (also called acid deposition) is a form of pollution depletion in which pollutants are transferred from the atmosphere to soil or water. It is possible to measure the acidity of the deposition by electrometric methods (American Society for Testing and Materials, 1995, ASTM D 5015). Acid rain is, in fact, precipitation that incorporates anthropogenic acids and acidic materials. However, the deposition of acidic materials on the land surface occurs in both wet and dry forms as rain, snow, fog, dry particles, and gases. Acid precipitation, strictly defined, contains a greater concentration of hydrogen (H^+) than of hydroxyl (OH^-) ions, resulting in a solution pH less than 7. Under this definition, nearly all precipitation is acidic but the phenomenon of acid rain is regarded as resulting from human activity.

Atomic oxygen is stable primarily in the thermosphere, where the atmosphere is so rarefied that the three-body collisions necessary for the chemical reaction of atomic oxygen seldom occur (the third body in this kind of three-body reaction absorbs energy to stabilize the products).

2.3 Nitrogen

The 78% v/v nitrogen contained in the atmosphere constitutes an inexhaustible reservoir of that essential element, either in the elemental form (N_2) or in combination with other elements (Parrish and Buhr, 1993). The nitrogen cycle and nitrogen fixation by microorganisms are essential atmosphere-biosphere interactions. A small amount of nitrogen is thought to be fixed in the atmosphere by lightning and some is also fixed by combustion processes, as in the internal combustion engine.

Unlike oxygen, which is almost completely dissociated to the monatomic form in higher regions of the thermosphere, molecular nitrogen is not readily dissociated by ultraviolet radiation. However, at altitudes exceeding ~62.5 miles (~100 km), atomic nitrogen is produced by photochemical reactions:

$$N_2 + h\nu = N + N$$

2.4 Other Gases

Although only about 0.035% v/v (350 ppm) of air consists of carbon dioxide, it is a species of some concern. Carbon dioxide, along with water vapor, is primarily responsible for the absorption of infrared energy reemitted by the land such that some of this energy is radiated back to the land surface. Current evidence suggests that change in the atmospheric carbon dioxide level will substantially alter the climate through the greenhouse effect (Pickering and Owen, 1994).

The most obvious factor contributing to increased atmospheric carbon dioxide is consumption of carbon-containing fossil fuels. In addition, release of carbon dioxide from the biodegradation of biomass and uptake by photosynthesis are important factors determining overall carbon dioxide levels in the atmosphere. The role of photosynthesis shows a seasonal cycle in carbon dioxide levels in the northern hemisphere.

Forests have a much greater influence than other vegetation because forests carry out more photosynthesis. Furthermore, forests store enough fixed, but readily oxidizable carbon in the form of wood (Goldstein, 1987) and humus (Stevenson, 1994) to have a marked influence on atmospheric carbon dioxide content. During the summer months, forests carry out enough photosynthesis to markedly reduce the atmospheric carbon dioxide content. During the winter, metabolizing biota, such as bacterial decay of humus, release a significant amount of carbon dioxide. The current worldwide trend toward destruction of forests and conversion of forest lands to agricultural uses will contribute substantially to a greater overall increase in atmospheric carbon dioxide levels.

Chemically and photochemically, carbon dioxide is a comparatively insignificant species because of its relatively low concentrations and low photochemical reactivity. However, calculations based on known photochemical reactions, carbon dioxide levels, and ultraviolet radiation intensity indicate that photodissociation of carbon dioxide by solar ultraviolet radiation should occur in the upper atmosphere:

$$CO_2 + hv = CO + O$$

Water circulates through the atmosphere in the hydrologic cycle (Chapter 4). The water vapor content of the troposphere is normally within a range of 1-3% v/v and the global average is about 1% v/v. However, air can contain as little as 0.1% v/v or as much as 5% v/v water. The percentage of water in the atmosphere decreases rapidly with increasing altitude.

Water vapor absorbs infrared radiation even more readily than does carbon dioxide, thus greatly influencing the earth's heat balance. Clouds formed from water vapor reflect light from the sun and have a temperature-lowering effect. On the other hand, water vapor in the atmosphere acts as a blanket, insofar as it retains heat from the land surface by absorption of infrared radiation.

Water vapor in the upper atmosphere is involved in the formation of hydroxyl ($HO^{\cdot}$) and hydroperoxyl ($HOO^{\cdot}$) radicals. Condensed water vapor in the form of very small droplets is of considerable concern in atmospheric chemistry (Cantrell et al., 1993; O'Brien and Hard, 1993). The harmful effects of some air pollutants (e.g., the corrosion of metals by acid-forming gases) require the presence of water, which may come from the atmosphere. Water vapor interacting with pollutant particulate matter in the atmosphere may reduce visibility to undesirable levels through the formation of aerosol particles (fog).

When ice particles in the atmosphere change to liquid droplets or when these droplets evaporate, heat is absorbed from the surrounding air. Reversal of these processes results in heat release to the air (as latent heat). This may occur many miles from the place where heat was absorbed and is a major mode of energy transport in the atmosphere. It is the predominant type of energy transition involved in thunderstorms, hurricanes, and tornadoes.

On a global basis, rivers drain only about one-third of the precipitation that falls on the continents. This means that two-thirds of the precipitation is lost as combined evaporation and transpiration. During the summer, losses by evaporation and transpiration may exceed precipitation because of the large quantities of water stored in the root zone of the soil. In some cases, the evaporation and transpiration furnish atmospheric water vapor necessary for cloud formation and precipitation. It is probable, therefore, that large-scale deforestation, soil damage (such as plowing up grasslands in semiarid areas), and irrigation could have an effect on regional climate and rainfall.

The cold tropopause serves as a barrier to the movement of water into the stratosphere. The main source of water in the stratosphere is the photochemical oxidation of methane:

$$CH_4 + 2O_2 + h\nu = CO_2 + 2H_2O$$

The water thus produced serves as a source of stratospheric hydroxyl radical ($HO^{\cdot}$):

$$H_2O + h\nu = HO^{\cdot} + H^{\cdot}$$

2.5 Particulate Matter

Particles are common components of the atmosphere (American Society for Testing and Materials, 1995, ASTM D 1704), particularly the troposphere. Colloidal-sized particles in the atmosphere are called aerosols. Most aerosols from natural sources have a diameter of <0.1 micron. These particles originate in nature from sea spray, smoke, dust, and the evaporation of organic materials from vegetation. Other typical particles of natural origin in the atmosphere are bacteria, fog, pollen grains, and volcanic ash.

Important atmospheric phenomena that involve aerosol particles are electrification phenomena, cloud formation, and fog formation. Particles help determine the heat balance of the atmosphere by reflecting light. Probably the most important function of particles in the atmosphere is their action as nuclei for the formation of ice crystals and water droplets. Current efforts at rain making are centered around the addition of condensing particles to atmospheres supersaturated with water vapor. Dry ice was used in early attempts; now silver iodide, which forms huge amounts of very small particles, is used.

Particles are involved in many chemical reactions in the atmosphere (Figure 5.2). Neutralization reactions, which occur most readily in solution, may take place in water droplets suspended in the atmosphere. Small particles of metal oxides and carbon have a catalytic effect on oxidation reactions. Particles may also participate in oxidation reactions induced by light.

3.0 Dispersion Effects

The dispersion of chemical pollutants in the atmosphere is determined by mean wind flow conditions and by atmospheric turbulence. Turbulence results from such factors as the friction of the land surface, physical obstacles to wind flow, and the vertical temperature profile of the lower atmosphere.

Stability class refers to the degree of turbulence in the atmosphere. For air-quality purposes, stability usually refers to the lower layers of the atmosphere, where pollutants are emitted. The idea of discrete stability classes is a simplification of the complex nature of the atmosphere, but has proved useful in predictive studies.

A stable atmosphere is marked by air that is cooler at the ground than aloft, by low wind speeds, and consequently, by a low degree of turbulence. A pollutant plume released into a stable lower layer of the atmosphere can remain relatively intact for long distances. On the other hand, an unstable atmosphere is marked by a high degree of turbulence. A visible plume released into an unstable atmosphere may exhibit a characteristic looping appearance produced by turbulent eddies. An intermediate turbulence class between stable and unstable conditions is the "neutral" stability class. A visible plume released into a neutral stability condition may display a cone-like appearance as the edges of the plume spread out in a V shape.

The term inversion refers to a layer in the atmosphere where temperature increases with height rather than decreases, as is usually the case. This inversion layer serves as a cap and restricts chemical pollutants from any further upward dispersal. As a result, pollutant levels will increase below the cap and an extended period of such a pollutant buildup is usually referred to as a *stagnation episode* during which smog (Chapter 10) can occur.

In order to reduce the emissions at a given locale, there has been a tendency to build larger stacks. Pollutants emitted from a source with a tall stack tend to

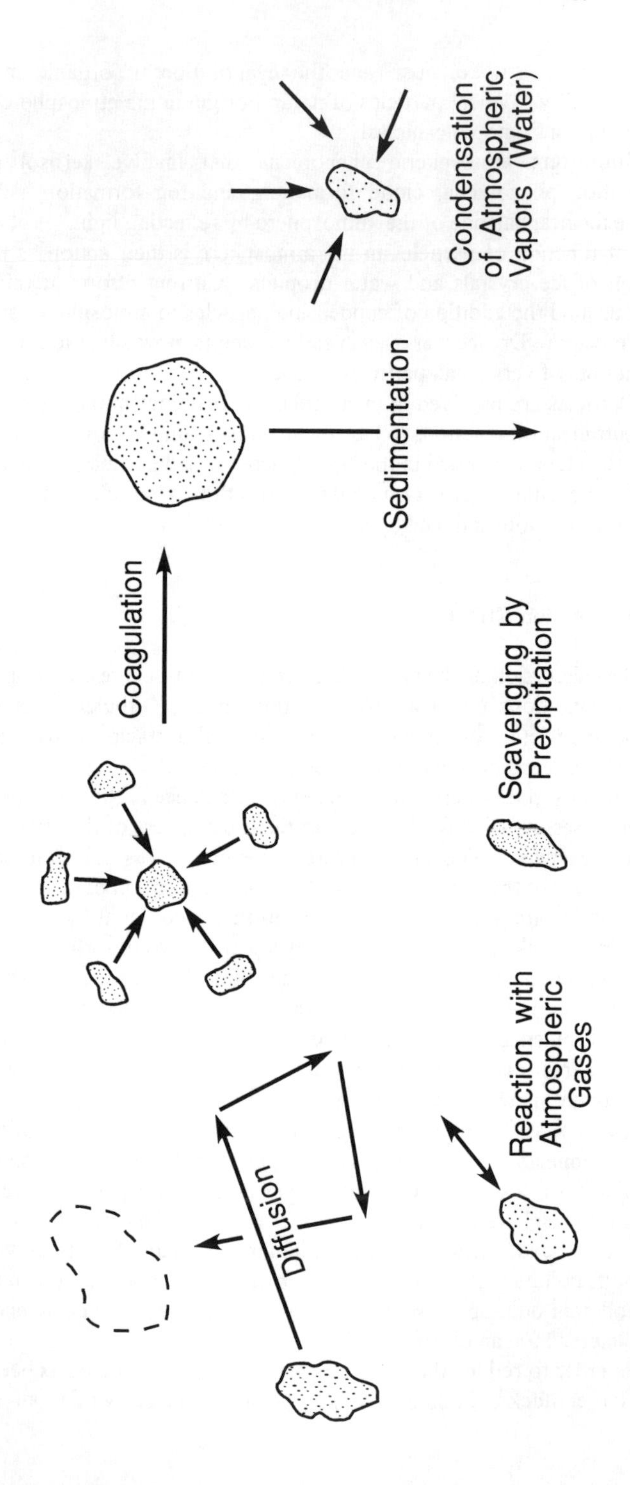

Figure 5.2: Particle behavior in the atmosphere

produce lower ground-level concentrations than the same amount of pollutants emitted from a source with a short stack. Sources with the same stack height can produce different impacts, depending on the plume rise above the stack (Figure 5.3), which depends on the exit velocity of the emissions, the temperature of the emissions, and the atmospheric conditions. The combination of the physical stack height and plume rise above the stack is referred to as the effective stack height.

Pollutants emitted or formed in the atmosphere are eventually depleted. In addition to chemical transformation, which acts as a depletion mechanism for precursors, two other common depletion mechanisms are dry deposition and washouts. *Dry deposition* refers to the removal of both particles and gases as they come into contact with the land surface. *Washouts* refer to the uptake of particles and gases by water droplets and snow, and their removal from the atmosphere when rain and snow fall on the ground.

The actual removal of pollutants from the atmosphere by *wet deposition* requires the formation of precipitation within the clouds. Without cloud elements greater than about 100 microns in diameter, the pollutant mass remains in the air, largely in association with the relatively small cloud drops, which have negligible rates of descent.

Since most precipitation in midlatitude storms is initiated by the formation of ice particles in the cold upper reaches of the clouds, pollutant fractionation between water phases inhibits the transfer of cloud water acidity to large particles, which do precipitate readily. Partly because of such microphysical phenomena and partly because of nonuniform distributions of pollutants within the clouds, the acidity of precipitation tends to be substantially less than that of the cloud water that remains aloft.

The outcome of all of these effects is that the atmosphere can be seriously polluted, as life in some inner city areas has attested. Determination of the pollutants in the atmosphere, through strict sampling protocols, is a major issue necessitating accuracy and reliability (Lodge, 1988).

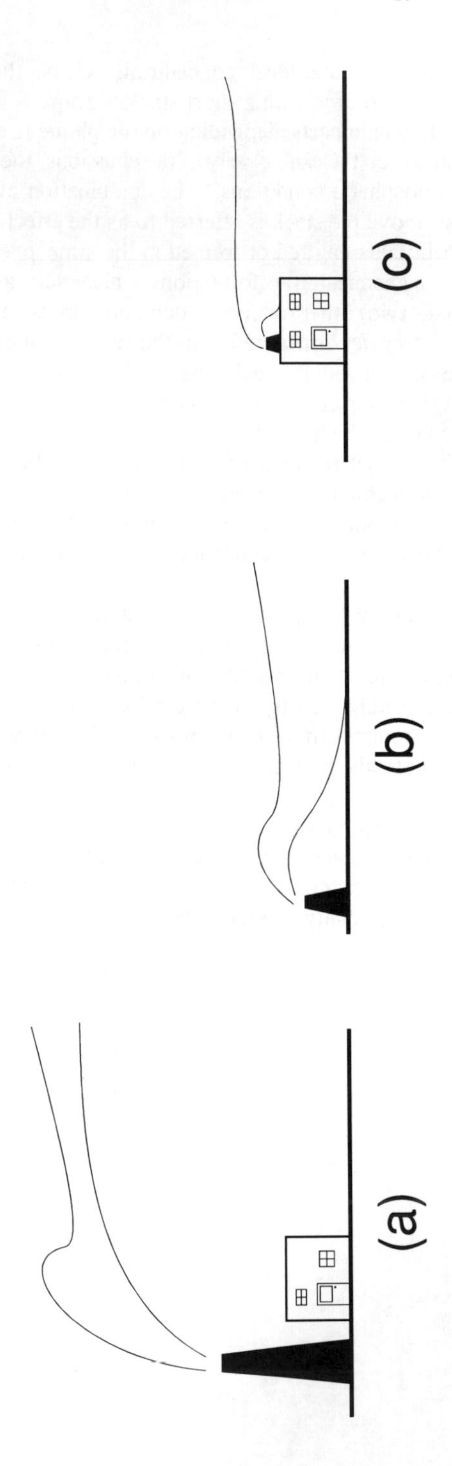

Figure 5.3: Structure of a smoke stack plume depending upon wind condition and location of the stack

4.0 References

American Society for Testing and Materials 1995. *Annual Book of ASTM Standards*. Vol. 11.03. American Society for Testing and Materials, Philadelphia, Pennsylvania.

American Society for Testing and Materials. 1995. ASTM D 1704. Amount of Particulate Matter in the Atmosphere. *Annual Book of ASTM Standards*. Vol. 11.03. American Society for Testing and Materials, Philadelphia, Pennsylvania.

American Society for Testing and Materials. 1995. ASTM D 5015. 1995. pH of Atmospheric Wet Deposition Samples. *Annual Book of ASTM Standards*. Vol. 11.03. American Society for Testing and Materials, Philadelphia, Philadelphia, Pennsylvania.

Batta, L.J. 1975. *Atmospheric Chemistry: Problems and Scope*. National Academy of Sciences, Washington, D.C.

Birks, J.W., Calvert, J.G., and Sievers, R.W. (eds.). 1993. *The Chemistry of the Atmosphere: Its Impact on Global Change*. American Chemical Society, Washington, D.C.

Cahill, T.A., and Wakabayashi, P. 1993. Chap. 7 in *Measurement Challenges in Atmospheric Chemistry*. Edited by L. Newman. Advances in Chemistry Series No. 232. American Chemical Society, Washington, D.C.

Cantrell, C.A., Shetter, R.A., McDaniel, A.H., and Calvert, J.G.. 1993. Chap. 11 in *Measurement Challenges in Atmospheric Chemistry*. Edited by L. Newman Advances in Chemistry Series No. 232. American Chemical Society, Washington, D.C.

Charlson, R.J., and Wigley, T.M.L. 1994. *Scientific American* 270(2): 48.

Colbeck, I., and MacKenzie, A.R. Chap. 4 in *Environmental Oxidants*. Edited by J.O. Nriagu and M.S. Simmons. John Wiley & Sons Inc., New York.

Dasgupta, P.K. 1993. Chap. 2 in *Measurement Challenges in Atmospheric Chemistry*. Edited by L. Newman. Advances in Chemistry Series No. 232. American Chemical Society, Washington, D.C.

Flagan, R.C. 1993. Chap. 6 in *Measurement Challenges in Atmospheric Chemistry*. Edited by L. Newman. Advances in Chemistry Series No. 232. American Chemical Society, Washington, D.C.

Goldstein, I.S. 1987. In *Encyclopedia of Science and Technology*. Vol. 19, p. 498. Edited by S.P. Parker. McGraw-Hill, New York.

Hidy, G.M., Mueller, P.K., Grosjean, D., Appel, B.R., and Wesolowski, J.J. 1979. *The Character and Origins of Smog Aerosols*. John Wiley & Sons Inc., New York.

Kyte, W.S. 1991. *Desulphurisation 2: Technologies and Strategies for Reducing Sulphur Emissions*. Institute of Chemical Engineers, Rugby, Warwickshire, England.

Lodge, J.P., Jr. (ed.). 1988. *Methods of Air Sampling and Analysis. 3rd ed*. Lewis Publishers, Chelsea, Michigan.

Lunine, J.I. 1995. *Chemical and Engineering News* 73(3): 40.

Maier, H. 1990. *VGB Kraftwerkstechnik* 70(10): 749.

Newman, L. (ed.). 1993. *Measurement Challenges in Atmospheric Chemistry*. Advances in Chemistry Series No. 232. American Chemical Society, Washington, D.C.

Nilsson, K. 1991. *UNEP Industry and Environment*. p. 73.

O'Brien, R.J., and Hard, T.M. 1993. Chap. 12 in *Measurement Challenges in Atmospheric Chemistry*. Edited by L. Newman. Advances in Chemistry Series No. 232. American Chemical Society, Washington, D.C.

Parrish, D.D. and Buhr, M.P. 1993. Chap. 9 in *Measurement Challenges in Atmospheric Chemistry*. Edited by L. Newman. Advances in Chemistry Series No. 232. American Chemical Society, Washington, D.C.

Pickering, K.T., and Owen, L.A. 1994. *Global Environmental Issues*. Routledge Publishers, New York.

Prinn, R.G. 1991. In *Energy and the Environment in the 21st Century*. p. 27 Edited by J.W. Tester, D.O. Wood, and N.A. Ferrari. MIT Press, Cambridge, Massachusetts.

Stevenson, F.J. 1994. *Humus Chemistry*. John Wiley & Sons Inc., New York.

Westberg, H., and Zimmerman, P. 1993. Chap. 10 in *Measurement Challenges in Atmospheric Chemistry*. Edited by L. Newman. Advances in Chemistry Series No. 232. Washington, American Chemical Society, Washington, D.C.

Zamaraev, K.I., M.I. Khramov, and V.N. Parmon. 1994. *Catalysis Reviews. Science and Engineering* 36: 617.

Zurer, P.S. 1995. *Chemical and Engineering News* 73(17): 10.

Part III Chemical Waste and Waste Management

Chapter 6

Chemical Waste

1.0 Introduction

The pollution of ecosystems, either inadvertently or deliberately, has been a fact of life for millennia (Pickering and Owen, 1994). In recent times, the evolution of industrial operations has led to issues related to the disposal of a wide variety of chemical contaminants (Table 6.1) (Chapter 2) (Easterbrook, 1995). Chemical wastes that were once exotic have become commonplace and hazardous (Tedder and Pohland, 1993, 1995). Recognition of this makes it all the more necessary that steps be taken to terminate the pollution, preferably at the source or before it is discharged into the environment. Several technologies are under development (United States Environmental Protection Agency, 1993). It is also essential that the necessary tests be designed to detect the pollution and its effect on living forms.

There is probably no one who can testify with any degree of accuracy (but with a high degree of uncertainty) as to when the earth was last pristine and unpolluted. Yet, to attempt to return the environment to such a mythical time might have a severe effect on the current indigenous life, perhaps a form of pollution in reverse!

The seepage of oil sand bitumen (a highly viscous petroleum-like material) (Speight, 1991) into the Athabasca River in northeastern Alberta (Canada) over the centuries is an example of inadvertent pollution. Yet the indigenous fish, a hardy northern perch, seem oblivious to the pollution. It would therefore be foolhardy to test the toxicity of the water of the Athabasca River on fish that had lived in a near-pristine system, although steps have been taken by the Alberta Government to reduce the amount of pollution flowing into the river.

On the other hand, in medieval Europe the deliberate (perhaps unknowingly, since raw sewage was often dumped into the street or any open space behind the house) dumping of raw sewage into rivers was the cause of many deaths from diseases such as typhoid (Cartwright and Biddiss, 1972; James and Thorpe, 1994). Realization that such practices caused disease and death brought on a new environmental awareness. It is now forbidden in many countries to deposit onto the land or introduce into the waterways any material not indigenous to the system or any material that, being indigenous to the system, would bring the quantities above the natural abundance. Yet, pollution still occurs.

Humans have been exposed to hazardous substances since prehistoric times, when they inhaled noxious volcanic gases or succumbed to carbon monoxide from inadequately vented fires in cave dwellings sealed against Ice Age cold. Slaves in

Table 6.1: Types of chemical waste

Source	Waste type
Chemical manufacturers	Strong acids and bases Spent solvents Reactive materials
Vehicle maintenance shops	Heavy-metal paints Ignitable materials Used lead-acid batteries Spent solvents
Printing industry	Heavy-metal solutions Waste ink Spent solvents Spent electroplating wastes Ink sludge containing heavy metals
Leather products	Waste toluene and benzene
Paper industry	Paint wastes containing heavy metals
Construction industry	Ignitable paint wastes Spent solvents Strong acids and bases
Cleaning agents and cosmetics manufacturing	Heavy-metal dusts Ignitable materials Flammable solvents Strong acids and bases
Furniture and wood manufacturing and refinishing	Ignitable materials Spent solvents
Metal manufacturing	Paint wastes containing heavy metals Strong acids and bases Cyanide wastes Sludge containing heavy metals

ancient Greece developed lung disease from weaving mineral asbestos fibers into cloth to make it more degradation-resistant. Some archaeological and historical studies have concluded that lead wine containers were a leading cause of lead poisoning in the more affluent ruling class of the Roman Empire. Lead poisoning caused erratic behavior, such as fixation on spectacular sporting events, chronic unmanageable budget deficits, poorly regulated financial institutions, and ill-conceived, overly ambitious military ventures in foreign lands. Alchemists who plied their art during the Middle Ages often suffered debilitating injuries and illnesses resulting from the hazards of their explosive and toxic chemicals. During the 1700s runoff from mine spoils began to create serious contamination problems in Europe.

As the production of dyes and other organic chemicals developed from the coal tar industry in Germany during the nineteenth century, pollution and poisoning from coal tar by-products were observed. By around 1900, the quantity and variety of chemical wastes produced each year were increasing sharply with the addition of spent steel and iron pickling liquor, lead battery wastes, chromium, petroleum refinery, radium, and fluoride from aluminum ore refining. As the century progressed into the World War II era, the wastes and hazardous by-products of manufacturing increased markedly from sources such as chlorinated solvents, pesticide synthesis, polymers, plastics, paints, and wood preservatives.

Any chemical substance, if improperly managed or disposed of, may pose a danger to living organisms, materials, structures, or the environment, by explosion or fire hazards, corrosion, toxicity to organisms, or other detrimental effects. In addition, many chemical substances, when released to the environment can be classified as hazardous or nonhazardous. Consideration must be given to the distribution of chemical wastes on land systems, in water systems, and in the atmosphere (Chapter 1).

Every industrial country in the world can admit to issues arising from the management of chemical wastes (American Chemical Society, 1994). Improper disposal of these waste streams in the past has created a need for very expensive cleanup operations. Efforts are under way internationally to remedy old problems caused by hazardous wastes and to prevent the occurrence of problems in the future.

In general terms, the origin of chemical wastes refers to their points of entry into the environment, which may consist of (1) deliberate addition to soil, water, or air by humans; (2) evaporation or wind erosion from waste dumps into the atmosphere; (3) leaching from waste dumps into groundwater, streams, and bodies of water; (4) leakage, such as from underground storage tanks or pipelines; (5) accidents, such as fire or explosion; and (6) release from improperly operated waste treatment or storage facilities.

2.0 Classification

There are three basic approaches to defining a chemical waste as hazardous: (1) a qualitative description of the waste by origin, type, and constituents; (2) classification by characteristics based upon testing procedures; and (3) classification as a result of the concentration of specific chemical substances. Various countries have different definitions of chemical waste. Usually the definition involves qualification of whether or not the material is hazardous. For example, in parts of Europe a hazardous waste is any material that is

> ...especially hazardous to human health, air, or water, or which are explosive, flammable, or may cause diseases." Poisonous waste is that material "....which is poisonous, noxious, or polluting and whose presence on the land is liable to give rise to an environmental hazard."

In more general terms:

> ...hazardous waste is waste material that is unsuitable for treatment or disposal in municipal.....treatment systems, incinerators or landfills and that therefore requires special treatment.

Somewhat paradoxically, measures taken to reduce air and water pollution may actually increase production of chemical wastes. As examples, most water treatment processes yield sludge or concentrated liquors that require stabilization and disposal (Cheremisinoff, 1995). Air scrubbing processes likewise produce sludge. Baghouses and precipitators used to control air pollution (Speight, 1993) yield significant quantities of solids, some of which are hazardous.

2.1 Hazardous Waste

A chemical waste is considered hazardous if it exhibits one or more of the following characteristics: *ignitability*, *corrosivity*, *reactivity*, and *toxicity*. Under the authority of the Resource Conservation and Recovery Act (RCRA), the United States Environmental Protection Agency (EPA) defines hazardous substances in terms of the above characteristics.

Ignitability: characteristic of substances that are liquids whose vapors are likely to ignite in the presence of ignition sources, nonliquids that may catch fire from friction or contact with water and which burn vigorously or are persistently ignitable compressed gases and oxidizers. Ignitable wastes can create fires under certain conditions; examples include solvents that readily catch fire and friction-sensitive substances.

Organic solvents are used widely and in large amounts in industries, laboratories, and homes (Table 6.2). They are released to the atmosphere as vapor and can pose a significant inhalation hazard. Improper storage, use, and disposal have resulted in the contamination of land systems as well as groundwater and drinking water (Barcelona et al., 1990).

Corrosivity: characteristic of substances that exhibit extremes of acidity or basicity or a tendency to corrode steel. Corrosive wastes are acidic and are/or capable of corroding metal such as tanks, containers, drums, and barrels.

Reactivity: violent chemical change (an explosive substance is an obvious example). Reactive wastes are unstable under normal conditions. They can create explosions, toxic fumes, gases, or vapors when mixed with water.

Toxicity: defined in terms of a standard extraction procedure followed by chemical analysis for specific substances. Toxic wastes are harmful or fatal when ingested or absorbed. When such wastes are disposed of on land, contaminated liquid may drain (leach) from the waste and pollute groundwater. Leaching of waste from landfills may be particularly evident when the area is exposed to acid rain. The acidic nature of the water may impart mobility to the waste by changing the chemical character of the waste or the character of the minerals to which the waste species are adsorbed.

Table 6.2: Effects of organic solvents.

Solvent	Affected parts of human body
Aliphatic hydrocarbons	
pentanes, hexanes, heptanes, octanes	central nervous system and liver
Halogenated aliphatic hydrocarbons	
methylene chloride	central nervous system, respiratory system
chloroform	liver
carbon tetrachloride	liver and kidneys
Aromatic hydrocarbons	
benzene	blood, immune system
toluene	central nervous system
xylene	central nervous system
Alcohols	
methyl alcohol (methanol and toxic metabolites)	optic nerve
isopropyl alcohol	central nervous system
Glycols	
ethylene glycol (and toxic metabolites)	central nervous system

2.1.1 Ignitable Materials

In a broad sense, a flammable substance is something that will burn readily, whereas a combustible substance requires relatively more persuasion to burn. Most chemicals that are likely to burn accidentally are liquids. Liquids form vapors that are usually more dense than air and thus tend to settle. The tendency of a liquid to ignite is measured by a test in which the liquid is heated and periodically exposed to a flame until the mixture of vapor and air ignites at the liquid's surface. The temperature at which this occurs is called the *flash point*.

There are several standard tests for determining the flammability of materials (American Society for Testing and Materials, 1995). For example, the upper and lower concentration limits for the *flammability* of chemicals and waste can be determined by standard tests (American Society for Testing and Materials, 1995, ASTM D 4982 and ASTM E 681) as can the *combustibility* and *flash point* (American Society for Testing and Materials, 1995, ASTM D 1310, ASTM E 176, and ASTM E 502). With these definitions in mind it is possible to divide ignitable materials into four subclasses.

A *flammable solid* is a solid that can ignite from friction or from heat remaining from its manufacture, or which may cause a serious hazard if ignited. Explosive materials are not included in this classification.

A *flammable liquid* is a liquid having a flash point below 37.8°C (100°F)

(American Society for Testing and Materials, 1995, ASTM D 92 and ASTM D 1310). A *combustible liquid* has a flash point in excess of 37.8°C, but below 93.3°C (200°F). Gases are substances that exist entirely in the gaseous phase at 0°C (32°F) and 1 atm (14.7 psi) pressure. A *flammable compressed gas* meets specified criteria for lower flammability limit, flammability range, and flame projection.

In considering the ignition of vapors, two important concepts are *flammability limit* and *flammability range*. Values of the vapor/air ratio below which ignition cannot occur because of insufficient fuel define the lower flammability limit. Similarly values of the vapor/air ratio above which ignition cannot occur because of insufficient air define the upper flammability limit. The difference between upper and lower flammability limits at a specified temperature is the flammability range (Table 6.3).

Finely divided particles of combustible materials are somewhat analogous to vapors with respect to flammability. One such example is a spray or mist of hydrocarbon liquid in which oxygen has the opportunity for intimate contact with the liquid particles. In this case the liquid may ignite at a temperature below its flash point.

Dust explosions (American Society for Testing and Materials, 1995, ASTM E 789) can occur with a large variety of solids that have been ground to a finely divided state. Many metal dusts, particularly those of magnesium and its alloys, zirconium, titanium, and aluminum can burn explosively in air. Coal dust and grain dust have caused many fatal fires and explosions in coal mines and grain elevators, respectively. Dusts of polymers such as cellulose acetate, polyethylene, and polystyrene can also be explosive. Thus, control of dust is essential (Mody and Jakhete, 1988).

Combustible substances are reducing agents that react with oxidizers (oxidizing agents or oxidants) to produce heat. Oxygen from air is the most common oxidizer. Many oxidizers are chemical compounds that contain oxygen (Table 6.4). However, whether or not a substance acts as an oxidizer depends upon the reducing strength of the material that it contacts. In addition, oxidizers can contribute strongly to fire hazards because fuels may burn explosively in contact with an oxidizer.

Substances that catch fire spontaneously in air without an ignition source are called *pyrophoric*. These include white phosphorus, the alkali metals and powdered forms of magnesium, calcium, cobalt, manganese, iron, zirconium and aluminum. Also included are some organometallic compounds, such as lithium ethyl (LiC_2H_5) and lithium phenyl (LiC_6H_5), and some metal carbonyl compounds, such as iron pentacarbonyl [$Fe(CO)_5$]. Another class of pyrophoric compounds consists of metal and metalloid hydrides, including lithium hydride (LiH). Moisture in air is often a factor in *spontaneous ignition*.

2.1.2 Corrosive Materials

Conventionally, corrosive substances are regarded as those that dissolve metals or cause oxidation of metals (Table 6.5). Again as with flammability, there are

Table 6.3: Flammabilities of selected organic liquids

Liquid	Flash point(°C)[a]	Volume percent in air LFL	UFL
Diethyl ether	-43	1.9	36
Pentane	-40	1.5	7.8
Acetone	-20	2.6	13
Toluene	4	1.3	7.1
Methanol	12	6.0	37
Gasoline (2,2,4-trimethylpentane)	--	1.4	7.6
Naphthalene	157	0.9	5.9

LFL, lower flammability limit.
UFL, upper flammability limit at 25°C (77°F)
[a] Closed-cup flash point test.

Table 6.4: Common oxidizing agents

Name	Formula	Gas/liquid/solid
Ammonium nitrate	NH_4NO_3	Solid
Ammonium perchlorate	NH_4ClO_4	Solid
Bromine	Br_2	Liquid
Chlorine	Cl_2	Gas (stored as liquid)
Fluorine	F_2	Gas
Hydrogen peroxide	H_2O_2	Solution in water
Nitric acid	HNO_3	Concentrated solution
Nitrous oxide	N_2O	Gas (stored as liquid)
Ozone	O_3	Gas
Perchloric acid	$HClO_4$	Concentrated solution
Potassium permanganate	$KMnO_4$	Solid
Sodium dichromate	$Na_2Cr_2O_7$	Solid

many tests that can be used to determine corrosivity (American Society for Testing and Materials, 1995, ASTM D1838 and ASTM D 2251). In a broader sense, corrosive substances cause deterioration of materials, including living tissue, that they contact. Most corrosive substances belong to at least one of the four following chemical classes: strong acids, strong bases, oxidants, or dehydrating agents.

Sulfuric acid is a prime example of a corrosive substance (American Society for Testing and Materials, 1995, ASTM C 694). As well as being a strong acid (American Society for Testing and Materials, 1995, ASTM E 1011), concentrated sulfuric acid is also a dehydrating agent and oxidant. The high affinity of sulfuric acid for water is illustrated by the heat generated when water and concentrated sulfuric acid are mixed. If this is done incorrectly by adding water to the acid, localized boiling and spattering can occur and result in personal injury. The major destructive effect of sulfuric acid on skin tissue is removal of water with accompanying release of heat. Sulfuric acid decomposes carbohydrates by removal of water. In contact with sugar, for example, concentrated sulfuric acid reacts to leave a charred mass.

Table 6.5: Examples of corrosive substances.

Name (formula)	Properties and effects
Nitric acid (HNO_3)	strong acid, strong oxidizer, corrodes metals, reacts with protein in tissue
Hydrochloric acid (HCl)	strong acid, corrodes metals, HCl gas damages respiratory tract
Hydrofluoric acid (HF)	corrodes metals, dissolves glass, causes bad burns
Alkali metal hydroxides (e.g., NaOH)	corrode zinc, lead, and NaOH and KOH; dissolve tissue; causes severe burns
Hydrogen peroxide (H_2O_2)	causes severe burns
Interhalogen compounds	corrosive irritants, dehydrate tissue
Halogen oxides (OF_2, Cl_2O, Cl_2O_7)	corrosive irritants, dehydrate tissue
Halogens (F_2, Cl_2, Br_2)	corrosive to mucous membranes, strong irritants

Contact of sulfuric acid with tissue results in tissue destruction at the point of contact. Inhalation of sulfuric acid fumes or mists damages tissues in the upper respiratory tract and eyes. Long term exposure to sulfuric acid fumes or mists has caused erosion of teeth, as well as destruction of other parts of the body!

2.1.3 Reactive Materials

Reactive substances are those that tend to undergo rapid or violent reactions under certain conditions. Such substances include those that react violently or form potentially explosive mixtures with water.

Explosives (Sudweeks et al., 1983; Austin, 1984) constitute another class of reactive substances (American Society for Testing and Materials, 1995). For regulatory purposes, those substances are also classified as reactive that react with water, acid, or base to produce toxic fumes, particularly hydrogen sulfide or hydrogen cyanide.

Heat and temperature are usually very important factors in reactivity (American Society Testing and Materials, 1995) since many reactions require energy of activation to get them started. The rates of most reactions tend to increase sharply with increasing temperature, and most chemical reactions give off heat. Therefore, once a reaction is started in a reactive mixture lacking an effective means of heat dissipation, the rate will increase exponentially with time (doubling with every 10° rise in temperature), leading to an uncontrollable event.

Other factors that may affect the reaction rate include the physical form of reactants (for example, a finely divided metal powder reacts explosively with oxygen, whereas a single mass of metal barely reacts), the rate and degree of mixing of reactants, the degree of dilution with nonreactive media (solvent), the presence of a catalyst, and pressure.

Some chemical compounds are self-reactive, in that they contain oxidant and reductant in the same compound. Many nitrogen containing-compounds are strong explosives (e.g., American Society for Testing and Materials, 1995, ASTM D 5143) insofar as they decompose spontaneously with a rapid release of a high amount of energy. As examples, pure nitroglycerin has such a high inherent instability that only a slight blow may be sufficient to detonate it. Trinitrotoluene (TNT) is also an explosive with a high degree of reactivity. However, it is inherently relatively stable in that some sort of detonating device is required to cause it to explode.

Many different classes of inorganic compounds are reactive. These include some of the halogen compounds of nitrogen (shock-sensitive nitrogen tri-iodide, NI_3, is an example), compounds with metal-nitrogen bonds, halogen oxides (ClO_2), and compounds with oxy-anions of the halogens. An example of the last group of compounds is ammonium perchlorate (NH_4ClO_4).

Explosives such as nitroglycerin that is a single compound containing both oxidizing and reducing functions in the same molecule are called redox compounds. Some redox compounds have more oxygen than is needed for a complete reaction and are said to have a positive balance of oxygen. Some have exactly the stoichiometric quantity of oxygen required (zero balance, maximum energy release), and others have a negative balance and require oxygen from outside sources to completely oxidize all components.

2.1.4 Toxic Materials

Toxicity is of the utmost concern in dealing with chemicals and their disposal (American Society for Testing and Materials, 1995, ASTM D 4447). This includes both long term chronic effects from continual or periodic exposures to low levels of toxic chemicals and acute effects from a single large exposure (Zakrzewski, 1991). Not all toxins are immediately apparent. For example, living organisms require certain metals for physiological processes. These metals when present at concentrations above the level of homeostatic regulation, can be toxic (e.g., American Society for Testing and Materials, 1995, ASTM E 1302). In addition, there are metals that are chemically similar to, but higher in molecular weight than, the essential metals (heavy metals). Metals can exert toxic effects by direct irritant activity, blocking functional groups in enzymes, altering the conformation of biomolecules, or displacing essential metals in a metallo-enzyme (American Society for Testing and Materials, 1995).

Agrochemicals represent an important group of chemicals used for the control of insects, acarids, weeds, and fungal diseases (Baker et al., 1995; Nelson et a., 1995). There are, however, subgroups of this general class of agrochemicals that must be controlled in their application lest serious effects on the environment are noted. In fact, it is because of the potential toxicity of the synthetic chemicals that a general class of chemicals known as bioregulators is being examined (Hedin, 1994; Hall et al., 1995).

The *pesticides* (American Society for Testing and Materials, 1995, ASTM E 609) represent an important group of materials (Kohn, 1983; Risebrough, 1992) that can enter an ecosystem as beneficial species but emerge/evolve as pollutants (Bourke et al., 1992). Pesticides are highly toxic, and many nontarget organisms can suffer harmful effects if misuse or unintended release occurs. The use of pesticides is not a modern development; the effects of chemicals on insects were recognized in pre-Christian era China and in Rome during the first century of the present era (James and Thorpe, 1994). The most diverse group of pesticides is the *insecticides* and *acaricides* (chemicals detrimental to mites). It has only been since World War II that synthetic insecticides have been widely used.

By way of a general description, an insecticide is a substance that kills or interferes in the life cycle of certain insects and is thus useful for reducing and controlling insect populations (Chenier, 1992; Clark, 1995). Examples of insecticides are the chlorinated hydrocarbon dichlorodiphenyl trichloroethane (DDT) and its analogs, the cyclodienes and related compounds, and benzene hexachloride (BHC). Beginning in the 1940s, chlorinated hydrocarbons were used extensively in mosquito control and agriculture. These compounds exert their toxic effect by interfering with the transmission of nerve impulses. Many are very persistent in the environment and undergo concentration (*biomagnification*; or *bioconcentration*) in food chains. Toxic effects on top predators such as birds and the contamination of human food supplies raised concern over their use. After the 1960s, most insecticides were replaced with other, less persistent compounds.

Organic phosphates inhibit the enzyme acetylcholinesterase and the result is the overstimulation and excitation of nerves. Because of their mode of action, organic phosphates are toxic to the target insects as well as the nontarget insects (bees and aquatic insects), birds, wildlife, and fish and other aquatic life. With few exceptions (Malathion), organic phosphates have not been considered toxic to mammals and humans. They are easily degraded and do not persist in the environment. However, because of their high toxicity to nontarget organisms, their use is limited to areas where undesired exposure can be minimized.

Carbamate insecticides exhibit neurotoxic action similar to the organic phosphates, by inhibiting acetylcholinesterase. However, this action is more readily reversible than that of the organic phosphates.

Botanical insecticides are natural and synthetic derivatives of toxic plant materials and have been used for hundreds of years. Nicotine and its analogs are neuroactive agents.

Rotenones are electron-transport inhibitors and are used as insecticides and pesticides (Chenier, 1992). *Pyrethroid insecticides* were originally derived from chrysanthemum plants; however, most pyrethroids in use today are synthetic derivatives of natural plant toxins. Pyrethroids are nerve poisons, acting through interference of ion transport along the axonal membrane. The pyrethroids are the most widely used insecticides, primarily because of their low toxicity to mammals. However, pyrethroids are extremely toxic to nontarget arthropods such as bees, aquatic insects, and crustaceans, and to non-mammalian vertebrates such as fish.

Fumigants are volatile substances used as soil pesticides and to control insects in stored products and scale insects on citrus. Common fumigants include ethylene dichloride, ethylene oxide, carbon disulfide, methyl bromide, hydrogen cyanide, phosphine, and chloropicrin. Because they are nonselective and toxic to humans, many fumigants are now restricted or banned.

Organic thiocyanate insecticides are mild general poisons and have been used as fly spays and fumigants that are also toxic to nontarget animals and plants. Dinitrophenols were important as early insecticides and acaricides and still have limited use as dormant acaricides. They are toxic to mammals and plants. Fluoroacetate derivatives are general toxic chemicals that form lethal metabolic products.

Inorganic *insecticides* are general toxic chemicals and have been largely replaced with synthetic organic insecticides. However, two classes of inorganic insecticides are being used, arsenic compounds and fluorides. Common arsenic compounds include lead and calcium arsenates and sodium arsenite. Sodium fluoride, sodium fluoroaluminate, and sodium fluorosilicate are common fluorides. Both groups are stomach poisons, increasing the toxicity with increasing metal content.

Herbicides are, supposedly, selectively toxic to plants. Examples include the growth stimulators 2,4-dichlorophenoxyacetic acid (2,4-D) and 2,4,5-trichlorophenoxyacetic acid (2,4,5-T), the protein synthesis-inhibiting Alachlor, the defoliant Paraquat, and the chlorophenolic contact herbicides. Although herbicides have in large part not been an environmental problem because of their selectivity and low persistence, some can be very toxic to nontarget organisms Paraquat, for example, can cause severe pulmonary symptoms.

Fungicidal compounds are used widely to treat seed grains and wood. Common ones are pentachlorophenol and the organic mercury compounds.

Polychlorinated biphenyls (*polychlorobiphenyls*) constitute a family of 209 chlorinated isomers of biphenyl. The term has been used to refer to the biphenyl molecule with one to ten chlorine substitutions. The parent compound biphenyl is formed from two benzene rings joined by a single bond external to both rings. The term *biphenyls* is often used for polychorinated biphenyls (PCBs), which have been implicated in several instances of food contamination (Sax, 1979).

Polychlorinated biphenyls were discovered in the late nineteenth century and have been used widely since that time for a variety of industrial tasks. The qualities that made polychlorobiphenyls attractive to industry were chemical stability, resistance to heat, low flammability, and high dielectric constant. The polychlorobiphenyl mixture is a colorless, viscous fluid, is relatively insoluble in water, and can withstand high temperatures without degradation. However, these characteristics are precisely the qualities that make polychlorinated biphenyls persistent in the environment.

Fluids contaminated with polychlorobiphenyls have been decontaminated by using chemical reagents to attack the chlorinated molecule. Generally, a metal, such as sodium, is used to remove the chlorine atoms from the biphenyl molecule. Microbial degradation of polychlorobiphenyl-contaminated soils has been pursued but there is little evidence of commercial-scale success.

Adsorption by solids is a key step in the degradation of a pesticide (Barney et al., 1992; Mullins et al., 1992). The degree of adsorption and the speed and extent of ultimate degradation are influenced by a number of factors, including solubility, volatility, charge, polarity, molecular structure, and size. Many of the properties and effects of solids in contact with water have to do with the sorption of solutes by solid surfaces. Surfaces in finely divided solids tend to have excess surface energy because of an imbalance of chemical forces among surface atoms, ions, and molecules. Surface energy level may be lowered by a reduction in surface area. Normally this reduction is accomplished by aggregation of particles or by sorption of solute species.

Under some circumstances, adsorption of polychlorobiphenyls by soil components retards degradation by separating the waste from the microbial enzymes that degrade it, whereas under other circumstances the reverse is true. Purely chemical degradation reactions may be catalyzed by adsorption. Loss of the pesticide by volatilization or leaching is diminished. The toxicity of a herbicide to plants may be strongly affected by soil sorption.

The degradation and eventual fate of pesticides on soil largely determine the ultimate environmental effects of the pesticides, and detailed knowledge of these effects are now required for licensing of a new pesticide. Among the factors to be considered are the sorption of the pesticide by soil; leaching of the pesticide into water and the potential for water pollution; effects of the pesticide on microorganisms and animal life in the soil; and any reactions that are capable of producing second-generation toxic products (Bourke et al., 1992).

The three primary ways in which pesticides are degraded in or on soil are: biodegradation (Karns, 1992; Massey et al., 1992), chemical degradation (Hapeman-Somich, 1992), and photochemical reactions (Matsumura and Katayama, 1992). Various combinations of these processes may operate in the degradation of a pesticide (Seiber, 1992). Although insects, earthworms, and plants may play roles in the degradation of pesticides, microorganisms have the most important role.

The sources, transport, interactions, and fates of pollutants in the geosphere involve a complex scheme (Chapter 1). The primary environmental concern regarding chemical wastes in the geosphere is the possible contamination of groundwater aquifers by waste leachates and leakage from wastes. An obvious source is leachate from landfills containing chemical wastes. In some cases, liquid materials are placed in lagoons, which can leak into aquifers. Leaking sewers can also result in contamination, as can discharge from septic tanks.

The transport of contaminants in land systems depends largely upon the hydrologic factors governing the movement of water underground and the interactions of waste constituents with geological strata, particularly unconsolidated earth materials. Groundwater contaminated with chemical wastes tends to flow as a relatively undiluted plug or plume along with the groundwater in an aquifer. Various technologies are under investigation to mitigate the effects of such plumes and involve a number of chemical, physical, and bioremediation methods (U.S. Department of Energy, 1995a).

Heavy metals are particularly damaging to groundwater and their movement through land systems is of considerable concern. Heavy-metal ions may be adsorbed by the soil, held by ion-exchange processes, interact with organic matter in soil, undergo oxidation-reduction processes leading to mobilization or immobilization, or even be volatilized as organometallic compounds formed by bacteria. A large number of factors affect heavy-metal mobility and attenuation in soil. These include acidity, temperature, cation-exchange capacity, the nature of soil mineral matter, and the kinds of soil organic matter present.

One of the most crucial aspects of fate and toxic effects of environmental chemicals is their accumulation by organisms from their surroundings. Many classes of organic compounds can be partially degraded by various microorganisms. These classes include nonhalogenated alkanes, halogenated alkanes (trichloroethane, dichloromethane) nonhalogenated aromatic compounds [benzene, naphthalene, benzo(a)pyrene], halogenated aromatic compounds (hexachlorobenzene, pentachlorophenol), phenols (phenol, cresol), polychlorinated biphenyls, phthalate esters, and pesticides (chlordane, parathion).

Coal-derived and petroleum-derived materials and by-products are also environmental pollutants (Speight, 1991, 1994; Loeher, 1992; Olschewsky and Megna, 1992). The world's economy is highly dependent on fossil fuels in energy production, industry, and transportation (Chapter 2). Widespread use has led to enormous releases to the environment of distillate fuels, crude oils, runoff from coal piles, exhaust from internal combustion engines, emissions from coal-fired power plants, industrial emissions, and emissions from municipal incinerators.

Point-source leaks and spills and non-point-source emissions have resulted in environmental contamination from millions of tons of petroleum hydrocarbons each year. Spills of crude oil and fuels have caused wide-ranging damage in the marine and freshwater environments. Oil slicks and tars in shore areas and beaches can ruin the esthetic value of entire regions.

The presence of naturally occurring polynuclear aromatic compounds is an interesting feature of soil organic matter, as many of these compounds are carcinogenic. Polynuclear aromatic compounds found in soil include fluoranthene, pyrene, and chrysene. The origin of polynuclear aromatic compounds in soil is unknown, although the most likely source is combustion. Terpenes also occur in soil organic matter.

The toxicity of polynuclear aromatic hydrocarbons is perhaps one of the most serious long-term problems associated with the use of petroleum. They comprise a large class of petroleum compounds containing two or more benzene rings. They are present in fossil fuels and are formed in the incomplete combustion of organic materials. Polynuclear aromatic hydrocarbons are formed in nature by long-term, low-temperature chemical reactions in sedimentary deposits of organic materials and in high-temperature events such as volcanoes and forest fires. The major source of this pollution is, however, human activity. Polynuclear aromatic hydrocarbons accumulate in soil, sediment, and biota. At high concentrations, they can be acutely toxic by disrupting membrane function. Many cause sunlight-induced toxicity in humans and fish and other aquatic organisms. In addition, long-term chronic toxicity has been demonstrated in a wide variety of organisms. Through metabolic activation, some polynuclear aromatic hydrocarbons form reactive intermediates that bind to deoxyribonucleic acid (DNA). For this reason, many of these hydrocarbons are *mutagenic* (tending to cause mutations), *teratogenic* (tending to cause developmental malformations), or *carcinogenic* (tending to cause cancer). They are also suspected of interfering in the reproduction of aquatic life. Low rates of reproduction and high rates of larval deformities and mortality have been observed in fish exposed to polynuclear aromatic hydrocarbons.

For regulatory and remediation purposes a standard test is needed to measure the likelihood of toxic substances getting into the environment and causing harm to organisms. The test required by the United States Environmental Protection Agency is the toxicity characteristic leaching procedure (TCLP), designed to determine the mobility of both organic and inorganic contaminants present in liquid, solid, and multiphasic wastes. For analysis of toxic species a solution is leached from the waste and is designated as the TCLP extract. If no significant solid material is present, the waste is filtered through a 0.60-8 micron (1 micron = 1 meter x 10^{-6}) glass fiber filter and designated as the TCLP extract. In mixed liquid-solid wastes the liquid is separated and analyzed separately.

Solid wastes to be extracted are required to have a surface area per gram of material equal to or greater than 3.1 cm^2 or to consist of particles smaller than 1 cm in their most narrow dimension. The kind of extraction fluid used on the solids is determined from the pH of a mixture of 5 g of the solids (reduced to approximately 1 mm in size if necessary) shaken vigorously with 96.5 mL of water. If the pH of the water after mixing is less than 5.0, the extraction fluid used is an acetic acid/sodium acetate buffer of pH 4.5 and if the pH is greater than 5.0, the extraction fluid is a dilute acetic acid solution with a pH of 2.9.

After the fluid to be used is determined, an amount of extraction fluid equal to 20 times the mass of the solid is used for the extraction which is carried out for 18 hours in a sealed container held on a device that rotates it end-over-end for 18 hours. After the TCLP extract is separated from the solids it is analyzed for 39 specified volatile organic compounds, semivolatile organic compounds, and metals to determine if the waste exceeds specified levels of these contaminants.

In addition to chemical spills, another source of toxic compounds is combustion. In fact, some of the greater dangers of fires are from toxic products and by-products of combustion. The most obvious of these is carbon monoxide (CO), which can cause serious illness or death because it forms carboxyhemoglobin with hemoglobin in the blood so that the blood no longer carries oxygen to body tissues. Toxic sulfur dioxide and hydrogen chloride are formed by the combustion of sulfur compounds and organic chlorine compounds, respectively.

A large number of noxious organic compounds such as aldehydes are generated as by-products of combustion. In addition to forming carbon monoxide, combustion under oxygen-deficient conditions produces polynuclear aromatic hydrocarbons consisting of fused-ring structures. Some of these compounds, such as benzo(a)pyrene are precarcinogenic compounds, insofar as they are acted upon by enzymes in the body to yield cancer-producing metabolites.

In summary, many specific chemicals in widespread use are hazardous because of their chemical reactivities, fire hazards, toxicities, and other properties (Ashford, 1994). In fact, a simple definition of a hazardous waste is that it is a chemical substance that has been discarded, abandoned, neglected, released, or designated as a waste material and that has the potential to be detrimental to the environment. Alternatively, a hazardous waste may be a chemical that may interact with other (chemical) substances to give a product that is hazardous to the environment.

2.1.5 Radioactive Materials

The use and subsequent disposal of radioactive waste are issues for any country with a significant nuclear power industry (Eister, 1983; Pickering and Owen, 1994). By way of historical perspective, the dawn of the nuclear age occurred in 1919 when Ernest Rutherford and his team at Cambridge University (England) converted a nucleus of nitrogen into a nucleus of hydrogen. Then in 1934, Enrico Fermi with Frederic and Irene Joliot-Curie created artificial radioactivity by bombarding a target with alpha particles. In 1938, Otto Hahn and Fritz Strassman split the uranium atom and recognized that nuclear fission had occurred. Research into nuclear physics and chemistry was given its real momentum in 1941, when President Roosevelt sanctioned the construction of an atomic bomb which was first used in 1945.

Atmospheric nuclear testing between 1945 and 1980 was responsible for putting a large, but still an unknown amount of nuclear fallout into the atmosphere. The Partial Test Ban Treaty of 1963 led to the cessation of most of the testing, but some nonsignatory countries, such as France, continued testing until after 1980. These continued tests deposited large quantities of additional radionuclides such as strontium-90 (^{90}Sr) and cesium-137 (^{137}Cs) which found their way, via precipitation, into the hydrosphere.

Radioactive waste is very much a legacy of the twentieth century. The main radioactive elements that are involved in polluting the environment and causing ill health and death are the isotopes of iodine, strontium, cesium, and ruthenium. Iodine accumulates in the thyroid gland. Strontium, which is chemically similar to calcium, is absorbed through the walls of the intestine and collects in the bones. Cesium behaves chemically in a manner similar to potassium and therefore can be distributed throughout the body in much the same way. Ruthenium has no chemical analog with a biological function.

In the United States, such radioactive wastes are regulated by the Nuclear Regulatory Commission (NRC) and the Department of Energy (DOE). However, if reports are true (New York Times International, 1994) there are occasions when nuclear waste is disposed of within the land systems without any form of protection, other than the depth of the overburden. Furthermore, special problems are posed by mixed waste containing both radioactive and chemical wastes (United States Department of Energy, 1995b).

Radon gas is a major component of the background radiation dose received by populations in certain geographic areas. For example it is the largest contributor to radiation exposure in Britain, where, from the total average annual radiation dose (2.5 millisieverts, 2.5 mSv) to the population, approximately 1.2 mSv comes from radon. There are tentative links between high incidences of cancer (e.g., lung cancer and leukemia) and high concentration of household radon levels. Epidemologists, however, argue that a direct link between radon and an increased risk of contracting cancer cannot be proved statistically.

There are many factors controlling the rate of migration and the concentration of radon in the environment. Concentrations of radon are generally highest in regions made up of rocks that contain the parent sources, namely uranium and thorium, and have effective pathways for the migration of the gas. Examples are granites and some sedimentary rocks such as certain types of shales and ironstones, particularly where there are phosphate nodules. The concentrations, however, cannot be directly correlated with the concentrations of uranium and thorium in rock because the uranium is present in different host minerals, some of which are more easily weathered, and so release higher concentrations of radon more effectively. The rate at which radon passes through the ground is important in controlling the concentration of radon at the surface or within the soil.

Rocks with higher permeability (interconnectedness of cavities), such as highly fissured and faulted rocks, allow radon, in the form of gas or dissolved in groundwater, to migrate at a faster rate into the human environment. Some of the highest concentrations of radon are often associated with springs rather than known uranium rocks.

During rainstorms, pore spaces within soils and rocks fill with water, prohibiting the migration of radon and allowing its buildup within the soil. When the ground dries, radon is released in high concentrations. Diurnal variations may occur because dew inhibits the release of radon from the soil at night and in the early morning, while during the day radon is released as the soil dries out. Also, reduced atmospheric pressure during a cyclonic depression increases the soil-atmosphere pressure gradient, which may result in an increase in the rate of release of radon from the soil and help concentrate it in the atmosphere.

Atmospheric radon, however, is easily dispersed by the wind; it is only when its dispersal is inhibited that it becomes a hazard. The most common way in which it is concentrated is by becoming trapped within a building, particularly a house. Most radon enters through the floor from the ground and accumulates beneath houses in cavities where there are gaps between the floor and the ground. The concentration of radon within a building is aided by the lower pressures indoors compared to the ambient atmospheric pressure, creating a pressure gradient that drags air into the building. This gradient is greatest in buildings with chimneys. Radon may also enter houses by the release of dissolved radon in shower and bath water, and from walls which have been constructed from rocks such as granite or gypsum boards with high uranium concentrations.

2.1.6 Other Classifications

In addition to classification by the above characteristics, the United States Environmental Protection Agency (U.S. EPA) designates more than 450 wastes that are specific substances or classes of substances known to be hazardous. Each such substance is assigned a U.S. EPA hazardous waste number in the format of a letter followed by 3 numerals, where a different letter is assigned to substances from each of the following list:

1. F-type: wastes from nonspecific sources;
2. K-type: wastes from specific sources;
3. P-type: acute hazardous wastes which are mostly specific chemical species such as fluorine; and
4. U-type: generally hazardous wastes that are predominantly specific compounds.

The Comprehensive Environmental Response, Compensation, and Liability Act (CERCLA) gives a broader definition of hazardous substances that includes the following:

1. any element, compound, mixture, solution, or substance, the release of which may substantially endanger public health, public welfare, or the environment;
2. any element, compound, mixture, solution, or substance in reportable quantities designated by CERCLA Section 102;
3. certain substances or toxic pollutants designated by the Water Pollution

Control Act;

4. any hazardous air pollutant listed under Section 112 of the Clean Air Act;
5. any imminently hazardous chemical substance or mixture that has been the subject of government action under Section 7 of the Toxic Substances Control Act (TSCA); and
6. with the exception of those suspended by Congress under the Solid Waste Disposal Act, any hazardous waste listed or having characteristics identified by the Resource Conservation Recovery Act.

In terms of quantity by weight, more wastes than all others combined are those from categories designated by hazardous waste numbers preceded by F and K. The "F" categories are those wastes from nonspecific sources (Table 6.6). The "K-type" hazardous wastes are those from specific sources produced by industries such as the manufacture of inorganic pigments, organic chemicals, pesticides, explosives, iron and steel, and nonferrous metals, and from processes such as petroleum refining or wood preservation (Table 6.7).

2.2 Nonhazardous Waste

It is appropriate to consider nonhazardous chemical waste (solid waste, the municipal refuse and garbage produced by human activities) along with hazardous waste because it may not be nonhazardous in all situations and may interact with hazardous wastes or produce a hazardous waste by chemical reaction. Thus a waste that is currently classed as nonhazardous can, depending upon the circumstances, be reclassed as hazardous.

Some wastes that might exhibit a degree of hazard are exempt from the Resource Conservation Recovery Act regulation by legislation and include the following:

1. fuel ash and scrubber sludge from power generation by utilities;
2. oil field and gas field drilling muds;
3. by-product brine from petroleum production;
4. cement kiln dust;
5. waste and sludge from phosphate mining and beneficiation; and
6. mining wastes from uranium and other minerals.

Eventual reclassification of these kinds of low-hazard wastes could increase the quantities of regulated wastes several-fold.

On a different note, the amounts of solid waste produced each year are so enormous that our capacity to deal with the problem is under severe strain. Currently, ~92% of the municipal refuse currently produced in the United States is placed into landfills. In fact, many landfills were little more than "dumping grounds" for waste material. As the total quantity of municipal waste has increased, the landfill capacity to handle waste has decreased to approximately 20% of the original space. As a result, many municipalities have to transport municipal refuse many miles to a new disposal site, thereby increasing the disposal costs.

Incineration of municipal refuse can reduce waste mass by 75% and volume

Table 6.6: Chemical wastes designated as "F" category wastes

Number	Waste material
F001	spent halogenated solvents chlorinated fluorocarbons; sludge from solvent recovery processes
F004	spent nonhalogenated solvents (cresols, nitrobenzene); still bottoms from solvent recovery operations
F007	spent solution from electroplating operations
F010	sludge from metal heat-treating operations

Table 6.7: Chemical wastes designated as "K" category wastes

Number	Waste material
K001	sediment/sludge from wastewater treatment from wood-preserving processes (especially creosote and pentachlorophenol sediment/sludge)
K002	wastewater treatment sludge from chrome yellow and orange pigments
K020	residue from vinyl chloride distillation
K034	2,6-dichlorophenol waste
K047	pink/red water from trichloroethane manufacture
K049	waste oil/emulsion/solids from petroleum refining
K060	ammonia lime still sludge
K067	electrolytic sludge from zinc production

by 90%. However, environmental concern about organic pollutants (particularly dioxin) in stack emissions and heavy metals in incinerator ash have slowed municipal incinerator development. Management of the ash waste material requires attention to the applicability of the current federal regulations and the type of test to apply to characterize the ash (Clarke, 1992; Goodwin, 1993).

Recycling (Chapter 9) can certainly reduce quantities of solid waste, perhaps by as much as 50%, but it is not the panacea claimed by its most avid advocates. The overall solution to the solid waste problem must involve several kinds of measures, particularly (1) reduction of wastes at the source, (2) recycling as much waste as is practicable, (3) reducing the volume of remaining wastes by measures such as incineration, (4) treating residual material as much as possible to render it immobile and innocuous, and (5) placing the residual material in landfills, properly protected from leaching or release by other pathways.

3.0 References

American Chemical Society. 1996. *Laboratory Waste Management: A Guidebook*. ACS Task Force on Laboratory Waste Management. American Chemical Society, Washington, D.C.

American Society for Testing and Materials. 1995. *Annual Book of ASTM Standards*. American Society for Testing and Materials, Philadelphia, Pennsylvania.

American Society for Testing and Materials. 1995. ASTM C 694. Test Method for Weight Loss (Mass Loss of Sheet Steel) During Immersion in Sulfuric Acid Solution. *Annual Book of ASTM Standards*. Vol. 02.02. American Society for Testing and Materials, Philadelphia, Pennsylvania.

American Society for Testing and Materials. 1995. ASTM D 92. Test Method for Flash and Fire Points by Cleveland Open Cup. *Annual Book of ASTM Standards*. Vol. 05.01. American Society for Testing and Materials, Philadelphia, Pennsylvania.

American Society for Testing and Materials. 1995. ASTM D 1310. Test Method for Flash Point and Fire Point of Liquids by Tag Open-cup Apparatus. *Annual Book of ASTM Standards*. Vol. 05.01. American Society for Testing and Materials, Philadelphia, Pennsylvania.

American Society for Testing and Materials. 1995. ASTM D 1838. Test Method for Copper Strip Corrosion by Liquefied Petroleum Gases. *Annual Book of ASTM Standards*. Vol. 05.01. American Society for Testing and Materials, Philadelphia, Pennsylvania.

American Society for Testing and Materials. 1995. ASTM D 2251. Test Method for Metal Corrosion by Halogenated Organic Solvents and Their Admixtures. *Annual Book of ASTM Standards*, Vol. 15.05. American Society for Testing and Materials, Philadelphia, Pennsylvania.

American Society for Testing and Materials. 1995. ASTM D 4447. Guide for the Disposal of Laboratory Chemicals and Samples. *Annual Book of ASTM Standards*. Vol. 11.04. American Society for Testing and Materials, Philadelphia, Pennsylvania.

American Society for Testing and Materials. 1995. ASTM D 4982. Method for Flammability Potential Screening Analysis of Waste. *Annual Book of ASTM Standards*. Vol. 11.04. American Society for Testing and Materials, Philadelphia, Pennsylvania.

American Society for Testing and Materials. 1995. ASTM D 5143. Test Method for Analysis of Nitroaromatic and Nitramine Explosive in Soil by High Performance Liquid Chromatography. *Annual Book of ASTM Standards*. Vol. 04.08. American Society for Testing and Materials, Philadelphia, Pennsylvania.

American Society for Testing and Materials. 1995. ASTM E 176. Terminology Relating to Fire Standards. *Annual Book of ASTM Standards*. Vol. 04.07. American Society for Testing and Materials, Philadelphia, Pennsylvania.

American Society for Testing and Materials. 1995. ASTM E 502. Test Method forSelection and Use of ASTM Standards for the Determination of Flash Point of Chemicals by Closed Cup Methods. *Annual Book of ASTM Standards*. Vol. 14.02. American Society for Testing and Materials, Philadelphia, Pennsylvania.

American Society for Testing and Materials. 1995. ASTM E 609. Standard Definitions of Terms Relating to Pesticides. *Annual Book of ASTM Standards*. Vol. 11.04. American Society for Testing and Materials, Philadelphia, Pennsylvania.

American Society for Testing and Materials. 1995. ASTM E 681. Test Method for Limits of Flammability of Chemicals. *Annual Book of ASTM Standards*. Vol. 14.02. American Society for Testing and Materials, Philadelphia, Pennsylvania.

American Society for Testing and Materials. 1995. ASTM E 789. Test Method for Pressure and Rate of Pressure Rise for Dust Explosions in a Closed Vessel. *Annual Book of ASTM Standards*. Vol. 14.02. American Society for Testing and Materials, Philadelphia, Pennsylvania.

American Society for Testing and Materials. 1995. ASTM E 1011. Specification for Sulfuric Acid. *Annual Book of ASTM Standards*. Vol. 15.05. American Society for Testing and Materials, Philadelphia, Pennsylvania.

American Society for Testing and Materials. 1995. ASTM E 1302. Guide for Acute Animal Toxicity of Water-miscible Metalworking Fluids. *Annual Book of ASTM Standards*. American Society for Testing and Materials, Philadelphia, Pennsylvania.

Ashford, R.D. 1994. *Ashford's Dictionary of Industrial Chemicals: Properties, Production, Uses*. Wavelength Publications, New York.

Austin, G.T. 1984. Chap. 22. In *Shreve's Chemical Process Industries*. 5th ed. McGraw-Hill, New York.

Baker, D.R., Fenyes, J.G., and Basarab, G.S. (eds.). 1995. *Synthesis and Chemistry of Agrochemicals IV*. Symposium Series No. 584. American Chemical Society, Washington, D.C.

Barcelona, M., Wehrmann, A., Keeley, J.F. and Pettyjohn, J. 1990. *Contamination of Ground Water*. Noyes Data Corp., Park Ridge, New Jersey.

Barney, G.S., Lueck, K.J., and Green, J.W. 1992. Chap. 3 in *Environmental Remediation*. Edited by G.F. Vandegrift, D.T. Reed, and I.R. Tasker. Symposium Series No. 509. American Chemical Society, Washington, D.C.

Bourke, J.B., Felsot, A.S., Gilding, T.J., Jensen, J.K., and Seiber, J.N. (eds.). 1992. *Pesticide Waste Management*. Symposium Series No. 510. American Chemical Society, Washington, D.C.

Cartwright, F.F., and Biddiss, M.D. 1972. *Disease and History*. Dorset Press, New York.

Cheremisinoff, P. 1995. *Handbook of Water and Wastewater Treatment Technology*. Marcel Dekker Inc., New York.

Chenier, P.J. 1992. *Survey of Industrial Chemistry*. 2nd ed. VCH Publishers, New York.

Clark, J.M. (ed.) 1995. *Molecular Action of Insecticides on Channels*. Symposium Series No. 591. American Chemical Society, Washington, D.C.

Clarke, L. 1992. *Applications for Coal-use Residues*. Report No. IEACR/62. International Energy Agency Coal Research, London, England.

Easterbrook, G. 1995. *A Moment on the Earth: The Coming Age of Environmental Optimism*. Viking Press, New York.

Eister, W.K. 1983. Chap. 24 in *Riegel's Handbook of Industrial Chemistry*. Edited by J.A. Kent. Van Nostrand Reinhold, New York.

Goodwin, R.W. 1993. *Combustion Ash Residue Management: An Engineering Perspective*. Noyes Data Corp., Park Ridge, New Jersey.

Hall, F.R., and Barry, J.W. (eds.). *Biorational Pest Control Agents: Formulation and Delivery*. Symposium Series No. 595. American Chemical Society, Washington, D.C.

Hapeman-Somich, C.J. 1992. Chap. 13 in *Pesticide Waste Management*. Edited by J.B. Bourke, A.S. Felsot, T.J. Gilding, T.J.K. Jensen, and J.N. Seiber. Symposium Series No. 510. American Chemical Society, Washington, D.C.

Hedin, P.A. (ed.). 1994. *Bioregulators for Crop Protection and Pest Control*. Symposium Series No. 557. American Chemical Society, Washington, D.C.

James, P., and Thorpe, N. 1994. *Ancient Inventions*. Ballantine Books, New York.

Karns, J.S. 1992. Chap. 12 in *Pesticide Waste Management*. Edited by J.B. Bourke, A.S. Felsot, T.J. Gilding, T.J.K. Jensen, and J.N. Seiber. Symposium Series No. 510. American Chemical Society, Washington, D.C.

Kohn, G.K. 1983. Chap. 21 in *Riegel's Handbook of Industrial Chemistry*. Edited by J.A. Kent. Van Nostrand Reinhold, New York.

Loeher, R.C.. 1992. In *Petroleum Processing Handbook*. p. 190. Edited by J.J. McKetta. Marcel Dekker Inc., New York.

Massey, J.H., Lavy, T.L., and Fitzgerald, M.A. 1992. Chap. 18 in *Pesticide Waste Management*. Edited by J.B. Bourke, A.S. Felsot, T.J. Gilding, T.J.K. Jensen, and J.N. Seiber. Symposium Series No. 510. American Chemical Society, Washington, D.C.

Matsumura, F., and Katayama, A. 1992. Chap. 17 in *Pesticide Waste Management*. Edited by J.B. Bourke, A.S. Felsot, T.J. Gilding, T.J.K. Jensen, and J.N. Seiber. Symposium Series no. 510. American Chemical Society, Washington, D.C.

Mody, V., and Jakhete, R. 1988. *Dust Control Handbook*. Noyes Data Corp., Park Ridge, New Jersey.

Mullins, D.E., Young, R.W., Hertzel, G.H., and Berry, D.F. 1992. Chap. 14 in *Pesticide Waste Management*. Edited by J.B. Bourke, A.S. Felsot, T.J. Gilding, T.J.K. Jensen, and J.N. Seiber. Symposium Series No. 510. American Chemical Society, Washington, D.C.

Nelson, J.O., Karu, A.E., and Wong, R.B. (eds.) 1995. *Immunoassays of Agrochemicals: Emerging Technologies*. Symposium Series No. 586. Amercian Chemical Society, Washington, D.C.

New York Times International. 1994. Nuclear Roulette in Russia: Burying Uncontained Waste. November 21.

Olschewsky, D., and Megna, A. 1992. In *Petroleum Processing Handbook*. p. 179. Edited by J.J. McKetta. Marcel Dekker Inc., New York.

Pickering, K.T., and Owen, L.A. 1994. *Global Environmental Issues*. Routledge Publishers, New York.

Risebrough, R.W. 1993. In *Encyclopedia of Environmental Science and Engineering*. 3rd ed. p. 395. Edited by S.P. Parker and R.A. Corbitt. McGraw-Hill, New York.

Sax, N.I. (ed.) 1979. *Dangerous Properties of Industrial Materials*. 5th ed. Van Nostrand Reinhold, New York.

Seiber, J.N. 1992. Chap. 11 in *Pesticide Waste Management*. Edited by J.B. Bourke, A.S. Felsot, T.J. Gilding, TJ.K. Jensen, and J.N. Seiber. Symposium Series No. 510. American Chemical Society, Washington, D.C.

Speight, J.G. 1990. Part II in *Fuel Science and Technology Handbook*. Edited by J.G. Speight. Marcel Dekker Inc., New York.

Speight, J.G. 1991. *The Chemistry and Technology of Petroleum*. 2nd ed. Marcel Dekker Inc., New York.

Speight, J.G. 1993. *Gas Processing: Environmental Aspects and Methods*. Butterworth Heinemann, Oxford, England.

Speight, J.G. 1994. *The Chemistry and Technology of Coal*. 2nd ed. Marcel Dekker Inc., New York.

Sudweeks, W.B., Larsen, R.D., and Balli, F.K. 1983. In *Riegel's Handbook of Industrial Chemistry*. p. 700. Edited by J.A. Kent. Van Nostrand Reinhold, New York.

Tedder, D.W., and Pohland, F.G. (eds.). 1993. *Emerging Technologies for Hazardous Waste Management III*. Symposium Series No. 518. Washington, D.C.: American Chemical Society, Washington, D.C.

Tedder, D.W., and Pohland, F.G. (eds.). 1995. *Emerging Technologies for Hazardous Waste Management V*. Symposium Series No. 607. Washington, D.C.: American Chemical Society, Washington, D.C.

United States Department of Energy. 1995a. *Contaminant Plumes: Containment and Remediation Focus Area*. Report No. DOE/EM-0248. Office of Environmental Management. United States Department of Energy, Washington, D.C.

United States Department of Energy. 1995b. *Mixed Waste Characterization, Treatment, and Disposal Focus Area*. Report No. DOE/EM-0252. Office of Environmental Management. United States Department of Energy, Washington, D.C.

United States Environmental Protection Agency. 1993. *Superfund Innovative Technology Evaluation Program. Technology Profiles*. 6th ed. Office of Research and Development, United States Environmental Protection Agency, Washington, D.C.

Zakrzewski, S.F. 1991. *Principles of Environmental Toxicology*. American Chemical Society, Washington, D.C.

Chapter 7

Chemical Waste Management and Biodegradation of Waste

1.0 Introduction

Waste management refers to an organized system for waste handling in which chemicals pass through appropriate pathways leading to elimination or disposal in ways that protect the environment (Kocurek and Woodside, 1994). The choice of an appropriate technology to treat a chemical waste is dictated by the nature of the waste, i.e., whether the waste is nonhazardous or hazardous (Tedder and Pohland, 1995). This choice is strongly influenced (in the United States) by the requirements of the Resource Conservation and Recovery Act (RCRA) (Chapter 12). In this legislation a *solid waste* is defined as

> ...garbage, refuse, sludge from a waste treatment plant, water supply treatment plant, or air pollution control facility and other discarded material, including solid, liquid, semisolid, or contained gaseous material resulting from industrial, commercial, mining, and agricultural operations, and from community activities.

This definition does not include the solid or dissolved materials in domestic sewage, or solid or dissolved materials in irrigation return flows or industrial discharges that are subject to permits. From this same legislation, a *hazardous waste* is defined as

> ...a solid waste, or combination of solid wastes, which because of its quantity, concentration, or physical, chemical, or infectious characteristics may: (1) cause or significantly contribute to an increase in mortality or an increase in serious irreversible, or incapacitating reversible, illness; or (2) pose a substantial present or potential hazard to human health or the environment when improperly treated, stored, transported, or disposed of or otherwise managed.

Thus, in terms of a simple classification system, a hazardous waste is a subset of the broader chemical or solid waste category.

Within this simple classification system, volatile organic compounds (VOCs) (Table 7.1), semivolatile organic compounds (SVOCs) (Table 7.2), metals (Table 7.3), radioactively contaminated materials, or a mixture of any or all of these types of chemicals can also be classified as hazardous, depending upon their properties and effects on flora, fauna, and the ecosystem in general.

Table 7.1: Representative volatile organic compounds

1,1,1-Trichloroethane	Cis-1,2-Dichlorethylene
1,1,2,2-Tetrachloroethane	Cis-1,3-Dichloropropene
1,1,2-Trichloroethane	Dibromochloromethane
1,1-Dichloroethane	Dibromochloropropane
1,1-Dichloropropylene	Dibromomethane
1,2-Dichloroethane	Dichlorodifluoromethane
1,2-Dichloropropane	Dichloroethane
1,2-Transdichloroethene	Dichloroethylene
1,2,3-Trichloropropane	Dichloromethane
1,3-Dichloropropane	Dichloropropene
1,3-Trichloropropane	Ethyl ether
1,4-Declare-2-butane	Ethyl methacrylate
2-Butanone (MEK)	Ethylbenzene
2-Chloroethyl vinyl ether	Iodomethane
2-Chloropropane	Isopropanol
2-Hexanone	M-psa
3-Hexanone	M-xylene
4-Methyl-2-pentanone	Methane
Acetone	Methylene
Acrolein	Methylene chloride
Acrylonitrile	o-Xylene
Benzene	p-Xylene
Bromodichloroethane	Polyvinyl chloride
Bromodichloromethane	Styrene
Bromoform	Tetrachloroethene
Bromomethane	Tetrachloroethylene
Carbon disulfide	Toluene
Carbon tretrachloride	Trans-1,2-dichloroethene
Chlorobenzene	Trans-1,3-dichoropropene
Chloroethane	Trichloroethene
Chloroform	Trichlorofluoromethane
Chloromethane	Vinyl acetate
Cis-1,2-dichlorethane	Vinyl chloride

For the purposes of this chapter (and the text in general), a waste will generally be referred to as chemical and will only be classed as hazardous when the nature of the text permits. The general uses of the descriptor *hazardous* to classify wastes is often lacking in specificity and ignores the purpose of the definition. The indiscriminate use, and twisting, of words to describe a chemical also cannot escape criticism.

Table 7.2: Representative semi-volatile organic compounds

1,2,3-Trichlorobenzene	4-Aminobiphenol
1,2,4,5-Tetrachlorobenzene	4-Bromophenyl phenyl ether
1,2,4-Trichlorobenzene	4-Chloro-3-methylphenol
1,2-Dichlorobenzene	4-Nitroaniline
1,2-Diphenylhydrazine	4-Nitrophenol
1,3-Dichlorobenzene	7,12-Dimethylbenz(a)anthracene
1,4-Dichlorobenzene	a,a-Dimethyl-b-phenylethlamine
1-Chloroaniline	Acenanthrene
1-Naphthylamine	Acenaphthene
2,2-Dichlorobenzidine	Acenaphthylene
2,3,4,5-Tetrachlorophenol	Acetoaphthylene
2,4,5-Trichlorophenol	Aldrin
2,4,6-Trichlorophenol	Alpha-bhc
2,4-Dichlorophenol	Amiben
2,4-Dichlorotoluene	Aniline
2,4-Dimethylphenol	Anthracene
2,4-Dinitrophenol	Benzidine
2,4-Dinitrotoluene	Benzol(a)anthracene
2,6-Dichlorophenol	Benzo(a)pyrene
2,6-Dinitorotoluene	Benzo(b)fluoranthene
2-Chloronaphthalene	Benzo(ghi)perylene
2-Chlorophenol	Benzo(k)fluoranthene
2-Naphthylamine	Benzo(j)fluoranthene
2-Nitroaniline	Benzo(k)pyrene
2-Nitrophenol	Benzoic acid
2-Picoline	Benzothiazole
3-Methylchlolanthrene	Benzyl alcohol
3-Methylphenol	Bis(2-Chloroethoxy)methane
3-Nitroaniline	Bis(2-chlorethyl)ether
4,4-DDD	Bis(ethylhexyl)phthalate
4,4-DTE	Butyl benzyl phthalate
4,4-DDT	Chlordane
4,6-Dinitro-o-cresol	Chrysene

As an example, oxygen is necessary for life and thus an absence, or relative scarcity, of oxygen will cause cessation of life functions. It might be stated that an atmosphere of nitrogen containing 3 ppm (0.0003%) oxygen (instead of the usual 20% oxygen) in an atmosphere of nitrogen will not support human life. However, this can be twisted to note that 3 parts per million (ppm) oxygen in an atmosphere of nitrogen is harmful to humans insofar as it (the proportion of oxygen) is inadequate

Table 7.3: Representative metals

Aluminum	Molybdenum
Antimony	Nickel
Arsenic	Plutonium
Barium	Potassium
Beryllium	Radium
Boron	Selenium
Cadmium	Silicon
Calcium	Silver
Cesium	Sodium
Chrome	Strontium
Chromite	Technetium
Chromium	Thallium
Cobalt	Thorium
Copper	Tin
Iron	Titanium
Lead	Uranium
Magnesium	Vanadium
Manganese	Zinc
Mercury	Zirconium

to support aerobic life.

The migration of chemical waste is a very necessary consideration in waste management as well as treatment. In fact, the migration of chemical waste through the land, water, and atmosphere is largely a function of the physical properties of the waste as well as the physical properties of the surrounding matrix and the physical conditions to which the waste is subjected.

In addition, chemical factors play an important role. There is real potential for the chemical waste to react with an outside agent (i.e., a chemical not originally in the waste). The changes that occur during the migration of petroleum from the source rock to the reservoir rock are documented (Speight, 1991). By inference, changes occurring to chemical waste during migration must be anticipated.

Highly volatile chemicals are more amenable to transportation through the atmosphere, and more soluble chemicals are more likely to be carried by water. Volatile wastes are more mobile under hot, windy conditions, and soluble ones during periods of heavy rainfall. Chemicals (especially nonabsorbable materials) will migrate further in porous formations (e.g., sandstone) than they will in dense formations (e.g., clay or soil), even if adsorptive effects were absent from all media. Chemicals that are more reactive (in the chemical and biochemical sense) will not migrate to any great extent before they are converted to other products.

The distribution of chemical waste constituents between the atmosphere and the land and water systems is largely a function of compound volatility. Usually, the tendency of water and soil to retain the chemical is a factor in the mobility of the chemical. Physical interactions such as hydrogen bonding (in water) and adsorption (on the soil) can retain the chemical at temperatures above the boiling point. The physical interactions with the soil constituents may also be weaker than hydrogen bonding forces in the water, resulting in a more ready release from the soil than from water.

The ultimate concern with the release of chemicals is related to the toxic nature of the chemical. A chemical may not necessarily be classed as toxic (Chapter 6) to cause a marked effect on an ecosystem. For example, salt is accepted by many as a normal dietary (taste) additive. However, there are few (if any) humans who can tolerate more than several grams of salt at any given moment. Therefore, consider the effects of a few grams of a chemical such as salt on animals, plants, microbes, and any other part of an ecosystem!

By this example, it must be inferred that virtually all chemicals are poisonous beyond certain limits, some extremely so (Chapter 6). The toxicity of a chemical is a function of many factors, including (1) the properties of the chemical; (2) the species exposed to the chemical; and among other circumstances, (3) the time of exposure.

Many chemicals are also corrosive to materials such as steel and can have a direct or indirect effect on the environment. As an example of an indirect effect, a steel drum containing an extremely toxic chemical waste that is exposed to a corrosive chemical can release the toxic material to the environment.

Aerial oxygen can cause corrosion, and oxygen can cause certain types of waste to ignite. On the other hand, the spontaneous ignition of coal (Speight, 1994) is also a matter for consideration. Caution in the definition of terms is warranted here before aerial oxygen is classified as a corrosive chemical!

Soil exposed to chemical waste can be severely damaged by alteration of its physical and chemical properties and ability to support plants. For example, soil continuously exposed to brine from petroleum production will be unable to support plant growth and will become extremely susceptible to erosion. Contaminated soils can be treated in a variety of ways to eliminate contaminants (Just and Stockwell, 1993). It is possible in principle to treat contaminated soils biologically in place by pumping oxygenated, nutrient-enriched water through the soil in a recirculating system.

The fate of a chemical waste in water is a function of solubility, density, and biodegradability, as well as chemical reactivity, and the chemistry of water plays a major role (Benefield et al., 1982). Dense, water-immiscible liquids may simply sink to the bottom of a water system and accumulate. Chemicals that accumulate are readily taken up by organisms, exchangeable cationic materials adsorb on/in sediments, and organophilic materials may be adsorbed by organic matter in sediments.

Because of the ways in which chemicals can disperse throughout the land and water systems as well as in the atmosphere, there is a growing acceptance of the desirability of using waste management to solve the issues related to chemical waste. The main aspects of waste management involve source reduction, treatment, and recycling (or disposal).

Source reduction consists of the reduction or elimination of chemical waste at the source, usually within a process. Source reduction measures include process modification, feedstock substitutions, improvements in feedstock purity, changes in housekeeping and management practices, increases in the efficiency of equipment, and recycling within a process.

Treatment (see also Chapter 8) refers to any method, technique, or process that changes the physical, chemical, or biological character of any chemical waste so as to neutralize the effects of the constituents of the waste.

Recycling (Chapter 9) is the use or reuse of chemical waste as an effective substitute for a commercial product or as an ingredient or feedstock in an industrial process. It includes the reclamation of useful constituent fractions within a waste material or the removal of contaminants from a waste material to allow it to be reused. Thus waste treatment also includes efforts to recover energy or material resources from a waste or to render such waste nonhazardous, less hazardous, safer to manage, amenable for recovery, or amenable for storage, or to merely reduce the volume of the waste to require less storage volume.

Disposal (Chapter 9) is the discharge, deposit, injection, or placing of a waste usually onto or into a land facility. Disposal of waste onto a body of water so that the waste or any constituents may be discharged into the water is considered to be more risky and offers the opportunity for contamination of the associated water systems. A similar line of thinking can be applied to disposal of a chemical into the atmosphere.

A crucial part of the regulation of chemical wastes in many countries pertains to treatment, storage, and disposal facilities (Majumdar, 1993; Olsen, 1994). Treatment alters the physical, chemical, or biological character or composition of a

Table 7.4: Common impurities found in water and methods for removal

Impurities	Process
Iron, manganese, and hydrogen sulfide	Oxidation (aeration) and precipitation
Carbon dioxide, methane	Degasification
Oxygen, nitrogen, carbon dioxide	Vacuum degasification
Calcium and magnesium	Cation exchange
Bicarbonate and carbonate	pH adjustment
Ionized salts, acids, or bases	Deionization via ion-exchange resins
	Reverse osmosis
	Electrodialysis
Turbidity, organic matter, colloidal silica	Cold lime and hot lime-soda
	Coagulation
	Filtration: sand, diatomaceous earth, and carbon filters

waste to make it safer (Table 7.4). Storage refers to the holding of chemical wastes for a temporary period pending treatment or disposal. Disposal refers to the ultimate fate of chemical substances or their treatment products.

The ideal treatment process reduces the quantity of chemical waste to a small fraction of the original amount and converts it to a nonhazardous form. However, most treatment processes yield material, such as sludge from wastewater treatment or incinerator ash, which requires disposal and which may be hazardous to some extent. In addition, more emphasis in treatment is being placed on recovery of recyclable materials and production of innocuous by-products (Bishop, 1995).

The technologies to manage chemical waste can be subdivided into five general categories: chemical, biological, physical (Chapter 8), thermal (Chapter 8), and recycling/disposal methods (Chapter 9) (United States Environmental Protection Agency, 1994). The effectiveness of the application of each of these methods to a specific waste varies depending on (1) the type of waste, including the concentration and mixture of individual chemicals in the waste; (2) the physical phase (solid or liquid) of the waste material; (3) the desired level of treatment; and (4) the method of disposal of any residue. Another consideration in selecting a treatment technology is where the wastes are to be treated. Wastes may be treated in place (in situ), within the confines of the site, or at an off-site facility (ex situ). The major distinction is that the technologies currently being applied to chemical wastes focus on the destruction of specific contaminants within the waste, while solid waste management technologies focus primarily on volume reduction.

2.0 Chemical Methods

Chemical treatment processes alter the chemical structure of the constituents of the waste to produce either an innocuous or a less hazardous by-product. Chemical processes tend to produce minimal air emissions and can often be carried out on the site of the waste generator. In fact, some chemical waste treatment processes are available as mobile units.

The applicability of chemical treatment processes to waste management depends upon the chemical properties of the waste constituents and the products of the processes. Particularly relevant properties are (1) acid-base character, (2) oxidation and/or reduction behavior, (3) precipitation and/or the tendency to form complexes, (4) flammability and combustibility, (5) reactivity and corrosivity, and (6) compatibility with other chemical waste. Thus there are a variety of chemical-based operations commonly used in treating wastes (Long, 1995).

2.1 Neutralization

Neutralization (pH adjustment) is a process for reducing the acidity or alkalinity of a waste stream by mixing acids and bases to produce a neutral solution.

This has proven to be a viable waste management process. Moreover, neutralization may be required prior to the application of other waste treatment processes.

Acidic and basic chemicals are treated by neutralization:

$$H^+ + OH^+ = H_2O$$

Lime (calcium oxide, CaO) and slaked lime [calcium hydroxide, $Ca(OH)_2$] are widely used as bases for treating acidic wastes. Because of lime's limited solubility, solutions of excess lime do not reach extremely high pH values. Sulfuric acid (H_2SO_4) is a relatively inexpensive acid for treating alkaline wastes. However, addition of too much sulfuric acid can produce highly acidic products; for some applications, acetic acid (CH_3COOH) is preferable. Acetic acid is a weak acid and it is also a natural and biodegradable product.

Hydrolysis, in a simple sense, may also be considered a form of neutralization. One of the ways to dispose of chemicals that are reactive with water is to allow them to react with water under controlled conditions (hydrolysis). Inorganic chemicals that can be treated by hydrolysis include metals that react with water (metal carbides, hydrides, amides, alkoxides, and halides) and nonmetal oxyhalides and sulfides (Table 7.5).

2.2 Precipitation

Precipitation is a process for removing soluble compounds contained in a waste stream (Long, 1995). A specific chemical is added to coagulate the offending species and form a precipitate (Table 7.6).

Precipitation is used primarily for the removal of heavy-metal ions from water: the most widely used means of precipitating metal ions is by the formation of hydroxides. The source of hydroxide ion is a base (alkali), such as lime [$Ca(OH)_2$], sodium hydroxide (NaOH), or sodium carbonate (Na_2CO_3).

Heavy metal ions in soil contaminated by chemical wastes may be present in a co-precipitated form with insoluble iron (Fe^{3+}) and manganese (Mn^{4+}) oxides (Fe_2O_3 and MnO_2, respectively). These oxides can be dissolved by reducing agents, such as solutions of sodium dithionate/citrate or hydroxylamine. This results in the production of soluble iron and manganese and the release of heavy metal ions which are removed with the water.

2.3 Ion Exchange

Ion exchange is used to remove from solution ions derived from inorganic materials (Natansohn et al., 1992). The solution is passed over a resin bed, which exchanges ions for the inorganic substances to be removed. When the bed loses its capacity to remove the component, it can be regenerated with a caustic solution. The greatest use of ion exchange in chemical waste treatment is for the removal of low

Table 7.5: Typical hydrolysis reactions of selected chemicals.

Active metals (Ca)	$Ca + 2H_2O = H_2 + Ca(OH)_2$
Hydrides ($NaAlH_4$)	$NaAlH_4 + 4H_2O = 4H_2 + NaOH + Al(OH)_3$
Carbides (CaC_2)	$CaC_2 + 2H_2O = Ca(OH)_2 + C_2H_2$
Amides ($NaNH_2$)	$NaNH_2 + H_2O = NaOH + NH_3$
Halides ($SiCl_4$)	$SiCl_4 + 2H_2O = SiO_2 + 4HCl$
Alkoxides ($NaOC_2H_5$)	$NaOC_2H_5 + H_2O = NaOH + C_2H_5OH$

Table 7.6: Typical coagulation reactions of selected chemicals

$Na_2Al_2O_4 + Ca(HCO_3)_2 + 2H_2O = 2Al(OH)_3 + CaCO_3 + Na_2CO_3$
$Al_2(SO_4)_3 + K_2SO_4 + 3Ca(HCO_3)_2 = 2Al(OH)_3 + K_2SO_4 + 3CaSO_4 + 6CO_2$
$Al_2(SO_4)_3 + 6NaOH = 2Al(OH)_3 + 3Na_2SO_4$
$Al_2(SO_4)_3 + (NH_4)_2SO_4 + 3Ca(HCO_3)_2 = 2Al(OH)_3 + (NH_4)2SO_4 + 3CaSO_4 + 6CO_2$
$Al_2(SO_4)_3 + 3Na_2CO_3 + 3H_2O = 2Al(OH)_3 + 3Na_2SO_4 + 3CO_2$
$Al_2(SO_4)_3 + 3Ca(HCO_3)_2 = 2Al(OH)_3 + 3CaSO_4 + 6CO_2$
$Fe_2(SO_4)_3 + 3Ca(HCO_3)_2 = 2Fe(OH)_3 + 3CaSO_4 + 6CO_2$
$FeSO_4 + Ca(OH)_2 = Fe(OH)_2 + CaSO_4$
$4Fe(OH)_2 + O_2 + 2H_2O = 4Fe(OH)_3$

levels of heavy metal ions from wastewater (Natansohn et al., 1992; Long, 1995).

2.4 Oxidation-Reduction

Oxidation-reduction is a process that can be used for the treatment and removal of a variety of inorganic and organic wastes (Table 7.7) (Ollis, 1993). In fact, there is the general opinion that most (>90%) hazardous wastes are primarily aqueous, so many of the advanced oxidation technologies are for wastewater treatment (Baker and Warren, 1991; Fischer, 1991; Tedder and Pohland, 1992, 1993). Indeed, it is the wastewater that carries with it many pollutants from various domestic and industrial sources (Lacy, 1983).

Ozone is a strong oxidant that can be generated on site by an electrical discharge through dry air or oxygen. Ozone employed as an oxidant gas at levels of 12% w/w in air and 2-5% w/w in oxygen has been used to treat a large variety of oxidizable contaminants, effluents, and wastes including wastewater and sludge containing oxidizable constituents.

Table 7.7: Selected oxidation/reduction reactions for waste treatment

Oxidation of Organic Substances	
Organic matter	$[CH_2NOS] + [O] = CO_2 + H_2O + NO_x + SO_2$
Aldehyde	$CH_3CHO + [O] = CH_3COOH$
Oxidation of Inorganic Substances	
Cyanide	$2CN^- + 5OCl^- + H_2O = N_2 + 2HCO_3^- + 5Cl^-$
Iron (Fe^{2+})	$4Fe^{2+} + O_2 + 10H_2O = 4Fe(OH)_3 + 8H^+$
Sulfur dioxide	$2SO_2 + 2O_2 + H_2O = 2H_2SO_4$
Reduction of Inorganic Substances	
Chromate	$2CrO_4^- + 3SO_2 + 4H^+ = Cr_2(SO_4)_3 + 2H_2O$
Permanganate	$MnO_4^- + 3Fe^{2+} + 7H_2O = MnO_2(s) + 3Fe(OH)_3(s) + 5H^+$

Oxidation by using catalytic agents (Shaw et al., 1993; United States Environmental Protection Agency, 1993) or photolysis (Blystone et al., 1993) can be used to destroy a number of chemical wastes. In such applications it is most useful in breaking chemical bonds in refractory organic compounds.

Electrolysis is a process in which one species in solution (usually a metal ion) is reduced by electrons at the cathode and another gives up electrons to the anode and is oxidized there. In chemical waste applications, electrolysis is most widely used in the recovery of cadmium, copper, gold, lead, silver, and zinc. Metal recovery by electrolysis is made more difficult by the presence of cyanide ion, which stabilizes metals in solution as the cyanide complexes, such as nickel tetracyanide $[Ni(CN)_4]$.

2.5 Extraction and Leaching

While not strictly a chemical-based method for waste treatment, *extraction and leaching* is included here, since the procedure depends to a large extent on the chemical properties of the waste material. More simply, extraction and leaching is a process for the selective solubilization and removal of waste from the surrounding medium.

In addition, chemical extraction or chemical leaching is the removal of a constituent by chemical reaction with an extractant in solution. Poorly soluble heavy-metal salts can be extracted by reaction of the salt anions with acid (H^+). Acids also dissolve basic organic compounds such as amines and aniline. Extraction with acids should be avoided if cyanides or sulfides are present to prevent formation of toxic hydrogen cyanide or hydrogen sulfide:

$$X\text{-}CN + \text{acid} = HCN$$
$$X\text{-}S + \text{acid} = H_2S$$

Nontoxic weak acids are usually safer and include acetic acid (CH_3COOH) and monosodium phosphate (NaH_2PO_4).

Chelating agents are also employed to remove heavy metal contaminants from soil by forming soluble species with the metal salts (Peters and Shem, 1992). Chelating ion-exchange resins can also be employed for removal of heavy metals from liquid streams (Kwon et al., 1992).

3.0 Bioremediation

Bioremediation is the use of living organisms (primarily microorganisms) to degrade pollutants previously introduced into the environment or to prevent pollution through treatment of waste streams before they enter the environment. Bioremediation is emerging as one of several alternate technologies for removing pollutants from the environment, restoring contaminated sites, and preventing further pollution (Ladisch and Bose, 1992; Eckenfelder and Norris, 1993; Atlas, 1995).

Biodegradation of waste is the conversion of waste materials by biological processes to simple inorganic molecules and, to a certain extent, to biological materials. The complete bioconversion of a substance to inorganic species such as carbon dioxide, ammonia, and phosphate is called mineralization.

Biological waste treatment is a generic term applied to processes that use microorganisms to decompose organic wastes either into water, carbon dioxide, and simple inorganic substances, or into simpler organic substances, such as aldehydes and acids.

The purpose of a biological treatment system is to control the environment for microorganisms so that their growth and activity are enhanced, and to provide a means for maintaining high concentrations of the microorganisms in contact with the wastes. Since biological treatment systems do not alter or destroy inorganic substances, and high concentrations of such materials can severely inhibit decomposition activity, chemical or physical treatment may be required to extract inorganic materials from a waste stream prior to biological treatment.

Detoxification refers to the biological conversion of a toxic substance to a less toxic species, which may still be relatively complex, or biological conversion to an even more complex material. An example of detoxification is the enzymatic conversion of paraoxon (a highly toxic organophosphate insecticide) to p-nitrophenol, which has only about l/200 the toxicity of the parent compound.

Biodegradation is usually carried out by the action of microorganisms, particularly bacteria. However, microorganisms such as bacteria are not panaceas for chemical cleanup. Their function can vary depending upon the specific bacterium and the metabolic processes of the bacteria (Fenchel and Finlay, 1994).

Biotransformation is the conversion of a substance through metabolization, thereby causing an alteration to the substance by biochemical processes in an organism. Metabolism is divided into the two general categories of catabolism, which is the breaking down of more complex molecules, and anabolism, which is the building up of life molecules from simpler materials. The substances subjected to biotransformation may be naturally occurring or anthropogenic (made by human activities). They may consist of xenobiotic molecules that are foreign to living systems.

An important biochemical process that occurs in the biodegradation of many chemical waste materials is co-metabolism. This does not serve a useful purpose to an organism in terms of providing energy or raw material to build biomass, but occurs concurrently with normal metabolic processes. An example of co-metabolism of chemical waste is provided by the white rot fungus (Bishop, 1993; Shah and Aust, 1993) which degrades a number of kinds of organic chlorine compounds (including polychlorobiphenyls, cyanides, and chlorodioxins) under the appropriate conditions. The enzyme system responsible for this degradation is one that the fungus uses to break down lignin in plant material under normal conditions.

Biodegradation of chemical waste that can be metabolized takes place whenever the wastes are subjected to conditions conducive to biological processes. The most common type of biodegradation is that of organic compounds in the presence of air, that is, *aerobic* processes. However, in the absence of air, *anaerobic* biodegradation may also take place. Furthermore, inorganic species are subject to both aerobic and anaerobic biological processes.

Although biological treatment of chemical waste is normally regarded as degradation to simple chemical species such as carbon dioxide, water, sulfates, and phosphates, the possibility must always be considered of forming more complex (in some cases hazardous) chemical species. An example of the latter is the production of volatile, soluble, toxic methylated forms of arsenic and mercury from inorganic species of these elements by bacteria under anaerobic conditions.

Physical-chemical and biological treatment processes are employed as for wastewater treatment. In addition, chemicals are introduced for precipitation of nutrients, followed by coagulation and filtration for removing solids remaining after biological treatment. In some cases, granular activated carbon or membrane filtration (Chiarizia et al., 1992; Noble and Stern, 1995; Tiberi and Brose, 1995) or a combination of membrane-assisted solvent extractions (Hutter and Vandegrift, 1992) is for additional purification of the groundwater streams and waste streams. This higher level of treatment is advisable because any visual traces of chemical waste can damage the appearance of the waters. In addition, the treatment may combat the potential eutrophic effect that the nutrients phosphorus and nitrogen can have on the receiving water.

For the most part, anthropogenic compounds resist biodegradation much more strongly than do naturally occurring compounds. This is generally due to the absence of enzymes that can bring about an initial attack on the compound. Physical and chemical characteristics of a compound that are involved in its amenability to biodegradation are hydrophobicity, solubility, volatility, and affinity for lipids. Some organic structural groups that impart particular resistance to biodegradation are branched carbon chains, ether linkages, meta-substituted benzene rings, chlorine, amines, methoxy groups, sulfonates, and nitro groups.

Actinomycetes are microorganisms that are morphologically similar to both bacteria and fungi. They are involved in the degradation of a variety of organic compounds, including degradation-resistant alkanes and lignocellulose. Other compounds attacked include pyridines, phenols, nonchlorinated aromatics, and chlorinated aromatics. Fungi are particularly noted for their ability to attack long-chain and complex hydrocarbons and are more successful than bacteria in the initial attack on polychlorobiphenyls.

Phototrophic microorganisms, which include algae, photosynthetic bacteria, and cyanobacteria (blue-green algae) tend to concentrate organophilic compounds in their lipid stores and induce photochemical degradation of the stored compounds.

Usually the products of biodegradation are molecular forms that tend to occur in nature and are in greater thermodynamic equilibrium with their surroundings than the starting materials.

Microbial bacteria and fungi possessing enzyme systems required for biodegradation of wastes are usually best obtained from populations of indigenous microorganisms at a chemical waste site where they have developed the ability to degrade particular kinds of molecules. Although it has some shortcomings in the degradation of complex chemical mixtures, biological treatment offers a number of significant advantages and has considerable potential for the degradation of chemical wastes, even in situ.

The biodegradability of a compound is influenced by its physical characteristics, such as solubility in water and vapor pressure, and by its chemical properties, including molecular mass, molecular structure, and presence of various kinds of functional groups, some of which provide a "biochemical handle" for the initiation of biodegradation. With the appropriate organisms and under the right conditions, even substances such as phenol that are considered to be biocidal to most microorganisms can undergo biodegradation.

Properties of chemical wastes and their media can be altered to increase biodegradability. This can be accomplished by adjustment of conditions to optimum temperature, pH (usually in the range of 6-9), stirring, oxygen level, and the amount of the material. Biodegradation can be aided by removal of toxic organic and inorganic substances, such as heavy-metal ions.

Aerobic processes for the treatment of waste utilize aerobic bacteria and fungi that require molecular oxygen. These processes are often favored by microorganisms, in part because of the high energy yield obtained when molecular oxygen reacts with organic matter. Aerobic and anaerobic digestion processes are well adapted to the use of an activated sludge process (Goodloe et al., 1993). These treatments can be applied to wastes such as chemical process wastes and landfill leachates. Some systems use powdered activated carbon as an additive to absorb nonbiodegradable organic wastes.

Anaerobic processes for the treatment of waste are those processes in which microorganisms degrade wastes in the absence of oxygen. This process can be practiced on a variety of organic wastes (Goodloe et al., 1993). Compared to the aerated activated sludge process, anaerobic digestion requires less energy; yields less sludge by-product, generates hydrogen sulfide which precipitates toxic heavy metal ions; and produces methane which can be used as an energy source.

Activated sludge is the biologically active sediment produced by the repeated aeration and settling of sewage and/or organic wastes. The dissolved organic matter acts as food for the growth of aerobic flora. These species produce a biologically active sludge which is usually brown in color and which destroys the polluting organic matter in the sewage and waste. Briefly, this activated sludge process (Figure 7.1) is a versatile and effective waste treatment process. Microorganisms in the aeration tank convert organic material in wastewater to microbial biomass and carbon dioxide. Organic nitrogen is converted to ammonium ion or nitrate and organic phosphorus is converted to orthophosphate. The microbial cell matter formed as part of the waste degradation processes is normally kept in the aeration tank until the microorganisms are past the log phase of growth, at which point the cells flocculate and separate from the liquid. These solids settle out in a settler and a fraction is discarded. Part of the solid, the return sludge, is recycled to the head of the aeration tank and comes into contact with fresh sewage. The combination of a high concentration of microorganisms in the return sludge and a rich food source in the in-flowing sewage provides optimum conditions for the rapid degradation of organic matter.

However, in terms of pollution by humans, sewage sludge is not the beneficial organic fertilizer that many believe it to be. Sewage sludge spread on land may contaminate water by release of a variety of organic and inorganic contaminants. The spread of disease in ancient and medieval times due to contamination of water supplies will attest to this (Cartwright and Biddiss, 1991). Similarly, materials leached from landfills by (acid) rain can also cause serious contamination of land and water systems. In addition, leachates from unlined pits and lagoons containing chemical liquids/sludge may cause a specific pollution (drinking water) problem or a more general pollution problem.

Soil is a natural medium for a number of living organisms that may have an effect upon biodegradation of chemical wastes. Of these, the most important are bacteria. Wastes that are amenable to land treatment are biodegradable organic substances. However, in soil contaminated with chemical wastes, bacterial cultures may develop that are effective in degrading normally recalcitrant compounds through acclimation over a long period of time. Land treatment is most often used for petroleum refining wastes and is applicable to the treatment of fuels and wastes from leaking underground storage tanks.

The regulations concerning storage tanks, especially aboveground tanks, are sufficiently strict that precautions to counteract spillage must be taken to avoid fines of several thousand dollars per day (Thompson Publishing Group, 1995).

Land treatment can also be applied to biodegradable organic chemical wastes, including some organic halogen compounds. Land treatment is not suitable for the treatment of wastes containing acids, bases, toxic inorganic compounds, salts, heavy metals, and organic compounds that are excessively soluble, volatile, or flammable.

Composting of chemical wastes is the biodegradation of solid or solidified materials in a medium other than soil. Bulking material, such as plant residue, paper, municipal refuse, or sawdust may be added to retain water and enable air to penetrate to the waste material. Successful composting of chemical waste depends upon a number of factors, such as the selection of the appropriate microorganism or

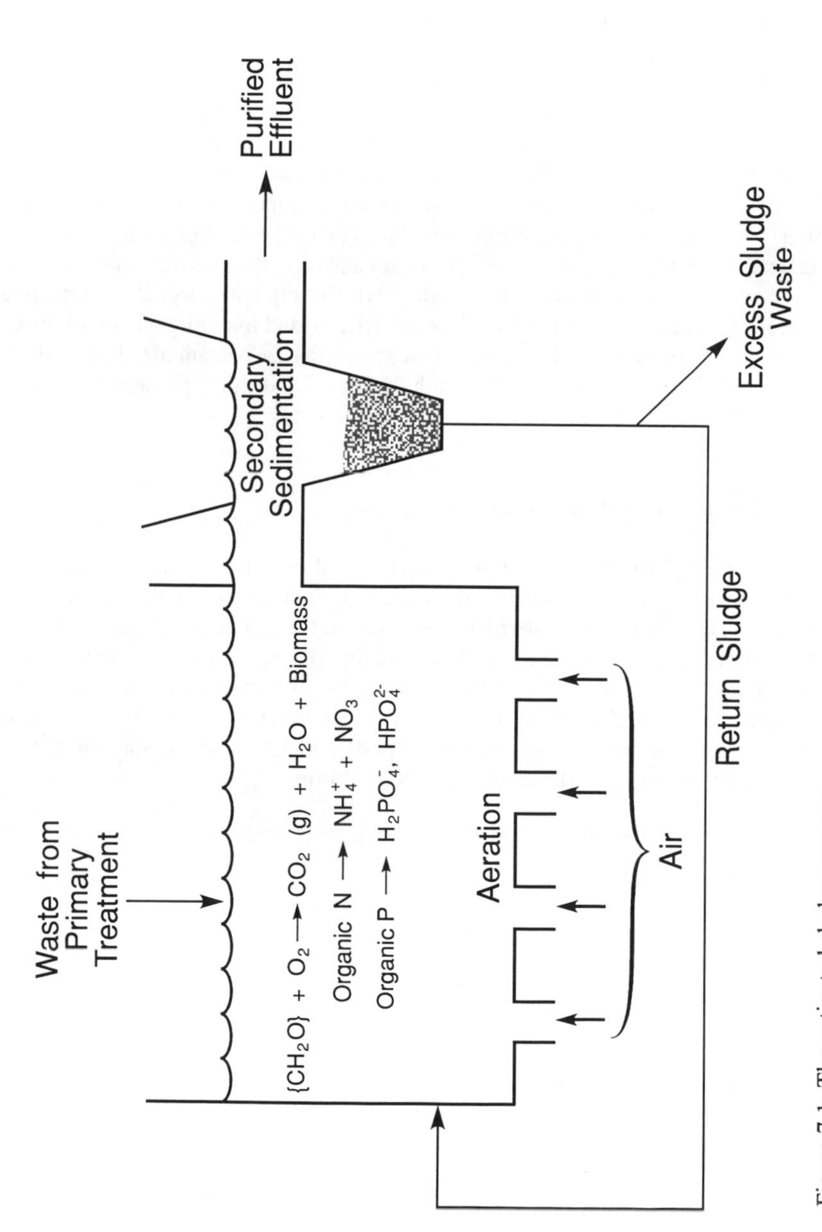

Figure 7.1: The activated sludge process

inoculum. Once a successful composting operation is under way, a good inoculum is maintained by recirculating spent compost to each new batch.

Other parameters that must be controlled include oxygen supply, moisture content (which should be maintained at a minimum of about 40%), pH (usually around neutral), and temperature. The composting process generates heat, so if the mass of the compost pile is sufficiently high, it can be self-heating under most conditions. Some wastes are deficient in nutrients, such as nitrogen, which must be supplied from commercial sources or from other wastes.

Bioreactors have been used for wastewater treatment processes for decades. The reactors may be either fixed film or slurry phase. Fixed-film reactors are similar to the traditional trickling filters or rotating biological contactors of the wastewater industry. In either case, the microorganisms are supported on the medium of the filter. The wastes are passed over the filter (or in the case of rotating biological contactors, the filter is passed over the waste) allowing the microorganisms to come into contact with the wastes and break down the organic material. Slurry phase reactors are tanks into which the wastes, nutrients, and microorganisms are placed. The tank is mixed and may be aerated. In many instances, contaminated groundwater is used to create the waste slurry. Both fixed-film and slurry phase treatments are either batch or continuous mode.

3.1 Solid Phase Bioremediation

Solid phase bioremediation, often referred to as land farming, treats wastes using conventional soil management practices to enhance the microbial degradation of the wastes. The wastes are placed directly on the ground or in shallow tanks, if required by RCRA restrictions. Nutrients and microorganisms are normally added to the wastes which are routinely tilled during the treatment process. This tilling improves aeration and the contact of the organisms with the wastes. While treatment may occur throughout the upper 3 to 5 feet (0.9 to 1.5 m) of the soil, most occurs within the top foot, called the zone of incorporation.

3.2 Soil Heaping

Soil heaping is piling wastes in heaps several feet high on an asphalt or concrete pad. Nutrients, microorganisms, and air are provided through perforated piping placed throughout the pile. The pile is covered to contain volatile organic compounds, to stabilize the microorganism's environment, and to control soil erosion. The volatile organic compounds can be further controlled by applying a vacuum to the pile and treating the exhaust.

3.3 Composting

Composting is another application of bioremediation. In this process the wastes are normally mixed with a structurally firm bulking material such as chopped hay and wood chips. As with the other bioremediation technologies, nutrients, air, and microorganisms must be added. The three major types of composting are open windrow, static windrow, and reactor systems. In the open windrow system, the compost piles are open to the air whereas in the static windrow system the air is mechanically forced into the compost piles. When reactors are used, the compost is mechanically mixed to ensure aeration.

3.4 In Situ Bioremediation

One of the advantages of bioremediation is that it can be effectively applied to treat wastes in place (Rittman, 1993; Rittman et al., 1994). The process usually entails introduction of nutrients, microorganisms, and air into the soil/waste through a series of injection wells or infiltration trenches. The term *bioventing* has also been applied to this technology, although the term could just as easily be applied to composting or to soil heaping. Whatever the name, the technology has shown some success for hydrocarbon degradation (Newman et al., 1993).

If the soil does not have sufficient moisture content, water may have to be added. In situ bioremediation is often applied in conjunction with groundwater pump and treat systems and soil flushing activities.

There is also the concept of gene manipulation as a means degrading polynuclear aromatic hydrocarbons (Zylstra et al., 1992). The concept offers promise for many sites (such as town gas sites where wastes containing polynuclear aromatic hydrocarbons are evident). The degradation products from such interactions may require cleanup, but it is quite possible that the degradation products are easier to clean than the original polynuclear aromatic hydrocarbons. There is also the concept of using biodegradation on wastes that have been reduced to residual saturation by flushing technologies (Johnson and Leuschner, 1992). A final flushing to remove the biodegraded material would be necessary.

4.0 References

Atlas, R.M. 1995. *Chemical and Engineering News* 73(14): 32.

Baker, R.D., and Warren, J.L. 1991. In Preprints. *2nd Topical Pollution Prevention Conference*. August 20-21. American Institute of Chemical Engineers, New York.

Benefield, L.D., Judkins, J.F., and Weand, B.L. 1982. *Process Chemistry for Water and Wastewater Treatment*. Prentice-Hall, Englewood Cliffs, New Jersey.

Bishop, J. 1993. *Hazmat World* 6(12): 32.

Bishop, J. 1995. *Environmental Solutions* 8(8): 31.

Blystone, P.G., Johnson, M.D., Haag, W.R., and Daley, P.F. 1993. Chap. 18 in *Emerging Technologies for Hazardous Waste Management III*. Edited by D.W. Tedder and F.G. Pohland. Symposium Series No. 518. American Chemical Society, Washington, D.C.

Cartwright, F.F., and Biddiss, M.D. 1972. *Disease and History*. Dorset Press, New York.

Chiarizia, R., Horwitz, E.P., and Hodgson, K.M. 1992. Chap. 2 in *Environmental Remediation*. Edited by G.F. Vandegrift, D.T. Reed, and I.R. Tasker. Symposium Series No. 509. American Chemical Society, Washington, D.C.

Eckenfelder, W.W., and Norris, R.D. 1993. Chap. 8 in *Emerging Technologies for Hazardous Waste Management III*. Edited by D.W. Tedder and F.G. Pohland. Symposium Series No. 518. American Chemical Society, Washington, D.C.

Fenchel, T., and Finlay, B.J. 1994. *American Scientist* 82(1): 22.

Fischer, L.M. 1991. In Preprints. *2nd Topical Pollution Prevention Conference*. August 20-21. American Institute of Chemical Engineers, New York.

Goodloe, J.C., Kitsos, H.M., Meyers, A.J. Jr., Hubbard, J.S., and Roberts, R.S. 1993. Chap. 12 in *Emerging Technologies for Hazardous Waste Management III*. Edited by D.W. Tedder and F.G. Pohland. Symposium Series No. 518. American Chemical Society, Washington, D.C.

Hutter, J.C., and Vandegrift, G.F. 1992. Chap. 4 in *Environmental Remediation*. Edited by G.F. Vandegrift, D.T. Reed, and I.R. Tasker. Symposium Series No. 509. American Chemical Society, Washington, D.C.

Johnson, L.A. Jr., and Leuschner, A.P. 1992. Chap. 20 in *Hydrocarbon Contaminated Soils and Groundwater*. Vol. 2. Edited by E.J. Calabrese and P.T. Kostecki. Lewis Publishers, Ann Arbor, Michigan.

Just, S.R., and Stockwell, K.J. 1993. Chap. 13 in *Emerging Technologies for Hazardous Waste Management III*. Edited by D.W. Tedder and F.G. Pohland. Symposium Series No. 518. American Chemical Society, Washington, D.C.

Kocurek, D.S., and Woodside, G. 1994. *Resources and References: Hazardous Waste and Hazardous Materials Management*. Noyes Data Corp., Park Ridge, New Jersey.

Kwon, K.C., Jermyn, H., and Mayfield, H. 1992. Chap. 12 in *Environmental Remediation*. Edited by G.F. Vandegrift, D.T. Reed, and I.R. Tasker. Symposium Series No. 509. American Chemical Society, Washington, D.C.

Lacy, W.J. 1983. In *Riegel's handbook of industrial chemistry*. p. 14. Edited by J.A. Kent. Van Nostrand Reinhold, New York.

Ladisch, M.R., and Bose, A. (eds.). 1992. *Harnessing Biotechnology for the 21st Century*. American Chemical Society, Washington, D.C.

Long, R.B. 1995. *Separation Processes in Waste Minimization*. Marcel Dekker Inc., New York.

Majumdar, S.B. 1993. *Regulatory Requirements for Hazardous Materials*. McGraw-Hill, New York.

Natansohn, S., Rourke, W.J., and Lai, W.C. 1992. Chap. 10 in *Environmental Remediation*. Edited by G.F. Vandegrift, D.T. Reed, and I.R. Tasker. Symposium Series No. 509. American Chemical Society, Washington, D.C.

Newman, W., Martinson, M., Smith, G., and McCain, L. 1993. *Hazmat World* 6(12): 34.

Noble, R.S., and Stern, A. 1995. (eds.). *Membrane Separations Technology: Principles and Applications*. Elsevier, New York.

Ollis, D.F. 1993. Chap. 2 in *Emerging Technologies for Hazardous Waste Management III*. Edited by D.W. Tedder and F.G. Pohland. Symposium Series No. 518. American Chemical Society, Washington, D.C.

Olsen, L.R. 1994. *Environmental Solutions* 7(10): 53.

Peters, R.W., and Shem, L. 1992. Chap. 6 in *Environmental Remediation*. Edited by G.F. Vandegrift, D.T. Reed, and I.R. Tasker. Symposium Series No. 509. American Chemical Society, Washington, D.C.

Rittman, B.E. 1993. *In Situ Bioremediation: When Does It Work?* National Research Council, Washington, D.C.

Rittman, B.E., Seagren, E., Wrenn, B.A., Valocchi, A.J., Ray, C., and Raskin, L. 1994. *In Situ Bioremediation*. Noyes Data Corp., Park Ridge, New Jersey.

Shah, M.M., and Aust, S.D. 1993. Chap. 10 in *Emerging Technologies for Hazardous Waste Management III*. Edited by D.W. Tedder and F.G. Pohland. Symposium Series No. 518. American Chemical Society, Washington, D.C.

Shaw, H., Wang, Y., Yu, T-C., and Cerkanowicz, A.E. 1993. Chap. 17 in *Emerging Technologies for Hazardous Waste Management III*. Edited by D.W. Tedder and F.G. Pohland. Symposium Series No. 518. American Chemical Society, Washington, D.C.

Speight, J.G. 1991. *The Chemistry and Technology of Petroleum*. 2nd ed. Marcel Dekker Inc., New York.

Speight, J.G. 1994. *The Chemistry and Technology of Coal*. 2nd ed. Marcel Dekker Inc., New York.

Tedder, D.W., and Pohland, F.G. (eds.). 1991. *Emerging Technologies for Hazardous Waste Management*. Symposium Series No. 422. American Chemical Society, Washington, D.C.

Tedder, D.W., and Pohland, F.G. 1992. Chap. 1 in *Emerging Technologies for Hazardous Waste Management II*. Edited by D.W. Tedder and F.G. Pohland. Symposium Series No. 468. American Chemical Society, Washington, D.C.

Tedder, D.W., and Pohland, F.G. 1993. Chap. 1 in *Emerging Technologies for Hazardous Waste Management III*. Edited by D.W. Tedder and F.G. Pohland. Symposium Series No. 518. American Chemical Society, Washington, D.C.

Tedder, D.W., and Pohland, F.G. (eds.). 1995. Chap. 1 in *Emerging Technologies for Hazardous Waste Management V*. Symposium Series No. 607. American Chemical Society, Washington, D.C.

Tiberi, T.P. and Brose, D.J. 1995. *Environmental Solutions* 8(2): 42.

Thompson Publishing Group. 1995. *Aboveground Storage Tank Guide*. Vols. 1 and 2. Thompson Publishing Group, Washington, D.C.

United States Environmental Protection Agency. 1993. *Evaporation-Catalytic Oxidation Technology*. Report No. EPA/540/AR-93/506. Superfund Innovative Technology Evaluation. United States Environmental Protection Agency, Washington, D.C.

United States Environmental Protection Agency. 1994. *The Superfund Innovative Technology Program*. Annual Report to Congress. Report No. EPA/540/R-94/518. United States Environmental Protection Agency, Washington, D.C.

Zylstra, G.J. , Wang, X.P., and Didolkar, V.A. 1992. In *Harnessing Biotechnology for the 21st Century*. p. 68. Edited by M.R. Ladisch and A. Bose. American Chemical Society, Washington, D.C.

Chapter 8

Physical and Thermal Methods of Waste Management

1.0 Introduction

The effects of chemical waste on the environment are reflected by the effects on organisms and on the overall ecosystem. Virtually all chemical wastes are poisonous to a degree, some being extremely so. Chemical wastes eventually reach a state of physical and chemical stability and equilibrium with the environment, although it may take many centuries. In many cases, the behavior (or reactivity) and ultimate fate of a chemical waste are functions of its physical properties and surroundings. Since it is not usually possible to wait for the chemical to reach an equilibrium with its surroundings, methods of remediation are essential (Borchardt, 1995; Tedder and Pohland, 1995).

The first step in considering the appropriate technology to treat a specific waste is to determine whether the waste is hazardous or nonhazardous (Chapter 6). The definition of a hazardous/nonhazardous waste is determined by the relevant legislation (Chapter 12). There is the general tendency to think of a chemical waste as some obscure and ill-defined sludge discarded from a process. However, chemical waste can also be in the form of volatile organic compounds, semivolatile organic compounds, metals, radioactively contaminated materials, or a mixture of any or all of these.

The ideal treatment process reduces the quantity of chemical waste to a small fraction of the original amount and converts it to a nonhazardous form (Chapter 6), if such a conversion is possible. Another consideration in selecting a treatment technology is the location where the wastes are to be treated (Wise and Trantolo, 1994). For example, wastes may be treated in place (in situ), within the confines of the site, or at an off-site facility (ex situ).

There are various alternative waste treatment technologies (United States Environmental Protection Agency, 1994), such as chemical treatment and biological treatment (Chapter 7). Physical treatment processes (which also include thermal methods such as incineration) as well as solidification or stabilization are the subject of this chapter. These processes are used to (1) recycle the waste (Chapter 9) and reuse waste materials, (2) reduce the volume and toxicity of the waste stream, or (3) produce a final residual material that is suitable for disposal. The effectiveness of the application of each of these technology groups to a specific waste varies depending on the type of waste, the concentration of the individual components in the waste, the physical phase of the material, the desired level of treatment, and the final method of disposal of any remaining residue.

The waste characteristics can include such properties as volatility (gases, volatile solutes in water, gases or volatile liquids held by solids, such as catalysts), liquid phase materials (wastewater, organic solvents), dissolved or soluble materials (water-soluble inorganic species, water-soluble organic species, compounds soluble in organic solvents), semisolid materials (sludge, grease), and solid materials (dry solids, including granular solids with a significant water content, such as dewatered sludge, as well as solids suspended in liquids). The chemistry that occurs in water (aquatic) systems (Chapter 4) can be extremely complex and requires detailed knowledge of the species in the aquatic system (Huang et al., 1995).

Waste treatment may occur at three major levels: primary, secondary, and tertiary (polishing). *Primary waste treatment* is generally regarded as preparation for further treatment, although it can result in the removal of by-products and reduction of the quantity and hazard of the waste. *Secondary waste treatment* detoxifies, destroys, and removes hazardous constituents. *Polishing* usually refers to treatment of a waste product for safe discharge, for example recycled water that is removed from wastes so that it may be safely discharged. In addition, especially in the case of water treatment, the processes employed can be generally categorized as *external treatment* or *internal treatment* (Benefield et al., 1982; Cheremisinoff, 1995).

External treatment uses processes such as flotation and clarification to remove material, including suspended or dissolved solids, hardness, and dissolved gases. Following this basic treatment, the water may be divided into different streams, some to be used without further treatment and the rest to be treated for specific applications. Internal treatment is, for example, the addition of chemicals to water to neutralize or change the state of the pollutants in the water.

2.0 Physical Methods

In contrast to the chemical methods of waste treatment (Chapter 7), these are processes that separate components of a waste stream or change the physical form of the waste without altering the chemical structure of the constituent materials (Long, 1995). These processes are very useful for separating hazardous materials from an otherwise nonhazardous waste stream, so that they may be treated in a more concentrated form. Various chemical components are separated for different treatment processes, and a waste stream is prepared for ultimate destruction in a biological or thermal treatment process.

Various unit operations for waste treatment that are based upon knowledge of the physical behavior of wastes (Long, 1995; United States Department of Energy, 1995a). These operations include phase separation, phase transfer (extraction, sorption), phase transition (distillation, evaporation, precipitation), and membrane separation (reverse osmosis, hyperfiltration, and ultrafiltration).

The important aspect of physical methods of remediation is to isolate the contaminant, so that it can be recovered/destroyed while being contained within a specific area. In terms of containment, there are several technologies available that involve construction of a containment wall (usually concrete) around the contaminated area on the side where leakage into nearby water systems can occur.

One aspect of remediation that is evolving is the application of soil-mix wall technology (Yang and Takeshima, 1995), which consists of mixing soils in situ with cement grout using multiple shaft augers to construct overlapped cement columns. The use of augers makes it possible to define the treatment zone clearly and to confine the contaminant(s).

2.1 Phase Separation

The most straightforward means of physical treatment involves separation of components of a mixture that are already in two different phases. In many cases the separation must be aided by mechanical means, particularly sedimentation, screening, centrifugation, flotation, and filtration.

Separation is used to divide the wastes into two or more distinct waste streams based upon size, density, or material type. Normally accomplished by either manual or mechanical means, separation allows for a more efficient operation of the subsequent technologies while reducing the quantities of material to be treated.

Sedimentation is usually accomplished by providing sufficient time and space in special tanks or holding ponds for settling. Chemical coagulating agents are often added to encourage the settling of fine particles.

Screening is a process for removing particles from waste streams, and it is used to protect downstream pretreatment processes. Screening is the most common technology employed for separation. Four general categories of screens are available: (1) grizzly screens, sets of parallel bars used for the removal of coarse material; (2) revolving screens, a cylindrical frame covered with wire cloth; (3) vibrating screens, normally used when higher capacities are required; and (4) oscillating screens, used at lower speeds than vibrating screens used for separating particles by grain size.

Flotation is a process for removing solids from liquids and moving the particles to the surface by using tiny air bubbles. Flotation is useful for removing particles too small to be removed by sedimentation.

In the process of dissolved air flotation, air is dissolved in the suspending medium under pressure and, when the pressure is released, comes out of the solution as minute air bubbles attached to suspended particles, which causes the particles to float to the surface. Flotation can also be used to remove hydrocarbon-type waste from in-situ locations without the more destructive dig-and-move procedures. For example, a chemical waste might be removed by a relatively simple (and perhaps standard) flushing technique (Figure 8.1) or by a somewhat more innovative technique that involves the principles of laminar flow and can be applied to lighter-than-water liquids as well as to heavier-than-water liquids (Figure 8.2) (Johnson and Leuschner, 1992).

Centrifugation is a process for separating solid and liquid components of a waste stream by rapidly rotating a mixture of solids and liquids inside a vessel. Centrifugation is most often used to dewater sludge.

An important and often difficult waste treatment step is emulsion breaking, in which colloidal-sized emulsions are caused to aggregate and settle from suspension.

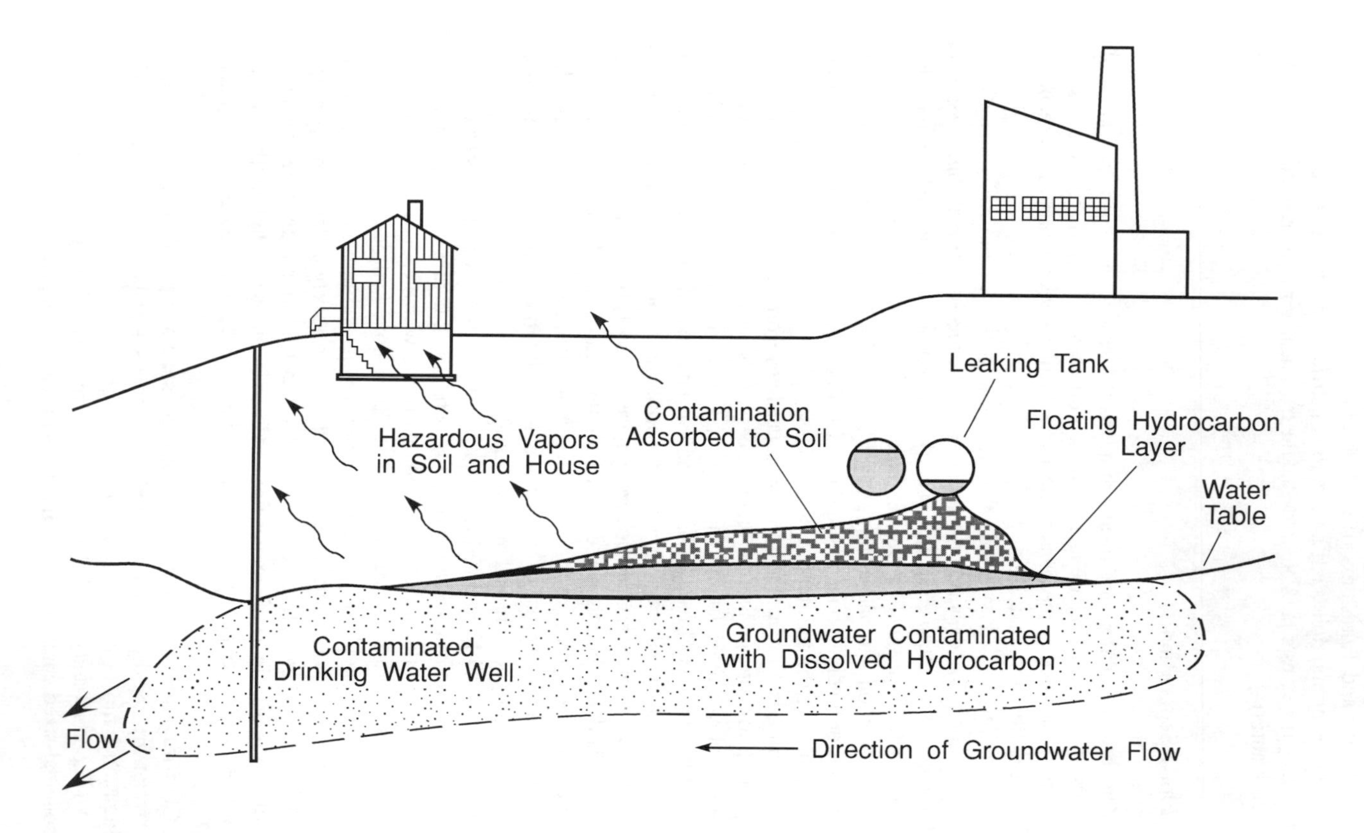

Figure 8.1: Flow of a contaminant plume in groundwater

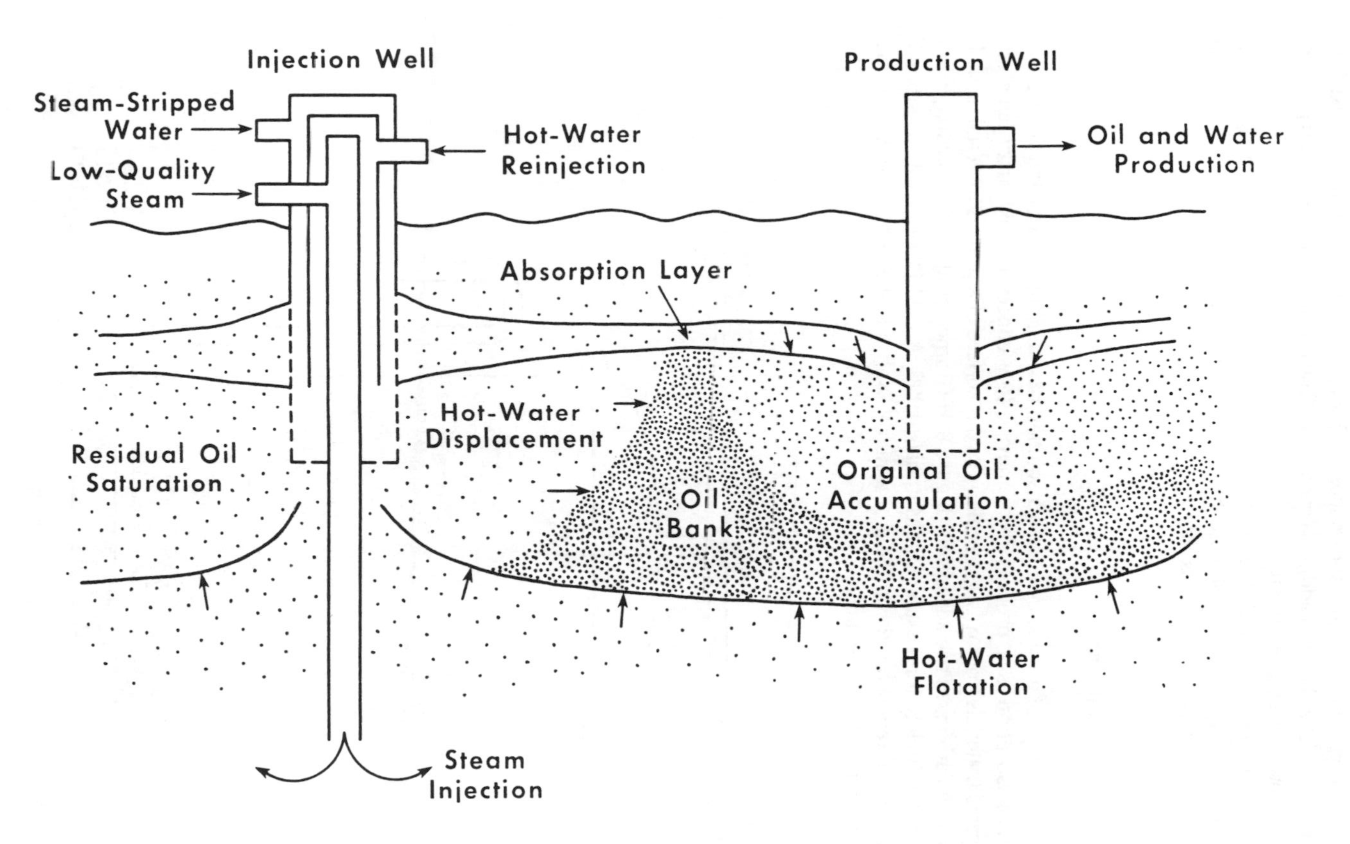

Figure 8.2: Subsurface cleanup of organic waste by the CROW^{TM} (Contained Recovery of Oil Waste) process

Agitation, heat, acid, and the addition of coagulants consisting of organic polyelectrolytes or inorganic substances, such as an aluminum salt, may be used for this purpose. The chemical additive acts as a flocculating agent to cause the particles to stick together and settle out.

Filtration is an older process that still finds wide application for waste treatment. The general principles involve the use of an impassible barrier that collects solids but allows liquids to pass through.

2.2 Phase Transfer

Transfer of a substance from a solution to a solid phase is called adsorption. Thus adsorption is a process for removing low concentrations of organic and inorganic materials from waste streams using the surface of a porous material, usually activated carbon, as the adsorbent (Figure 8.3) (Barney et al., 1992; Suzuki, 1993; American Society for Testing and Materials, 1995, ASTM D 3922). The carbon is replaced and regenerated with heat or a suitable solvent when its capacity to attract organic substances is reduced.

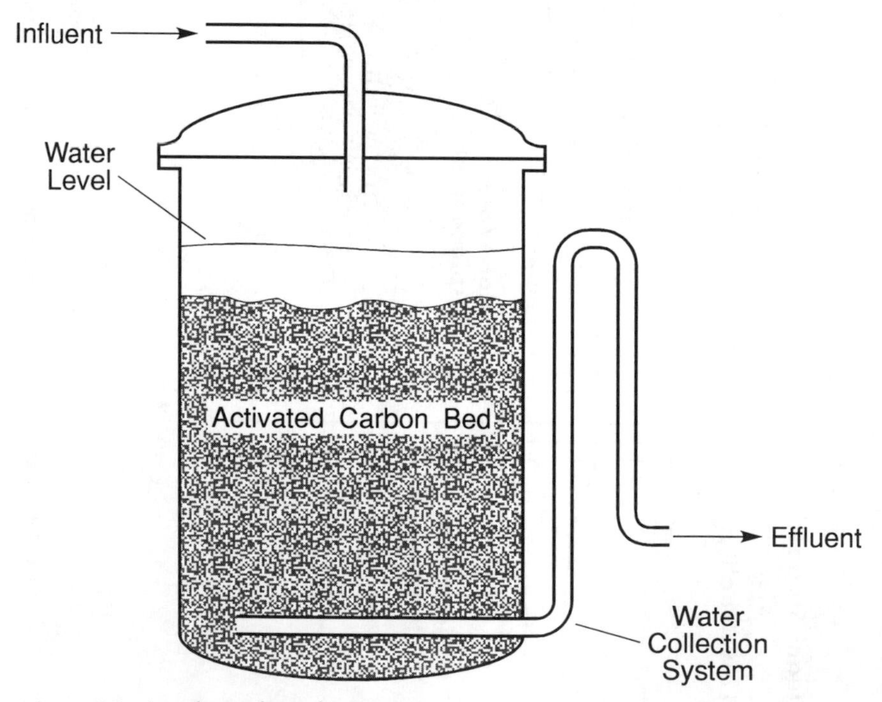

Figure 8.3: A carbon adsorption system

Effluents from other treatment processes, such as biological treatment of degradable organic solutes in water, can be polished with activated carbon. Activated carbon sorption is most effective for removing from water those chemicals that are poorly water soluble and that have high molecular masses, such as xylene, naphthalene, cyclohexane, chlorinated hydrocarbons, phenol, aniline, dyes, and surfactants. Activated carbon does not work well for organic compounds that are highly water soluble or polar.

Solids other than activated carbon that can be used for sorption of contaminants from liquid wastes are synthetic resins composed of organic polymers and mineral substances. For example, clay and regenerable adsorbents (Chovit and McMillen, 1994) are employed to remove impurities from waste lubricating oils (not yet a Resource Conservation Recovery Act, RCRA, waste) in some oil recycling processes.

Solvent extraction is a process for separating liquids by mixing the stream with a solvent that is immiscible with part of the waste but which will extract certain components of the waste stream (Long, 1995). The extracted components are then removed from the immiscible solvent for reuse or disposal. One of the more promising approaches to solvent extraction and leaching of chemical waste constituents is the use of supercritical fluids, most commonly carbon dioxide, as extraction solvents. A supercritical fluid is one that has characteristics of both liquid and gas and consists of a substance above its supercritical temperature and pressure (31.1°C/88°F and 73.8 atm/1085 psi/7.5 MPa, respectively, for carbon dioxide) (Kiran and Brennecke, 1993).

After a substance has been extracted from a waste into a supercritical fluid at high pressure, the pressure can be released, resulting in separation of the substance extracted. The fluid can then be compressed again and recirculated through the extraction system.

Some possibilities for treatment of chemical wastes by extraction with supercritical carbon dioxide include removal of organic contaminants from wastewater, extraction of organic halogen pesticides from soil, extraction of oil from emulsions used in aluminum and steel processing, and regeneration of the spent carbon. Waste oils contaminated with polychlorobiphenyls (PCBs), metals, and water can be purified using supercritical ethane.

2.3 Phase Transition

A second major class of physical separation is that of phase transition, in which a material changes from one physical phase to another.

Distillation is a process for separating liquids with different boiling points (Long, 1995). The mixed-liquid stream is exposed to increasing amounts of heat, and the various components of the mixture are vaporized and recovered. The vapor may be recovered and reboiled several times to effect a complete separation of components. Distillation is used in treating and recycling solvents, waste oil, aqueous phenolic wastes, and xylene contaminated with other hydrocarbons such as ethylbenzene. Distillation produces distillation bottoms (still bottoms, residua in the petroleum and coal tar industries), which consist of solids, semisolid tar, and sludge.

Evaporation is a process for concentrating nonvolatile solids in a solution by boiling off the liquid portion of the waste stream (Long, 1995). Evaporation units are often operated under some degree of vacuum to lower the thermal energy required to boil the solution. Evaporation is usually employed to remove water from an aqueous waste to concentrate it. A special case of this technique is thin-film evaporation in which volatile constituents are removed by heating a thin layer of liquid or sludge waste spread on a heated surface.

Drying is the removal of solvent or water from a solid or semisolid (sludge) or the removal of solvent from a liquid or suspension. This is a very important operation because water is often the major constituent of waste products, such as sludge obtained from emulsion breaking. In freeze drying, the solvent, usually water, is sublimed from a frozen material. Solids and sludge are dried to reduce the quantity of waste, to remove solvent or water that might interfere with subsequent treatment processes, and to remove volatile constituents. *Dewatering* (drying) can often be achieved thermally (Welch and Joachim, 1994) and can be improved with addition of a filter, such as diatomaceous earth, during the filtration step.

Stripping is a means of separating volatile components from less volatile components in a liquid mixture by the partitioning of the more volatile materials to a gas phase of air or steam (air stripping or steam stripping) (Figure 8.4). The gas phase is introduced into the aqueous solution or suspension containing the waste in a stripping tower that is equipped with trays or packed to provide maximum turbulence and contact between the liquid and gas phases (Billet, 1995). The two major products are condensed vapors and a stripped bottom residue.

Examples of two volatile components that can be removed from water by air stripping are benzene and dichloromethane. Air stripping can also be used to remove ammonia from water that has been treated with a base to convert ammonium ions to volatile ammonia.

Precipitation is used here as a term to describe processes in which a solid forms from a solute in solution as a result of a physical change in the solution. This is in contrast to chemical precipitation (Chapter 7), in which a chemical reaction in solution produces an insoluble material (Long, 1995). The major changes that can cause physical precipitation are cooling the solution, evaporation of solvent, or alteration of solvent composition. The most common type of physical precipitation by alteration of solvent composition occurs when a water-miscible organic solvent is added to an aqueous solution, so that the solubility of a salt is lowered below its concentration in the solution.

2.4 Phase Conversion

Vitrification (also referred to as glassification) is a phase conversion process in which a chemical waste (or constituents of the waste) is melted at high temperature to form an impermeable capsule around the remainder of the waste. The process has

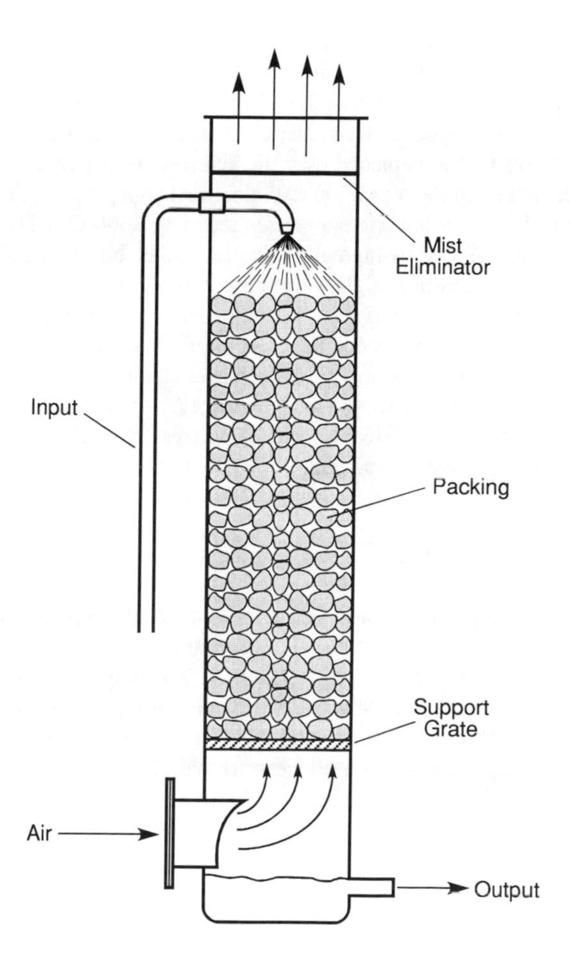

Figure 8.4: A stripping tower

a benefit over other processes insofar as it can be applied both in situ and ex situ.

The principles behind vitrification are the same as those applied to the production of glass. High-temperature electrodes are used to melt the wastes, and organic constituents are transformed by pyrolysis and either collected as product or destroyed in secondary processes. The inorganic components are immobilized in the resulting glass matrix.

In ex situ applications, the waste is introduced into the furnace along with the silica, soda, and lime. The organic materials are driven off, captured, and treated while the inorganic materials are incorporated into the glass. The plasma arc has also been shown to be suitable for ex situ vitrification (Tardy et al., 1994).

In situ vitrification involves insertion of large electrodes into the soil. Graphite is spread on the soil surface between the electrodes to complete the circuit. A negatively pressurized hood is placed over the site to collect any off gases for later treatment. High voltage (4160 V; a 3000-kW electrical source is required) is applied across the electrodes to produce temperatures reaching 3600°C (6510°F). The use of in situ vitrification is limited by high groundwater tables, buried metal objects, and the need for sufficient quantities of glass-forming material in the waste or in the soil.

The vitrification process has shown great promise for treating radioactive and mixed wastes, which are immobilized in the glass matrix and, supposedly, can then be stored until the radioactivity decays to a safe level (United States Department of Energy, 1995b). In the case of mixed wastes, vitrification drives off the nonradioactive components, allowing them to be treated as chemical waste while immobilizing the radioactive component.

2.5 Membrane Separations

Another major class of physical separation is molecular separation, often based upon membrane processes in which dissolved contaminants or solvent pass through a size-selective membrane under pressure (Cox, 1990; Huang, 1991; Chiarizia et al., 1992; Chemical & Engineering News, 1993; Long, 1995; Noble and Stern, 1995; Tiberi and Brose, 1995) (Figures 8.5 and 8.6). The products are relatively pure solvent phase (usually water) and a concentrate enriched in the solute impurities. Water and solutes of lower molecular weight under pressure pass through the membrane as a stream of purified permeate, leaving behind a stream of concentrate containing impurities in solution or suspension.

Dialysis is a process for separating components in a liquid stream by using a membrane. Components of a liquid stream will diffuse through the membrane if a stream with a greater concentration of the component is on the other side of the membrane. Dialysis is used to extract pure process solutions from mixed waste streams.

Electrodialysis is an extension of dialysis and is used to separate the components of an ionic solution by applying an electric current to the solution, which causes ions to move through the dialysis membrane (Figure 8.7). It is very effective for extracting acids and metal salts from solutions.

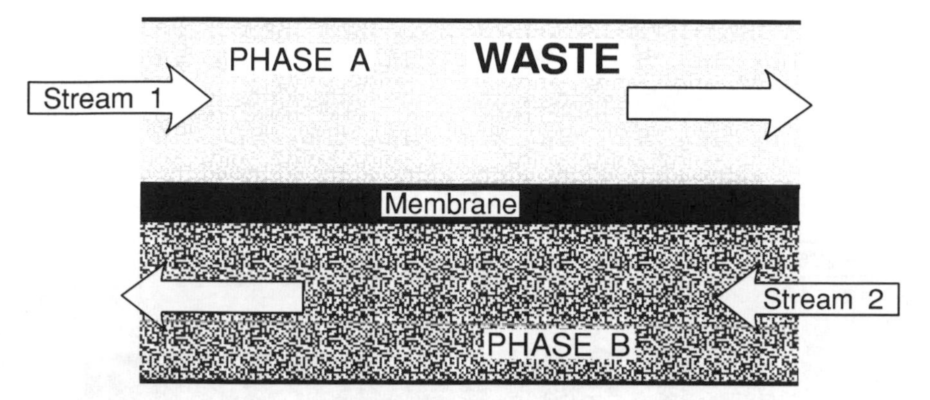

Figure 8.5: A membrane contactor system

The related process of electrolysis shows promise as a method for the dechlorination of polychlorinated biphenyls (Rusling and Zhang, 1994). Commercial polychorinated biphenyl mixtures in spiked soil samples have been reduced to biphenyls and hydro-biphenyls. The technique involves mixing the contaminated soil into a microemulsion that is used as the medium to carry out the catalytic electrolysis. The procedure certainly shows promise on the small scale and, if proven as a feasible/economic process on the large scale (especially as an in situ), might find considerable application.

Reverse osmosis separates components in a liquid stream by applying external pressure to one side of the membrane so that solvent will flow in the opposite direction (Figure 8.8). Reverse osmosis is the most widely used of the membrane techniques. Although superficially similar to ultrafiltration and hyperfiltration, it operates on a different principle, in that the membrane is selectively permeable to water and excludes ionic solutes. Reverse osmosis uses high pressures to force permeate through the membrane, producing a concentrate containing high levels of dissolved salts.

Ultrafiltration is similar to reverse osmosis, but the separation begins at higher molecular weights. The result is that dissolved components with low molecular weights will pass through the membrane with the bulk liquid, while compounds with higher molecular weights become concentrated through the loss of solvent. Ultrafiltration systems can handle much more corrosive fluids than reverse osmosis units.

Ultrafiltration and hyperfiltration are especially useful for concentrating suspended oil, grease, and fine solids in water. They also serve to concentrate solutions of large organic molecules and heavy metal ion complexes.

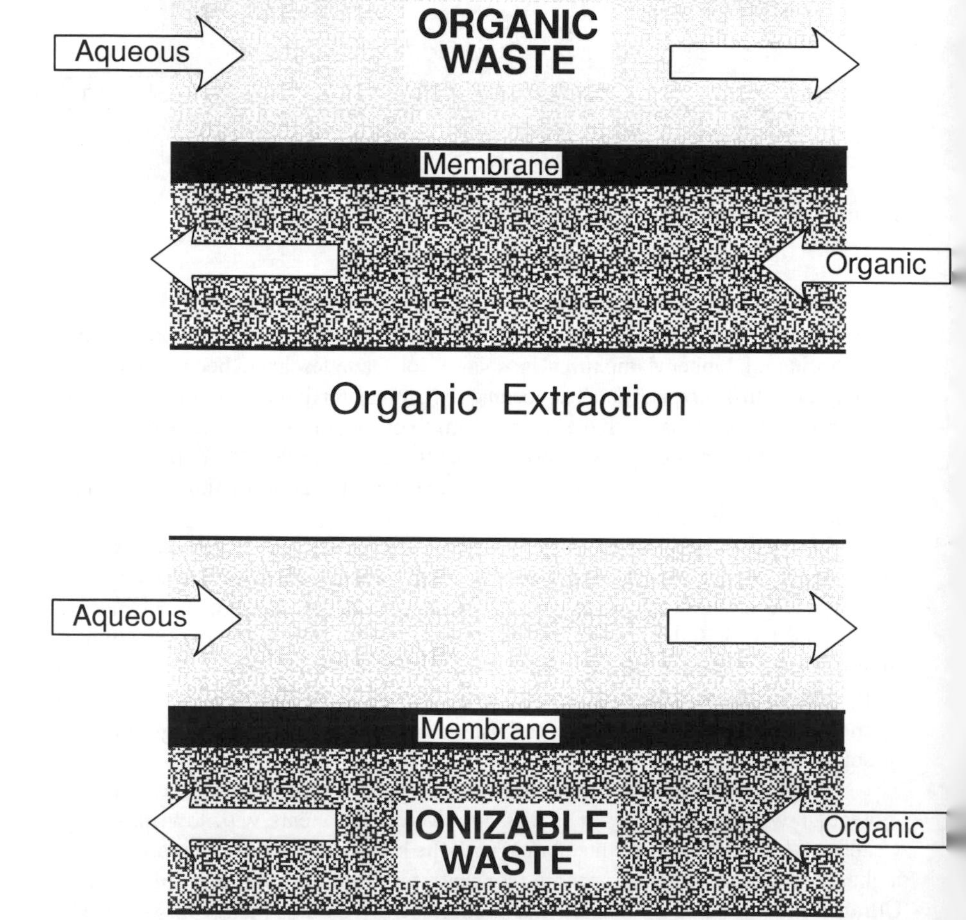

Figure 8.6: Cleanup of liquid streams using membranes

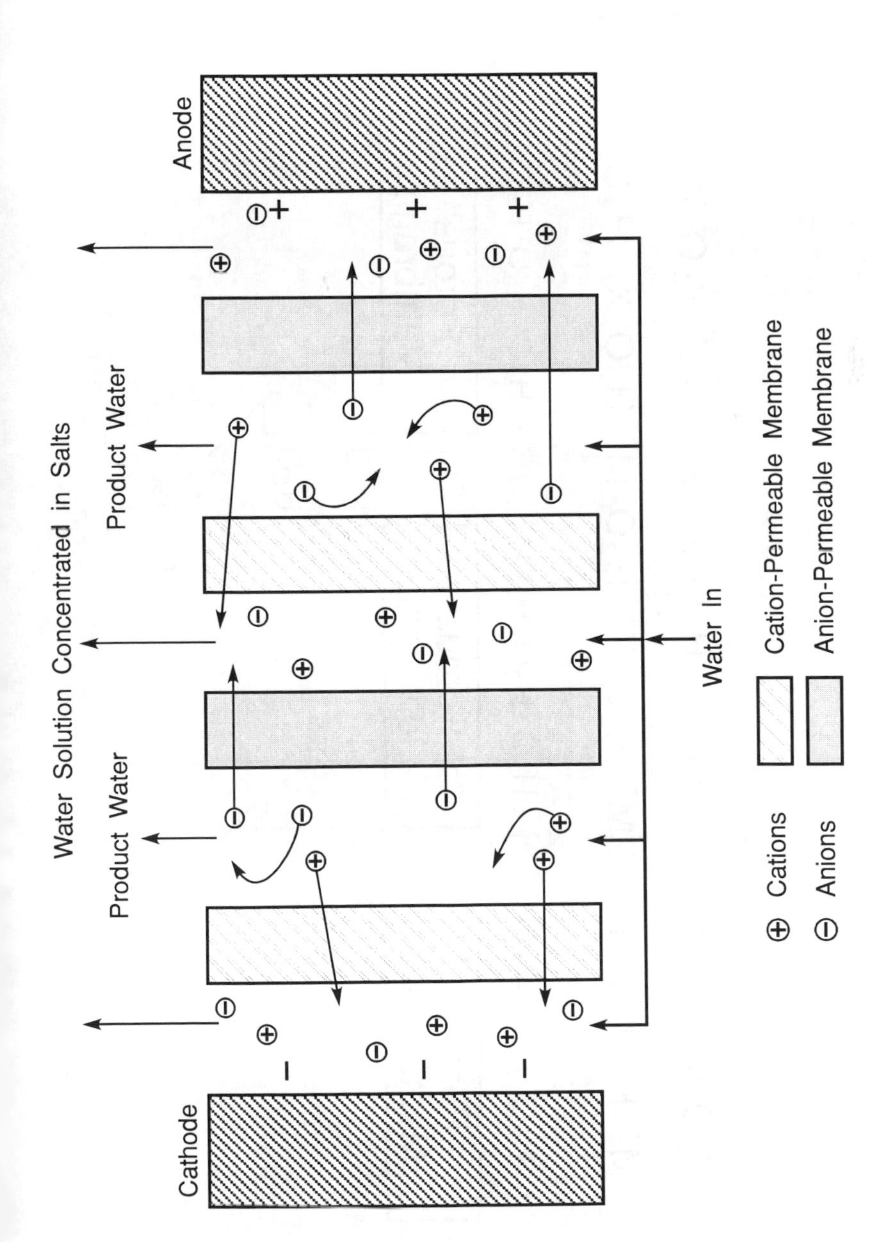

Figure 8.7: Electrodialysis

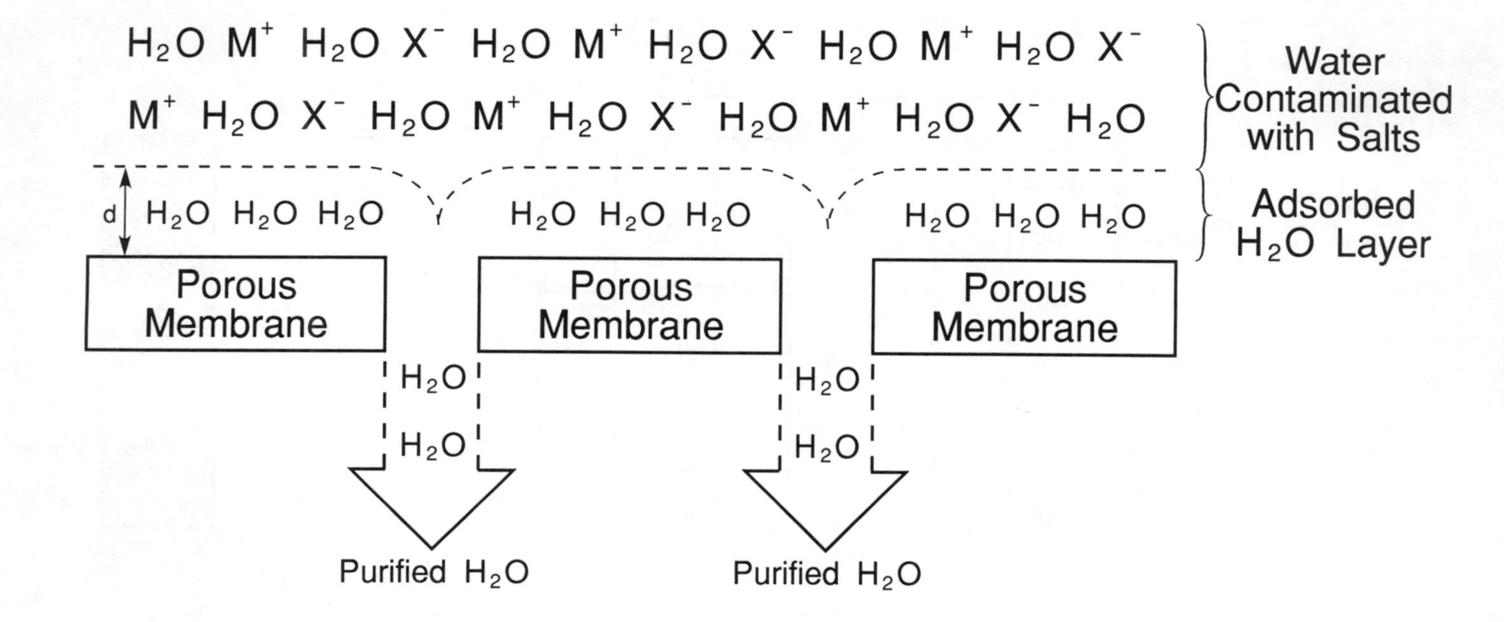

Figure 8.8: Reverse osmosis

3.0 Thermal Methods

Outside of the traditional containment methods (land filling and capping), application of some type of thermal process has, until recently, been the most common form of treatment for chemical waste.

Thermal treatments have lost some popularity recently due to the threat of emissions from incomplete combustion. Except for vitrification, thermal technologies are ex situ processes requiring the wastes to be transported to the processing unit. However, the use of a thermal destruction technology may be (depending upon the nature of the emissions) superior to wet scrubbing technologies for cleaning of emissions (Dieterman, 1995).

3.1 Incineration

Incineration is the controlled combustion of materials in an enclosed area. Thus, incineration of chemical waste and, for example, sewage sludge, is a process that involves exposure of the waste materials to oxidizing conditions at a high temperature, usually in excess of 900°C (1650°F) (Brunner, 1989; Lowe and Frost, 1990).

Incineration of solid and liquid chemical wastes, such as polychlorobiphenyl materials, is a common and highly efficient method of destruction. The destruction of polychlorobiphenyls commences at ca. 800°C (1470°F) and commercial incinerators operate at temperatures in excess of 1000°C (1800°F). Suitable for the purpose are stationary incinerators, cement kilns, incinerator ships, and smaller mobile thermal destruction units.

Emissions from incinerators may be gaseous (SO_x, NO_x, particulate matter) (Pershing, 1993), which can pollute the atmosphere or, through the formation of the constituents of acid deposition can, in turn, pollute the water systems and the land systems, and therefore control is necessary. Alternatively, these emissions might have a direct effect on the vegetation by deposition of the gases or particulate matter directly onto the land at a point downwind of the incinerator. In addition, solid waste (ash) from an incinerator can pollute the land and, as a result of leaching by rain (acidic or otherwise), pollute the water systems. Therefore attention must be focused on the disposal of the ash (Goodwin, 1993).

The air emission systems from these thermal reactors, as well as from the furnaces that burn solids emanating from wastewater plants, must be designed to prevent chemical and particulate air pollution. These air emission systems include electrostatic precipitators and afterburners.

In addition, sulfur-containing fuels emit sulfur dioxide and other sulfur-containing gases as a result of combustion (Chapter 2). These sulfur-containing gases have the potential for conversion to sulfates (sulfuric acid in the atmosphere), which can be deposited on the land as sulfate.

Nitrogen oxides (also produced during fuel combustion) (Chapter 2) are converted to nitrates in the atmosphere and the nitrates eventually are deposited on soil. Soil also adsorbs nitric oxide (NO) and nitrogen dioxide (NO_2) readily and these gases are oxidized to nitrate in the soil. Carbon monoxide is converted to carbon dioxide and possibly to biomass by soil bacteria and fungi. Elevated levels of heavy metals (such as lead) from mines and smelters are also found on in the soil near such facilities.

Thermal treatment of chemical waste can be used to accomplish most of the common objectives of waste treatment: (1) volume reduction; and (2) removal of volatile, combustible, mobile organic matter and destruction of toxic and pathogenic materials. Incineration utilizes high temperatures, an oxidizing atmosphere, and often turbulent combustion conditions to destroy wastes.

The effective incineration of wastes depends upon the combustion conditions, namely, (1) a sufficient supply of oxygen in the combustion zone; (2) thorough mixing of the waste, the oxidant, and any supplemental fuel; (3) combustion temperatures above 900°C (>1650°F); and (4) sufficient residence time to allow reactions to occur. Usually, the heat required for incineration comes from the oxidation of organically bound carbon and hydrogen contained in the waste material or in supplemental fuel:

$$C_{(organic)} + O_2 = CO_2 + \text{heat}$$
$$4H_{(organic)} + O_2 = 2H_2O$$

These reactions destroy organic matter and generate heat required for endothermic reactions, such as the breaking of C-Cl bonds in organic chlorine compounds.

Incineration detoxifies chemical waste by destroying organic compounds, reduces the volume of the waste, and converts liquid waste to a solid product by vaporizing any fluids present in the wastes. The primary use of incineration is for the destruction of volatile organic compounds and semi-volatile organic compounds. Incinerators have been extremely capable of destroying organic compounds in waste, routinely achieving removal efficiencies as high as 99.9999% (often referred to as the six-9s treatment level).

Incineration systems are designed to accept specific types of materials; they vary according to feed mechanisms, operating temperatures, equipment design, and other parameters. The main products from complete incineration include water, carbon dioxide, ash, and certain acids and oxides, depending upon the waste in question.

The ideal wastes for incineration are predominantly organic materials that will burn with a heating value of at least 5000 Btu/lb and preferably more than 8000 Btu/lb. Such heating values are readily attained with waste having a high content of organic constituents. In some cases, however, it is desirable to incinerate wastes that will not burn alone and that require supplemental fuel, such as methane and petroleum liquids. Examples of such wastes are nonflammable organic chlorine wastes, some aqueous wastes, or soil in which the elimination of a particularly troublesome contaminant is worth the expense and trouble of incinerating it. Inorganic matter, water, and organic heteroelement contents of liquid wastes are important in determining their susceptibility to incineration.

Many wastes, including hazardous waste, are burned to produce fuel for energy recovery in furnaces and boilers. This process (co-incineration) uses combustible waste material more for energy generation than for waste destruction. In addition to heat recovery from combustible waste, an existing on-site facility, rather than a separate waste incinerator, can be used for waste disposal.

Incinerators can be designed to handle wastes in any physical state and have proven effective in treating solids, liquids, sludge, slurries, and gases. The effectiveness of an incinerator depends on three factors: (1) temperature of the combustion chamber, (2) residence time in the chamber, and (3) amount of mixing of the material with air while in the chamber.

Normal combustion temperatures range between 900°C and 1500°C (1650°F and 2280°F); in some instances the temperatures employed are much higher. Many incinerators for hard-to-burn compounds employ two combustion chambers. The first chamber converts the compounds to gas and initiates the combustion process. In the second chamber, combustion of the gases is completed.

There are four major components of chemical waste incineration systems: preparation, combustion, pollutant removal, and ash disposal. Preparation for liquid wastes may require filtration, settling to remove solid material and water, blending to obtain the most appropriate mixture for incineration, or heating to decrease viscosity. Solids may require shredding and screening. Atomization is commonly used to feed liquid wastes. Several mechanical devices, such as rams and augers, are used to introduce solids into the incinerator.

The most common kinds of combustion chambers are liquid injection, fixed hearth, rotary kiln, and fluidized bed. Often the most complex part of a waste incineration system is the air pollution control system, which involves several operations, the most common of which are combustion gas cooling, heat recovery, quenching, particulate matter removal, acid gas removal, and treatment and handling of by-product solids, sludge, and liquids (Noyes, 1991).

The inert portion of the waste remains as ash after incineration. For liquid waste, the amount of ash remaining is generally minimal whereas for solid waste, the volume of ash can be as much as one-third by weight of the original material. If the ash contains metals or radioactive material, it must be further treated prior to disposal. The most frequently employed method of treating the ash remaining from the incineration process is solidification/stabilization.

Hot ash is often quenched in water and, prior to disposal, it may require dewatering and chemical stabilization. A major consideration with waste incinerators and the types of wastes that are incinerated is the disposal problem posed by the ash, especially in respect to potential leaching of heavy metals.

Waste incineration and pyrolysis systems include single-chamber liquid systems, rotary kilns, and fluidized-bed incineration systems. In a single-chamber liquid system a brick-lined combustion chamber contains liquids that are burned in suspension. In addition to being the primary parts of an incineration system, these units can be used as afterburners for rotary kilns.

A rotary kiln is a versatile large refractory-lined cylinder capable of burning virtually any liquid or solid organic waste, the unit is rotated to improve turbulence in the combustion zone. Fluidized-bed incineration uses a stationary vessel within which solid and liquid wastes are injected into a heated, extremely agitated bed of inert granular material; the process promotes rapid heat exchange and can be designed to scrub off the gases.

Waste incinerators may be divided among the following, based upon type of combustion chamber: (1) rotary kiln incinerators in which the primary combustion chamber is a rotating cylinder lined with refractory materials and an afterburner downstream from the kiln to complete destruction of the wastes; (2) liquid injection incinerators that burn liquid waste dispersed as small droplets; (3) fixed-hearth incinerators with single or multiple hearths upon which combustion of liquid or solid wastes occurs; and (4) fluidized-bed incinerators that have a bed of granular solid (such as sand) maintained in a suspended state by injection of air to remove pollutant acid gas and ash products.

Advanced design incinerators, including plasma incinerators, make use of an extremely hot plasma of ionized air injected through an electrical arc. Examples are electric reactors which use resistance-heated incinerator walls at around 2200°C (3990°F) to decompose the waste by radiative heat transfer. There are also infrared systems which generate intense infrared radiation by passing electricity through silicon carbide resistance heating elements. Molten salt combustors use a bed of molten salt, such as sodium carbonate, at ~900°C (~1650°F) to destroy the waste and retain gaseous emissions through chemical reaction. Finally, there are molten glass processes, which use a molten glass to transfer heat to the waste and to retain products in a nonleachable form.

Incinerators may be mobile, transportable, or stationary (fixed). Mobile incinerators are normally relatively small units that are mounted on a flat bed trailer and transported to the job site. Transportable incinerators are larger units that can be disassembled into manageable components and moved from one site to another by a caravan of trucks. Stationary/fixed incinerators are permanently erected at a site, and the wastes are brought to the site for treatment.

3.2 Thermal Desorption

Thermal desorption is the process of heating a waste in a controlled environment and thereby volatilizing any organic constituents. Thermal desorption works especially well for volatile organic compounds but can also be employed for semivolatile organic compounds. Removal efficiencies ranging from 65% to 99% have been achieved, depending upon the type of waste.

Prior to entering the thermal desorption unit, the wastes are screened to eliminate coarse pieces. If the wastes have a high moisture content, this excess moisture is also removed. The wastes are then passed to a furnace which operates at temperatures in the range 300-600°C (570-1110°F). Volatile organic compounds become gaseous in the process and are either collected on an adsorbent, such as activated carbon, for further treatment or passed through an incinerator connected in-line with the thermal desorption unit.

3.3 Pyrolysis

Pyrolysis is a chemical change brought about by the action of heat. If the waste is exposed to high temperatures in an oxygen-poor environment, the process is known as pyrolysis. The products of this process are simpler organic compounds, which may be recovered or incinerated, and char or ash. The process differs from incineration, which is the combustive destruction of a material in direct flame in the presence of oxygen. Pyrolysis can also be defined as destructive distillation in the absence of oxygen (or other oxidant) with the simultaneous removal of volatile products. Pyrolysis converts wastes containing organic material to combustible gas, charcoal, organic liquids, and ash/metal residues. In some instances, the organic liquid fraction produced during pyrolysis has the potential to produce constituents for synthetic crude oil.

The effectiveness of waste destruction by pyrolysis depends upon the residence time within the retort, the rate of temperature increase, the final temperature, and the composition of the feed material.

Plasma torch processes apply the principles of pyrolysis at temperatures in the range 5000°C to 15,000°C (9000°F to 27,000°F). The wastes are fed into the thermal plasma where they are dissociated into their basic atomic components which recombine in the reaction chamber to form carbon monoxide, nitrogen, and hydrogen, as well as methane and ethane. Acid gases that are removed from the emissions by scrubbers and any solid products are either incorporated into the molten bath at the bottom of the chamber or removed from exhaust gases by particulate scrubbers or filters.

Another process designed for the recovery of tank bottom wastes involves using a flash tank to recover products from waste oil as might occur after oil field drilling and recovery operations (Johnson et al., 1993). The waste oil is first preheated and pressurized after which the stream is fed to a flash tank. The hydrocarbon volatile components are purified and the nonvolatile components enter a pyrolyzing reator to produce more volatile products.

4.0 Solidification and Stabilization

Solidification and stabilization are treatment systems that move beyond the older concept of *dig and move* (Bishop, 1991; Bishop, 1994) and are designed to: (1) improve the handling and the physical characteristic of the waste; or (2) decrease the surface area across which transfer or loss of contained pollutants can occur; or (3) limit the solubility of, or detoxify, any chemical constituents in the waste.

Stabilization/solidification processes are being used to minimize the potential for groundwater pollution from land disposal of hazardous wastes. Many variations are used but most rely on pozzolanic reactions to chemically stabilize and physically solidify the waste. Portland cement alone or in combination with fly ash, cement kiln dust, lime, or other ingredients is the principal solidifying agent used (Wilk, 1995).

Stabilization and solidification processes are very effective at immobilizing most heavy metals present in sludge, contaminated soils and other wastes. They are not as effective at immobilizing toxic organic materials. Organically modified clays are now being evaluated as an additive to stabilization and/or solidification processes in order to adsorb and retain these organic pollutants in the solidified waste form.

The environmental acceptability of stabilization and solidification processes depends on the long-term ability of the waste form to retain contaminants. This will be governed by the chemical binding mechanisms involved and by the durability of the waste form.

Stabilization usually involves the addition of materials that ensure that the chemical constituents are maintained in their least soluble or least toxic form. In the solidification process, the results are obtained primarily, but not exclusively, via the production of a monolithic block of treated waste with high structural integrity. Stabilization techniques limit the solubility of, or detoxify, the waste contaminants even though the physical characteristics of the waste may not be changed. Both of these techniques might also have been classified under phase transition or phase conversion.

Encapsulation, another perhaps more specific form of waste stabilization, is a method by which mixed wastes, i.e., waste containing radioactive material and higher level radioactive wastes (Chapter 6), can be rendered less harmful to the environment (Poon, 1989). Polyethylene, chosen for its durability, has been suggested as a suitable encapsulating agent (Kalb et al., 1993).

5.0 References

American Society for Testing and Materials. 1995. ASTM D 3922. Estimating the Operating Performance of Virgin/Reactiviated Granular Activated Carbon (GAC) for Removal of Soluble Pollutants from Water. *Annual Book of ASTM Standards*. Vol. 15.01. American Society for Testing and Materials, Philadelphia, Pennsylvania.

Barney, G.S., Lueck, K.J., and Green, J.W. 1992. Chap. 3 in *Environmental Remediation*. Edited by G.F. Vandegrift, D.T. Reed, and I.R. Tasker. Symposium Series No. 509. American Chemical Society, Washington, D.C.

Benefield, L.D., Judkins, J.F., and Weand, B.L. 1982. *Process Chemistry for Water and Wastewater Treatment*. Prentice-Hall, Englewood Cliffs, New Jersey.

Billet, R. 1995. Packed Towers in *Processing and Environmental Technology*. VCH Publishers, Deerfield Beach, Florida.

Bishop, J. 1994. *Environmental Solutions* 7(10): 44.

Bishop, P.L. 1991. Chap. 15 in *Emerging Technologies for Hazardous Waste Management II*. Edited by D.W. Tedder and F.G. Pohland. Symposium Series No. 518. American Chemical Society, Washington, D.C.

Borchardt, J.K. 1995. *Today's Chemist at Work* 4(3): 47.

Brunner, C.R. 1989. *Handbook of Hazardous Waste Incineration*. TAB Books, Blue Ridge Summit, Pennsylvania.

Chemical and Engineering News. 1993. *Ion-selective membranes. Chemical and Engineering News* 73(19): 6.

Cheremisinoff, P. 1995. *Handbook of Water and Wastewater Treatment Technology*. Marcel Dekker, Inc., New York.

Chiarizia, R., Horwitz, E.P., and Hodgson, K.M. 1992. Chap. 2 in *Environmental Remediation*. Edited by G.F. Vandegrift, D.T. Reed, and I.R. Tasker. Symposium Series No. 509. American Chemical Society, Washington, D.C.

Chovit, K., and McMillen, T. 1994. *Hazmat World* 7(3): 57.

Cox, M. 1990. In *Effluent Treatment and Waste Disposal*. p. 287. Edited by D. Handley. Institute of Chemical Engineers, Rugby, Warwickshire, England.

Dieterman, J.R. 1995. *Chemical Processing* 58(1): 61.

Goodwin, R.W. 1993. *Combustion/Ash Residue Management: An Engineering Perspective*. Noyes Data Corp., Park Ridge, New Jersey.

Huang, C.P., O'Melia, C.R., and Morgan, J.J. (eds.). 1995. Aquatic Chemistry: Interfacial and Interspecies Processes. Advances in Chemistry Series No. 244. American Chemical Society, Washington, D.C.

Huang, R.Y.M. 1991. *Pervaporation Membrane Separation*. Elsevier, Amsterdam, The Netherlands.

Johnson, L.A. Jr., and Leuschner, A.P. 1992. Chap. 20 in *Hydrocarbon Contaminated Soils and Groundwater*. Vol. 2. Edited by E.J. Calabrese and P.T. Kostecki. Lewis Publishers, Ann Arbor, Michigan.

Johnson, L.A. Jr., Satchwell, R. M., Glaser, R.R., and Brecher, L.E. 1993. Process for Recovery of Tank Bottom Wastes. United States Patent No. 5,259,945. November 9.

Kalb, P.D., Heiser, J.H., and Colombo, P. 1993. Chap. 22 in *Emerging Technologies for Hazardous Waste Management III*. Edited by D.W. Tedder and F.G. Pohland. Symposium Series No. 518. American Chemical Society, Washington, D.C.

Kiran, E., and Brennecke, J.F. (eds.). 1993. *Supercritical Fluid Engineering Science*. Symposium Series No. 514. American Chemical Society, Washington, D.C.

Long, R.B. 1995. *Separation Processes in Waste Minimization*. Marcel Dekker Inc., New York.

Lowe, P., and Frost, R.G. 1990. In *Effluent Treatment and Waste Disposal*. p. 261. Edited by D. Handley. Institute of Chemical Engineers, Rugby, Warwickshire, England.

Noble, R.D., and Stern, S.A. (eds.). 1995. *Membrane Separations Technology: Principles and Applications*. Elsevier, New York.

Noyes, R. (ed.). 1991. *Handbook of Pollution Control Processes*. Noyes Data Corp., Park Ridge, New Jersey.

Pershing, D.W. 1993. *Energy & Fuels* 7: 782.

Poon, C. 1989. In *Environmental Aspects of Stabilization and Solidification of Hazardous and Radioactive Wastes*. p. 114. ASTM STP 1033. American Society for Testing and Materials, Philadelphia, Pennsylvania.

Rusling, J.F., and Zhang, S. 1994. *Environmental Science and Technology* 29: 1195.

Suzuki, M. 1993. *Fundamentals of Adsorption*. Elsevier, Amsterdam, The Netherlands.

Tardy, P., Labrot, M., and Pineau, D. 1994. *Environmental Solutions* 7(12): 30.

Tedder, D.W., and Pohland, F.G. (eds.). 1995. Emerging Technologies in Hazardous Waste Management V. Symposium Series No. 607. American Chemical Society, Washington, D.C.

Tiberi, T.P. and Brose, D.J. 1995. *Environmental Solutions* 8(2): 42.

United States Department of Energy. 1995a. *Efficient Separations and Processing Crosscutting Program*. Report No. DOE/EM-0249. Office of Environmental Management. United States Department of Energy, Washington, D.C.

United States Department of Energy. 1995b. *Mixed Waste Characterization, Treatment, and Disposal Focus Area*. Report No. DOE/EM-0252. Office of Environmental Management. United States Department of Energy, Washington, D.C.

United States Environmental Protection Agency. 1994. *The Superfund Innovative Technology Program*. Annual Report to Congress. Report No. EPA/540/R-94/518. United States Environmental Protection Agency, Washington, D.C.

Welch, T., and Joachim, M. 1994. *Environmental Solutions* 7(8): 58.

Wilk, C. 1995. *Environmental Solutions* 8(5): 47.

Wise, D.L., and Trantolo, D.J. 1994. *Remediation of Hazardous Waste Contaminated Soils*. Marcel Dekker Inc., New York.

Yang, D.S., and Takeshima, S. 1995. *Environmental Solutions* 8(3): 28.

Chapter 9

Waste Recycling and Waste Disposal

1.0 Introduction

In many cases the disposal of a waste material by a destructive process such as incineration (Chapter 8) is necessary. In other cases, there is the potential to produce a useable material from the waste, thereby protecting a resource from overuse and eventual depletion. Such technologies are called recycling and are an integral part of a disposal program (American Society for Testing and Materials, 1995; Rader et al., 1995).

It would not be possible to fully describe all of the concepts that fall within the recycling and disposal arenas. Simple examples will be used as a guide to the concepts employed in waste management. Wherever (and whenever) possible, recycling and reuse should be accomplished on-site to avoid having to move wastes and because a process that produces recyclable materials is often the most likely to have use for them.

2.0 Waste Recycling

Concepts of waste removal generally involve destruction of the waste (Chapters 7 and 8). However, another concept by which a chemical waste can be disposed involves recycling the material for other use(s) (Rogoff and Williams, 1994). There are four general areas in which valuable products may be obtained from waste materials:

1. Recycle unconsumed raw material in a process as feedstock to the process.
2. Recycle unconsumed raw material in a process as feedstock for another process.
3. Utilize a waste for pollution or emission control; an example is the gas processing industry (Speight, 1993) where waste alkali can be used to neutralize waste acid.
4. Recover energy from the incineration of combustible wastes (Chapter 8).

Recycling can suffer from the disadvantage that, even though many of the primary products are not hazardous, there is always the potential to produce a secondary product that is hazardous. As an example, waste oil from lubricants and hydraulic fluids is often re-used/recycled (Chovit and McMillen, 1994). The collection, recycling, treatment, and disposal of waste oil are all complicated by the fact that the oil is from diverse sources and contains several classes of chemical contaminants. Waste oil from lubricants and hydraulic fluids contains contaminants that are predominantly organic constituents (polynuclear aromatic hydrocarbons from lubricating oil and chlorinated hydrocarbons from transformer oil) and inorganic constituents (aluminum, chromium, and iron from wear of metal parts; barium and zinc from oil additives, lead from leaded gasoline). Mercury is another contaminant that requires disposal or recycling (United States Environmental Protection Agency, 1993). The metal (the seventh metal of antiquity) has been used since at least 1500 B.C. without recognition of its hazards. Currently, emphasis is on the minimal use of, and exposure to, this element, and recycling by distillation or retorting are practiced (Queneau and Smith, 1994).

Another area where recycling can play an important role is in the destruction of waste plastics from sources such as automobiles. In fact, plastics represent one segment of solid waste that can be recycled to decrease the mass of solids and yield value-added products.

The United States generates more than 160 million tons of municipal solid waste or garbage each year, of which plastics are ~7% of the total. Disposal of plastic wastes is becoming a serious problem because landfill site capacity is decreasing. In the last 10 years, the number of landfill sites has dropped from 18,500 to 6000. One, if not the only, practical solution to this problem is to decrease the volume of material being disposed of in landfill sites through rigorous conservation and recycling efforts.

Conventional approaches to address recycling of municipal waste plastics rely on secondary or tertiary recycling processes (Ehrig, 1992; Hegberg et al., 1992). Secondary recycling processes recycle the bulk plastics back to the plastics industry for the manufacture of new products. Tertiary recycling processes utilize plastic waste as a carbon-containing resource and thermally process the plastics to produce a distillate of low molecular weight.

Secondary recycling of waste plastic involves collection, sorting, reclamation, and reuse of the plastic material. There are several problems anticipated in the future associated with large-scale secondary recycling. The first problem is the separation of bulk waste plastic into individual polymer types for recycling, which is normally done manually and is costly. The second problem, critical to the health and safety of our society, is the possible contamination of the plastic with organic material associated with the plastic. This is particularly true for plastic wastes from the food service industry, chemical laboratories, and hospitals, where in some cases, the contamination associated with the plastic may be classed as hazardous wastes. A third problem is developing markets for recycled plastics. Current efforts have been directed at replacing virgin plastics with recycled plastics in the manufacture of more durable goods, such as park benches, curb stops, and building materials. For this reason, the market for recycled plastics may be limited and easily saturated if large-scale recycling becomes a reality. This will decrease the economic benefit of recycling and discourage this conservation effort.

An alternative philosophy is to convert the polymeric structure to smaller chemical molecules such as monomeric units and related chemical structures (tertiary recycling). This improves the market for the products because there are existing plants in the refining and petrochemical industries.

The low-molecular weight distillate from the thermal decomposition of plastics in the presence of various heavy oils has potential to produce a gasoline blending stock for the production of unleaded gasoline (Guffey et al., 1991). The average volume of gasoline imported over a seven year period (to the end of 1990) was 356 thousand barrels per day (United States Department of Energy, 1990). The distillate product from low-temperature thermal decomposition of waste plastics represents (assuming 11 x 10^6 tons of waste plastic each year) 40-50% of the gasoline imports into the United States.

Scrap tires from automobiles have commonly been disposed of in landfills, which is a procedure of much comment and debate (Reese, 1995). By the early 1960s, some 100 million tires were being discarded annually in the United States. By the mid-1970s, that number exceeded 200 million, and it continues to rise. As of mid-1994, approximately 90 million scrap tires per year were being disposed, but 160 million more needed outlets other than landfills (Clark, 1993; Reese, 1995). About one-third of the used tires were being reclaimed by depolymerization with various agents at high temperature and pressure and reprocessed into lower-grade rubber products. The rest were being placed into landfills or stockpiled. Rubber tires in landfills tend to work their way to the surface and collect water, creating a breeding place for insects, especially mosquitoes. Leachate from such fill material is potentially a concern but does not appear at this time to have caused any difficulties.

This problem can be alleviated by reusing the tires in such products as railroad crossing pads, flooring materials, sound insulation, carpet underlay, and rubber-modified asphalt. This latter use is of particular interest and has been included in national legislation. Section 1038 of the Intermodal Surface Transportation Efficiency Act (ISTEA) of 1991 directed states to use 20 lbs of scrap tire rubber per ton of asphalt hot mix in 5% of the federally funded asphalt highway paving in 1994, or lose federal funding equal to the mandated goal. The percentage rises 5% a year until 20% is reached in 1997. As it happens, Congress imposed a spending moratorium on section 1038 for 1994 and again for 1995.

Pyrolysis of scrap tires to produce marketable gaseous and liquid hydrocarbons and char has been investigated for at least 20 years, but to little avail commercially. During the past decade or so, some three dozen scrap tire pyrolysis projects have been designed, patented, licensed, and /or implemented as scale models, but no plants are operating commercially at this time. Technology is available, but the attendant economics does not suggest an immediate path to commercialization.

3.0 Waste Disposal

Ultimately, after all waste treatment has been completed (successful or not), there remains a residue that must be sent to disposal. If the method of waste treatment has been successful in the fullest sense of the word, the residue will be benign and easily disposable. However, this is not always (if ever) the case! Therefore, when any form of waste treatment is designed, there must always be at least one option for residue handling.

Thus, options for such disposal are:

1. Underground injection using steel-encased and concrete-encased sub-surface shafts, which are usually maintained under pressure.
2. Surface impoundment using natural depressions, engineered depressions, or diked areas for treatment, storage, or disposal of chemical waste. Such impoundments may be referred to as pits, ponds, lagoons, or basins.
3. Landfills where waste is placed in or on land (often in separate trenches depending upon the waste and to prevent contact of reactive waste materials) and which may/should be lined to prevent leakage and to prevent runoff of the contaminated surface water.
4. Land treatment whereby the chemical waste is incorporated into the soil and (natural) microbes in the soil break down or immobilize the constituents.
5. Waste piles are used to contain accumulations of solid waste and may be used for final disposal or for temporary storage.

Waste disposal requires some degree of immobilization to minimize the potential for leaching of the constituents thereby reducing the tendency of the waste constituents to migrate. And, immobilization is achieved by solidification and stabilization (Chapter 8).

As a recap, *solidification* may involve chemical reaction of the waste with the solidification agent, mechanical isolation in a protective binding matrix, or a combination of chemical and physical processes. *Stabilization* is the conversion of a waste from the original form to a physically and chemically more stable material. *Fixation* is a process that binds a waste in a less mobile and less toxic form and is generally included in the definition of *stabilization*. Adsorption (Chapter 8) may be used to convert liquids and semisolids to dry solids as well as to improve waste compatibility with substances (such as cement) used for solidification and setting. Vitrification (or glassification) (Chapter 8) consists of embedding wastes in a glass material. Molten glass can be used, or glass can be synthesized in contact with the waste by mixing and heating with glass constituents (silicon dioxide, SiO_2, sodium carbonate, Na_2CO_3, and calcium oxide, CaO). Glassification is used in conjunction with thermal waste destruction processes and can prevent mobilization of the constituents of the ash. Glassification is, however, not a panacea for waste disposal since some of the ash constituents can prevent the glass from fusing.

Encapsulation is used to coat waste with an impermeable material so that there is no contact between the waste constituents and the surroundings. A common method of encapsulation involves the use of molten thermoplastics, asphalt, and wax, all of which solidify when cooled. The use of asphalt, or other petroleum derivatives, allows for timed release of the encapsulated material (Moschopedis and Speight, 1974).

Chemical fixation is a process that controls a substance in a less mobile, less toxic form. Chemical reaction is involved and the chemical form of the waste constituents are changed. For example, humic acid wastes react with calcium ions in a solidification matrix to produce insoluble calcium humate.

Regardless of the destruction, treatment, and immobilization techniques used, there will always remain some waste constituents that require final disposal. Thus, there is the need to address the ultimate disposal of ash, liquids, solids and other residues that must be placed where their potential to harm the environment is minimized (Cook and Kocunik, 1993; Goodwin, 1993).

3.1 Surface Facilities

Landfill (a waste disposal surface facility) is, historically, the most common way of disposing of solid waste and, in some instances, liquid waste. However, the indiscriminate dumping of domestic and industrial waste materials (trash, garbage) onto the land has caused serious pollution problems.

In the early days of the evolution of cities, garbage was dumped just outside the city walls. This is not to deny the indiscriminate throwing of sewage out of the upper windows of buildings or just outside the door. The garbage in question included any large amounts of materials, and the occasional body or two! The term *Gehenna* of Biblical reference (late Latin: Gehenna, Geenna; Greek: Genna; Hebrew: Ge Hinnom, i.e., valley of Hinnom; and meaning *a place of misery*) was often used to refer to the refuse disposal area just outside a city wall.

Currently, refuse disposal can be controlled in a well-designed and operated landfill, and this can be a suitable process for certain small communities. However, land for the disposal of refuse for mid-size and large-size communities is being seriously reduced, and composting of refuse for reuse on the land may have limited application (United States Department of Energy, 1995). The appropriate system for the majority of communities appears to be incineration of the refuse, with recovery of energy from the combustion of organic materials.

A major environmental concern with landfill of hazardous wastes is the generation of leached material from infiltrating surface water and groundwater with resultant contamination of groundwater supplies. What is often overlooked is the character of the rainwater, it being generally assumed to be at, or near to, neutral in terms of acidity/alkalinity. With the current concern over acid rain, there is the need to ensure that the materials in the landfill are protected against acid leaching. In addition, most landfills now provide systems to contain, collect, and control any leached material.

In brief, a landfill should be placed on a compacted low-permeability medium, preferably clay, which is covered by a flexible-membrane liner consisting of watertight impermeable material. The liner can also be covered with granular material in which is installed a secondary drainage system. Next is another flexible-membrane liner above which is installed a primary drainage system for the removal of leached material. This drainage system is covered with a layer of granular filter medium, upon which the wastes are placed.

Different types of waste are separated in the landfill by berms consisting of clay or soil covered with liner material. When the fill is complete, the waste is capped to prevent surface water infiltration and covered with compacted soil. Capping is done to cover the wastes, prevent infiltration of excessive amounts of surface water, and prevent release of wastes to overlying soil and the atmosphere. Caps are usually multi-layered although they are still subject to the potentially problematic effects associated with settling and erosion. In addition, provision may be made for a system to treat evolved gases, particularly when methane-generating biodegradable materials are disposed in the landfill.

Many liquid wastes (including slurries and sludge) are placed in surface impoundments which usually serve for treatment and are often designed to be filled in eventually as a landfill disposal site. A surface impoundment may consist of an excavated pit, a structure formed with dikes, or a combination thereof. The construction is similar to that of a landfill in that the base and the walls need to be impermeable to liquids; provision must also be made for leached material collection.

Leached material can be treated by a variety of physical processes. In some cases, simple density separation and sedimentation can be used to remove water-immiscible liquids and solids. Filtration is frequently required, and flotation may be useful. Leached material solutes can be concentrated by evaporation, distillation, and membrane processes, including reverse osmosis, hyperfiltration, and ultrafiltration (Chapter 8).

3.2 Underground Disposal

Disposal of liquid waste in deep wells consists of pressurized injection of the waste to underground strata isolated from aquifers by impermeable rock strata. This method has been used by the petroleum industry to dispose of the brine (saline wastewater) produced with crude oil. The method has been employed by other industries for disposal of brine, acid, heavy-metal solutions, organic liquids, and a variety of other liquid waste.

Although there are several factors that must be addressed (such as the potential for chemical and physical interactions between the waste constituents and the mineral strata), most problems can be mitigated by waste pretreatment. However, a very serious concern is the potential for groundwater contamination through fractures, faults, and other wells. The disposal well itself can act as a route for contamination if it is damaged or not properly constructed and eased.

3.3 In situ Disposal

In situ treatment refers to waste treatment processes that can be applied to wastes in a disposal site by direct application of treatment processes and reagents to the wastes. In situ treatment is considered to be a preferential but problematic waste site remediation option. Early efforts at groundwater cleanup were costly, time-consuming and ineffective (Knox et al., 1985). Recent years have seen dramatic increases in the number of technologies being promoted for subsurface remediation, many representing simple innovations of existing procedures.

A broad categorization of remediation technologies could include (United States Environmental Protection Agency, 1990) the following:

1. Pump and treat: extraction of contaminated groundwater with subsequent treatment at the surface and disposal or reinjection.
2. Soil vacuum extraction: enhanced volatilization of compounds by applying a vacuum to the subsurface.
3. Soil flushing/washing: use of extractant solvents to remove contaminants from soils.
4. Containment: placement of physical, chemical, or hydraulic barriers to isolate contaminated areas (Chapter 8).
5. Bioremediation: enhanced biodegradation of contaminants by stimulating indigenous subsurface microbial populations (Chapter 7).

Most successful remediation schemes will require a combination of technologies, as it is more than likely that any one technology will not be the cleanup panacea for a site. However, for any technology to be effective, a thorough understanding of the processes governing the transport and ultimate fate of the target contaminants needs to be developed (Hemond and Fechner, 1994). With few exceptions, the above technologies are highly influenced by physicochemical and biological processes active in the subsurface environment.

In situ immobilization is used to convert waste constituents to insoluble forms that will not leach from the disposal site. Heavy-metal contaminants can be immobilized by chemical precipitation, but one disadvantage is the potential for contamination by the precipitating additive.

In situ solidification can be used as a remedial measure at waste sites. One approach is to inject, for example, soluble sodium silicate followed by calcium chloride (or lime, CaO) to form the solid calcium silicate.

If there are a limited number of contaminants in a disposal site, it may be practical to consider detoxification in situ. This approach may be very practical for cleanup of organic contaminants. Among the chemical and biochemical processes that can detoxify such materials are chemical and enzymatic oxidation, reduction, and hydrolysis.

Groundwater contaminated by soluble waste constituents can be treated by a permeable bed of material placed in a trench through which the groundwater must flow. An example is the use of limestone (contained in a permeable bed) to neutralize acid and to precipitate certain metals as hydroxides or carbonates. Another example is the use of ion exchange resins (in a permeable bed) to retain heavy metals (Natansohn et al., 1992). Activated carbon in a permeable bed will remove some organic compounds, especially less-soluble organic compounds of higher molecular weight. However, the permeable beds that have been effective in collecting waste materials may, themselves, be considered a hazardous waste, thereby requiring treatment and disposal.

Heating of wastes in situ can remove or destroy certain chemicals in the waste. Steam injection (analogous to petroleum recovery by steam injection) has been proposed, for this purpose and any volatile materials recovered at the surface can be collected as condensed liquid or can be adsorbed on activated carbon.

Extraction with water (containing additives) can be used to cleanse soil contaminated with chemical waste. When the soil is left in place and the water pumped into and out of it, the process is called *flushing*; when soil is removed and put in contact with liquid, the process is referred to as *washing*.

The water may be pure or it may contain acids (to leach out metals or neutralize alkaline soil contaminants), bases (to neutralize contaminant acids), chelating agents (to solubilize heavy metals), surfactants (to enhance the removal of organic contaminants from soil and improve the ability of the water to emulsify insoluble organic species), or reducing agents (to reduce oxidized species).

Soil contaminants may dissolve, form emulsions, or react chemically. Inorganic species commonly removed from soil by washing include heavy metal salts; lighter aromatic hydrocarbons, such as toluene and xylene; and lighter organic-halogen compounds, such as trichloroethylene or tetrachloroethylene.

4.0 Abandoned Disposal Sites

Improper disposal (by modern standards) of chemical waste at many sites has necessitated cleanup operations to restore the site to the original state. Cleaning abandoned disposal sites involves isolating and containing the contaminated material, removal and redeposit of contaminated sediments, and in-place and direct treatment of the hazardous wastes involved (Table 9.1).

As the state of the art for remedial technology improves (United States Environmental Protection Agency, 1994), there is a clear preference for processes (Chapters 7 and 8) that result in the permanent destruction of contaminants rather than the removal and storage of the contaminating materials.

5.0 Radioactive Waste Management

The main objective of the nuclear industry, in addition to the weapons production of the past decades, is the production of electrical or thermal energy for industrial and domestic consumers (Eister, 1983). Radioactive waste management is the treatment and containment of radioactive waste materials that arise from such an industry. These wastes originate almost exclusively in the nuclear fuel cycle and in the nuclear weapons program and their toxicity requires that they be isolated from the biosphere. The radioactivity of such wastes is commonly measured in curies (Ci). The curie (1 curie = 3.7×10^{10} becquerels) is the activity of 1 gram of radium-226 (^{226}Ra). The curie (based on toxicity) is considered to be a large unit of activity and a more appropriate unit is the microcurie (1 microCi = 10^{-6} Ci), but the nanocurie (1 nanoCi = 10^{-9} Ci) and picocurie (1 picoCi = 10^{-12} Ci) are also frequently used.

Radioactive wastes are classified in four major categories: *spent fuel elements* and *high-level waste*, *transuranic waste*, *low-level waste*, and *uranium mill tailings*. Other contamination such as *radioactive emissions* produced during reactor operation or from combustion of uranium-containing coal and contamination from uranium mine waters is also possible.

Table 9.1: Treatment technologies for abandoned disposal sites

Aqueous waste treatment	
	Activated carbon treatment
	Biological treatment
	Filtration
	Precipitation/flocculation
	Sedimentation technology
	Ion exchange and adsorptive resin
	Reverse osmosis
	Neutralization
	Gravity separation
	Air stripping
	Oxidation
	Chemical reduction
Solids treatment	
	Solids separation
	Dewatering
Solidification/stabilization	
	Cement-based solidification
	Silicate-based process
	Adsorbents
	Thermoplastic solidification
	Surface microencapsulation
	Vitrification
Gaseous waste treatment	
	Flaring
	Adsorption
Thermal destruction	
	Liquid injection
	Rotary kiln
	Multiple hearth
	Fluidized bed
	Mobile incineration

Spent fuel elements arise when uranium undergoes fission in a reactor to generate energy. The fuel elements needed for the production of 1 GW(e)-year of electrical energy contain 40 (metric) tons of uranium. The spent fuel contains 1 (metric) ton (2240 lb) of fission-product nuclides as well as transuranic nuclides (such as plutonium), which are produced by neutron capture by the uranium nuclei. Uranium mill tailings are the residues of the chemical extraction of uranium from the ore. Low-level waste is a very broad category of wastes, accounting for almost every form of radioactive waste not covered by the other categories.

In order to extract plutonium from nuclear fuel, the spent fuel must be chemically reprocessed. The resulting high-level waste contains most of the fission products and transuranic elements, including residual plutonium. Transuranic waste, arising mainly during this reprocessing, is now defined as solid material contaminated to greater than 100 nanoCi/g with certain alpha-emitting radioactive nuclides. Protection against the harmful radiation from radioactive nuclides must rely on their isolation from the biosphere until their radioactivity has decayed.

For the first 100 years, the toxicity is dominated by the beta-emitting and gamma-emitting fission products such as strontium-90 (^{90}Sr) and cesium-137 (^{137}Cs), with half-lives of approximately 30 years. Thereafter, the long-lived, alpha-emitting transuranium elements, such as plutonium-239 (^{239}Pu) with a half-life of 24,000 years and the radioactive decay product americium-241 (^{241}Am) with a half-life of 432 years, continue the toxicity.

The disposal of radioactive waste is one of the most sensitive environmental issues (Young and Yalow, 1995). Many radioactive wastes are extremely toxic, their radioactivity exponentially decaying through time to produce fission products. Thus nuclear accidents can contaminate the natural environment into the foreseeable future, and beyond!

Although the radioactivity of the transuranic wastes is considerably smaller than that of high-level waste or spent fuel, the high toxicity and long lifetimes of these wastes have required disposal in a geological repository.

Uranium is naturally radioactive, decaying in a series of steps to stable lead. It is currently a rare element, averaging between 0.1 and 0.2% in the mined ore. At the mill, the rock is crushed to fine sand, and the uranium is chemically extracted. The residues (uranium mill tailings), several hundred thousand cubic meters for the annual fuel requirements of a 1 GW(e) reactor, are discharged to the tailings pile. The tailings contain the radioactive daughters of the uranium.

The long-lived isotope thorium-230 (^{230}Th, half-life 80,000 years) decays into radium-226 (^{226}Ra, half-life 1600 years), which in turn decays to radon-222 (^{222}Rn, half-life 3.8 days). Radium and radon are known to cause cancer, the former by ingestion, the latter by inhalation. Radon is an inert gas and thus can diffuse out of the mill tailings pile and into the air.

Although the radioactivity contained in the mill tailings is very small relative to that of the high-level waste and spent fuel, it is comparable to that of the transuranic waste. It is mainly the dilution of the thorium and its daughters in the large volume of the mill tailings that reduces the health risks to individuals relative to those posed by the transuranium elements in the transuranic wastes. However, this advantage is offset by the great mobility of the chemically inert radon gas, which emanates into the atmosphere from the unprotected tailings.

New mill tailings piles will be built with liners to protect the groundwater and will be covered with earth and rock to reduce atmospheric release of the radon gas. None of these measures provides protection on the timescales required for the ^{230}Th to decay.

By definition, practically everything that does not belong to the first three major categories (above) is considered low-level waste. This name is misleading because some wastes, though low in transuranic content, may contain very high beta and gamma activity. The past method of low-level waste disposal was shallow-land burial, which is relatively inexpensive but provides less protection than a geologic repository. Prior to 1970, low-level waste was also disposed of by ocean dumping and, prior to 1983, was also injected with grout into fractured shale formations. However, these practices are no longer allowed.

Currently, nuclear power plants are an important category of nuclear facilities, since they contain the largest amounts of radioactive wastes. These wastes can be grouped in three classes: neutron-activated wastes, surface-contaminated wastes, and miscellaneous wastes. In addition, nuclear facilities have to be dismantled (decommissioned) at the end of their useful life and the accumulated radioactivity disposed. Delaying the dismantling of the facility, in a procedure called safe storage or entombment, allows much of the radioactivity to decay. However, the burden of complete disposal is shifted to future generations.

Radioactive waste can be introduced into water as liquid effluent discharged from two commercial nuclear-fuel reprocessing plants (one in the UK and the other in France). Long-lived nuclides such as strontium-90 (^{90}Sr), cesium-137 (^{137}Cs), and ruthenium-106 (^{106}Ru) are frequently discharged. Underground and undersea repositories for radioactive waste storage are being adopted by many governments and/or waste-disposal companies as the most acceptable option. Both options have serious drawbacks, but an undersea site would have been the worse choice, principally because it would have posed the greatest difficulties in monitoring the radioactive waste. Furthermore, the retrieval of such waste from below the seabed would be logistically much more difficult than if it were stored underground.

6.0 References

American Society for Testing and Materials. 1995. *Annual Book of ASTM Standards.* Vol. 11.04. American Society for Testing and Materials, Philadelphia, Pennsylvania.

Chovit, K., and McMillen, T. 1994. *Hazmat World* 7(3): 57.

Clark, C., Meardon, K., and Russell, D. 1993. *Scrap Tire Technology and Markets.* Noyes Data Corp., Park Ridge, New Jersey.

Cook, R.H., and Kocunik, D.C. 1993. Proceedings. *American Power Conference* 55: 1272.

Ehrig, R.J. (ed.). 1992. *Plastics Recycling.* Hanser Publishers, Munich, Germany.

Eister, W.K. 1983. Chap. 24 in *Riegel's Handbook of Industrial Chemistry.* Edited by J.A. Kent. Van Nostrand Reinhold, New York.

Goodwin, R.W. 1993. *Combustion/Ash Residue Management: An Engineering Perspective.* Noyes Data Corp., Park Ridge, New Jersey.

Guffey, F.D., Holper, P.A., and Hunter, D.E. 1991 *Summary of Laboratory Simulation Studies of the ROPE Process.* Western Research Institute, Laramie, Wyoming for the United States Department of Energy, Washington, D.C. Report in Press.

Hegberg, B.A., Brenniman, G.R., and Hallenbeck, W.H. . 1992. *Mixed Plastics Recycling Technology.* Noyes Data Corp., Park Ridge, New Jersey.

Hemond, H.F., and Fechner, E.J. 1994. *Chemical Fate and Transport in the Environment.* Academic Press, San Diego, California.

Knox, R.C., Canter, L.W., Kincannon, D.F., Stover, E.L., and Ward, C.H. 1985. Report No. EPA/600/S2-84/182. United States Environmental Protection Agency, Washington, D.C.

Moschopedis, S.E., and J.G. Speight, J.G. 1974. Preprints. *American Chemical Society Division of Fuel Chemistry* 19(2): 291.

Natansohn, S., Rourke, W.J., and Lai, W.C. 1992. Chap. 10 in *Environmental Remediation.* Edited by G.F. Vandegrift, D.T. Reed, and I.R. Tasker. Symposium Series No. 509. American Chemical Society, Washington, D.C.

Queneau, P.B., and Smith, L.A. 1994. *Hazmat World* 7(2): 31.

Rader, C.P., Baldwin, S.D., Cornell, D.D., Sadler, G.D., and Stockel, R.F. (eds.). 1995. Plastics, Rubber, and Paper Recycling. Symposium Series No. 609. American Chemical Society, Washintgon, D.C.

Reese, K. 1995. *Today's Chemist at Work* February, p. 75.

Rogoff, M.J., and Williams, J.F. 1994. *Approaches to Implementing Solid Waste Recycling Facilities.* Noyes Data Corp., Park Ridge, New Jersey.

Speight, J.G. 1993. *Gas Processing: Environmental Aspects and Methods.* Butterworth Heinemann, Oxford, England.

United States Department of Energy. 1990. *Annual Energy Review 1990.* Report No. DOE/EIA-0384(90). Energy Information Administration. United States Department of Energy, Washington, D.C.

United States Department of Energy. 1995. *Landfill Stabilization Focus Area.* Report No. DOE/EM-0251. Office of Environmental Management. United States Department of Energy, Washington, D.C.

United States Environmental Protection Agency. 1990. *Subsurface Contamination Reference guide.* Report No. EPA/540/2-90/011. Office of Energy and Remedial Response. United States Environmental Protection Agency, Washington, D.C.

United States Environmental Protection Agency. 1993. *Mercury and Arsenic Wastes: Removal, Recovery, Treatment, and Disposal.* United States Environmental Protection Agency, Washington, D.C.

United States Environmental Protection Agency. 1994. *Superfund Innovative Technology Evaluation Program. Technology Profiles.* Office of Research and Development. United States Environmental Protection Agency, Washington, D.C.

Young, J.P., and Yalow, R.S. (eds.). 1995. *Radiation and Public Perception: Benefits and Risks.* Advances in Chemistry Series No. 243. American Chemical Society, Washington, D.C.

Wash[illegible] Waste Disposal [illegible]

[illegible] J. Fuller [illegible] DOE/EM-0[illegible] Environmental Management [illegible] States Department of Energy, Washington, D.C.

United States Environmental Protection Agency. [illegible] Report No. EPA [illegible] Office of Emergency and Remedial Response, United States Environmental Protection Agency, Washington, D.C.

[illegible] Protection Agency. [illegible] United States Environmental Protection Agency, Washington, D.C.

United States General Accounting Office. August [illegible] GAO/RCED [illegible] Development [illegible] United States General Accounting Office, Washington, D.C.

[illegible] and [illegible] Washington, D.C.

Chapter 10

Sources and Effects of Gaseous Emissions

1.0 Introduction

Emissions into the atmosphere have been known, but perhaps not identified with the intensity that they are in modern times, since fire was first discovered and used by ancient man. The continued development of various societies saw the increased burning of coal in urban areas to the extent that there was pollution during medieval times (Galloway, 1882; James and Thorpe, 1994) (Chapter 1). The invention of the internal combustion engine and the ensuing popularity of motor vehicles in this century increased the emissions to the atmosphere.

Before any cleanup of air emissions can occur, it is necessary first to define not only the nature of the emissions but also any ancillary factors (Table 10.1). In addition, air emissions can be classified as either gaseous or particulate emissions (Chapter 2). Common gaseous emissions are carbon monoxide, sulfur dioxide, nitrogen dioxide, and ozone. The particulate matter found in the atmosphere can be made up of many different substances including inorganic and organic constituents.

Another important consideration in the classification of air emissions is the distinction between primary and secondary air pollutants (or emissions). *Primary emissions* or (*secondary pollutants*) are those directly emitted to the atmosphere. A common example is the carbon monoxide emitted from automobile exhausts. *Secondary emissions* or (*secondary pollutants*), on the other hand, are formed in the atmosphere as the result of various transformation mechanisms involving primary emissions or the secondary emissions. Another term used in the categorization of pollutants is the term *regulated pollutants*. Many pollutants can occur in the atmos-

Table 10.1: Necessary actions for cleanup of emissions

1. Identification of the emissions
2. Identification of the emission sources
3. Estimation of emission rates
4. Atmospheric dispersion, transformation, and depletion mechanisms
5. Emission control methods
6. Air-quality evaluation methods
7. Effects on stratospheric ozone
8. Regulations

sphere, but not all potential pollutants are specifically addressed in air pollution laws and regulations. The pollutants that have been singled out for regulatory control are sometimes referred to as regulated pollutants.

One of the atmospheric pollutants of most concern in both urban and rural environments is ozone (Chapter 5). Ozone is a secondary pollutant produced by the reaction of volatile organic compounds and oxides of nitrogen in the presence of sunlight (a photochemical reaction).

Carbon dioxide is also a by-product of respiration and fossil fuel combustion. Increasing amounts of carbon dioxide in the atmosphere have caused concern over a global warming trend (Smith, 1988; Smith and Thambimuthu, 1991; Easterbrook, 1995). Carbon dioxide also traps long-wave radiation near the surface, causing the temperature of the surface layer to increase (the *greenhouse effect*). The greenhouse effect, caused by the absorption of long-wave radiation, is affected by small amounts of water vapor, carbon dioxide, ozone, nitrous oxide, methane, and other minor constituents of air, and by clouds. Clouds absorb, on average, about one-fifth of the solar radiation striking them, but unless they are extremely thin, they are almost completely opaque to infrared radiation. The appearance even of cirrus clouds after a period of clear sky at night is enough to cause the surface air temperature to increase rapidly by several degrees because of long-wave radiation emitted by the cloud. Although the term *greenhouse effect* has generally been used for the role of the whole atmosphere (mainly water vapor and clouds) in keeping the land surface warm, it has been increasingly associated (erroneously, as many scientists believe) with perceived increases in the levels of carbon dioxide in the atmosphere.

It is very difficult to model the temperature effect accurately because of complicated atmospheric responses and unknown natural sources and sinks (Pickering and Owen, 1994). Nevertheless, it has been estimated that the global temperature may rise as much as 5°C (8°F) by the year 2040. This warming would result in significant environmental effects, such as a rise in mean sea level of a few meters due to the melting of ice sheets and a shift in vegetation patterns.

The energy released by anthropogenic activities is known to affect surface temperatures, particularly near urban centers. For example, temperatures may be several degrees warmer near the surface in urban areas than in nearby rural areas. However, the question is how real are predictions based on such phenomena?

Some years ago (in the late 1960s and early 1970s), it was predicted that the nighttime low during the winter could never fall as low as -40°C (-40°F) because of the increase in size of the city of Edmonton (Alberta, Canada) and the associated generation of heat. Within a winter or so, the temperature had dipped to below -40°C (-40°F), and has done so on other occasions.

Metabolic processes and natural combustion (such as forest fires and volcanoes) can release large amounts of gaseous emissions (Pickering and Owen, 1994). However, natural inputs are minor compared to atmospheric emissions due to human activity. Although most anthropogenic air emissions are produced by the various forms of transportation (*mobile emissions sources*), emissions occur from sources such as fuel combustion (*stationary sources*).

Sulfur oxides (SO_X where x = 2 or 3) produced by the combustion of sulfur-containing fuels (which can be biomass, in addition to the fossil fuels) are converted to acids in the atmosphere and precipitated in rain. Work is ongoing in technologies that have been proposed and tested for the conversion of biomass (Table 10.2).

Nitrogen oxides (NO_X, where x = 1 or 2) are formed during high-temperature combustion of fuels (which again may be biomass and/or polymer wastes):

$$N_{fuel} + O_2 = NO_X$$

As with sulfur oxides (SO_x):

$$SO_2 + H_2O = H_2SO_3$$
$$2SO_2 + O_2 = 2SO_3$$
$$SO_3 + H_2O = H_2SO_4$$

nitrogen oxides can form acid precipitation (Chapter 2):

$$2NO + H_2O = 2HNO_2$$
$$2NO + O_2 = 2NO_2$$
$$NO_2 + H_2O = HNO_3$$

They can also react with hydrocarbons (in sunlight) to form ozone and other constituents of smog (Section 4.3, below).

Table 10.2: Technologies for biomass conversion (gasification)

Technology	Notes	Comments
Pyrolysis	Low operating temperatures simple design	Char and tar produced tar removal may be difficult
Gasification (air)	Simple design less tar than pyrolysis	Gas diluted with nitrogen cannot cannot be used for synthesis low-Btu gas may have a reduced combustion efficiency
Gasification (oxygen)	Medium-Btu gas produced	Oxygen plant required.
Gasification (steam)	Medium-Btu gas produced	Product is methane-rich can be steam-reformed for methanol synthesis; can be used as synthetic natural gas

It must also be acknowledged that gasification and combustion of biomass can lead to the release of contaminant species (Hoerning et al., 1995; Moilanen and Kurkela, 1995) such as alkali metals, which in turn, can cause fouling and slagging in the combustors or which can occur as volatile matter with ensuing release into the atmosphere. Careful monitoring for such species is essential (Hald, 1995; Dayton and Milne, 1995).

2.0 Fossil Fuel Sources

Industrial sources (*stationary industrial emission sources*) of gases are often classified as either point-sources or fugitive sources. *Point-sources* involve air emissions through a confined vent or stack. The stack of a fuel combustion source is an example. *Fugitive emissions*, on the other hand, are those emissions that enter the atmosphere from an unconfined area (Greenberg, 1993). Fugitive emissions are often thought of as dust or particle emissions, but the term can apply equally well to many types of gaseous emissions. Examples of fugitive emissions are the dust particles stirred up by wind blowing over exposed storage piles and the vapors that escape from leaking pumps and valves.

Stationary source emissions at industrial facilities can result from fuel combustion and from various process operations (Austin, 1984; Speight, 1993). There are several activities that produce air emissions:

1. external combustion of coal, fuel oil, natural gas, and wood waste,
2. solid waste disposal (municipal waste incinerators, open burning, sludge incinerators),
3. internal combustion devices (combustion turbines, compressor engines, diesel-fired generator engines),
4. evaporation loss sources (dry cleaning; surface coating including can, fabric, automobile, and large appliance coating; magnetic tape manufacturing; distribution and marketing of petroleum liquids such as gasoline; solvent degreasing; printing),
5. chemical process industry (chlor-alkali, paint and varnish, pharmaceutical, phosphoric acid, soap and detergents, sulfuric acid, synthetic organic chemicals, synthetic fibers),
6. food and agriculture industries (coffee roasting, cotton ginning, grain elevators, meat smokehouses, phosphate fertilizers, bread baking, cattle feedlots),
7. metallurgical industry (iron and steel production; primary aluminum production; primary copper, lead, and zinc smelting; secondary aluminum, copper, lead, magnesium, and zinc processing; iron foundries),
8. mineral products industry (asphalt roofing, bricks and clay products, Portland cement, concrete batching, glass and glass fiber manufacturing, aggregate quarrying, metallic minerals mining and processing, coal mining),
9. petroleum industry (petroleum refining, natural gas processing),
10. wood products industry (chemical wood pulping, wood building products

such as plywood and particle board),

11. storage of organic liquids (gasoline, fuel oil, process organic chemicals).

Thus, by definition there are several types of gas that are produced during a variety of industrial processes (Austin, 1984; Probstein and Hicks, 1990), and strategies are necessary to protect the environment from the potential emissions (Frosch and Gallopoulos, 1989).

The composition of process gases is very dependent upon the process, be it coal gas manufacture, gases from coal carbonization or combustion, or the numerous process gases produced by a host of manufacturing industries (Chapter 2). When gases are produced as by-products of industrial processes, it is extremely likely that all of the by-product emissions will be considered as potential pollutants. The gas that tends to receive the most attention as a major pollutant is sulfur dioxide (Kyte, 1991).

2.1 Coal

The cumulus-like plumes of industrial gases, belching forth so prominently, and unabated, into the lives of many people in the industrialized urban areas of the world, were the true sign of the industrial complex. Industry has now embraced the concept of a clean environment, and a concerted effort is being made to ensure that emissions are of minimal, and preferably of no detriment to the environment. There must be a balance between product supply and environmental issues (Malin, 1990).

Fossil fuels play a major role in driving modern industry and therefore require serious consideration of the ways in which they are used. For example, coal is usually considered a dirty fuel but it is more the manner in which coal is used than the fuel itself.

Since at least the late days of the seventeenth century coal has been a source of gas (Chapter 2) for industrial use. In addition to specific tasks such as carbonization and gasification, there is also the recognition that coal specifications can have an impact on power plant performance (Skorupska, 1993). Thus there is the immediate inference that the type of coal (especially low-sulfur coal) used for power generation can also affect (and, in the case of low-sulfur coal, markedly reduce) the emissions (World Coal, 1995).

Manufactured gases, or coal gases, are the gaseous mixtures that are produced when coal is thermally decomposed under a variety of conditions. The processes consist, essentially, of heating coal to drive off the volatile products, which may be gases and others are liquids and tars, to leave a solid carbonaceous residue:

$$[C]_{coal} + heat = [C]_{char} + C_nH_{2n+2} + CO + CO_2 + H_2$$

or

$$coal + heat = char + liquid + gas$$

The residue/char is then treated under a variety of conditions to produce fuels varying which vary from a "purified" char to different types of gaseous mixtures.

The different processes by which these gaseous mixtures are produced are much more complex than the relatively simple chemical equations would indicate, coal being an extremely complex material (for example, see Berkowitz, 1979; Elliott, 1981; Meyers, 1981; Funk, 1983; Hessley et al., 1986; Hessley, 1990; Speight, 1994). Moreover, the presence of nitrogen, sulfur, and mineral matter in the coal (Davidson, 1994; Speight, 1994), as well as the complex nature of the thermal degradation process dictates the production of a gaseous mixture that is by no means pure and may even need adjustment of the relative amounts of the different constituents before further use:

$$[N]_{coal} = NO_x + HCN... \; x = 1 \text{ or } 2$$
$$[O]_{coal} = Co_x... \; x = 1 \text{ or } 2$$
$$[S]_{coal} = SO_x + H_2S... \; x = 1 \text{ or } 2$$

Combustion probably represents the oldest known use of coal. There are two major methods of coal combustion: fixed-bed combustion and combustion in suspension. The first fixed beds (e.g., open fires, fireplaces, domestic stoves) were simple in principle. Suspension burning of coal that began in the early 1900s with the development of pulverized coal-fired systems was in widespread use by the 1920s. Spreader stokers, which were developed in the 1930s, combined both methods by providing for the smaller particles of coal to be burned in suspension and larger particles to be burned on a grate.

A major concern in the present-day combustion of coal is the performance of the process in an environmentally acceptable manner through the use of either low-sulfur coal or postcombustion cleanup of the flue gases (Soud and Takeshita, 1994; United States Department of Energy, 1995; World Coal, 1995). Thus there is a marked trend to use more efficient methods of coal combustion. A desirable system would be a combustion system that is able to accept coal without the necessity of a postcombustion treatment or without emitting objectionable amounts of sulfur and nitrogen oxides and particulate matter.

In direct combustion, coal is burned (i.e., the carbon and hydrogen in the coal are oxidized into carbon dioxide and water) to convert the chemical energy of the coal into thermal energy:

$$[C]_{coal} + O_2 = CO_2$$
$$[2H]_{coal} + O_2 = H_2O$$
$$[C]_{coal} + H_2O = CO + H_2$$

after which the sensible heat in the products of combustion can be converted into steam that can be used for external work or converted directly into shaft horsepower (e.g., in a gas turbine). In fact, the combustion process actually represents a means of achieving the complete oxidation of coal.

On a more formal basis, the combustion of coal may be simply represented as the staged oxidation of coal carbon to carbon dioxide:

$$[2C]_{coal} + O_2 = 2CO$$
$$2CO + O_2 = 2CO_2$$

with any reactions of the hydrogen in the coal being considered to be of secondary importance. In reality, the combustion of coal, and for that matter other carbonaceous materials (which contain hydrogen and oxygen as well as carbon) involves a wide variety of reactions among reactants, intermediates, and products. The reactions occur simultaneously and consecutively (in both forward and reverse directions) and may at times approach a condition of equilibrium. Furthermore, there is a change in the physical and chemical structure of the fuel particle as it burns.

The conversion of nitrogen and sulfur, during coal combustion, to their respective oxides during combustion is a major issue:

$$[S]_{coal} + O_2 = SO_2$$
$$2SO_2 + O_2 = 2SO_3$$
$$[2N]_{coal} + O_2 = 2NO$$
$$2NO + O_2 = 2NO_2$$
$$[N]_{coal} + O_2 = NO_2$$

The sulfur dioxide that escapes into the atmosphere is either deposited locally or converted to sulfurous and sulfuric acids by reaction with moisture in the atmosphere:

$$SO_2 + H_2O = H_2SO_3$$
$$2SO_2 + O_2 = 2SO_3$$
$$SO_3 + H_2O = H_2SO_4$$

or

$$2SO_2 + O_2 + 2H_2O = 2H_2SO_4$$

Nitrogen oxides also contribute to the formation and occurrence of acid rain, in a similar manner to the production of acids from the sulfur oxides, yielding nitrous and nitric acids:

$$NO + H_2O = H_2 + NO_2$$
$$2NO + O_2 = 2NO_2$$
$$NO_2 + H_2O = HNO_3$$

or

$$2NO + O_2 + H_2O = 2HNO_3$$

Table 10.3: Constituents of coal ash.

	Percent
SiO_2	40-90
Al_2O_3	0-60
Fe_2O_3	5-25
CaO	1-15
MgO	<4
$Na_2O + K_2O$	1-4

In addition to causing objectionable stack emissions, coal ash (Table 10.3) and volatile inorganic material generated by thermal alteration of mineral matter in coal (Speight, 1993, 1994) will adversely affect heat transfer processes by fouling heat-absorbing and radiating surfaces and will also influence the performance of the combustion system by causing corrosion. Operating procedures must therefore provide for effective countering of all these effects.

Combustion systems vary in nature, depending upon the feedstock and the air needed for the combustion process. However, the two principal types of coal-burning systems are usually referred to as layer and chambered. The former refers to fixed beds while the latter is more specifically for pulverized fuel.

For fuel-bed burning on a grate, a distillation effect occurs and the result is that liquid components that are formed will volatilize before combustion temperatures are reached; cracking may also occur. The ignition of coal in a bed is almost entirely by radiation from hot refractory arches and from the flame burning of volatile components/products. In fixed beds the radiant heat above the bed can only penetrate a short distance into the bed.

Consequently, convective heat transfer determines the intensity of warm-up and ignition. In addition, convective heat transfer plays an important part in the overall flame-to-surface transmission. The reaction of gases is greatly accelerated by contact with hot surfaces, and while the reaction away from the walls may proceed slowly, reaction at the surface proceeds much more rapidly.

In the simplest terms, fluidized-bed combustion occurs in expanded beds. Reaction occurs at lower temperatures (925°C; 170O°F), but high convective transfer rates exist due to the bed motion. In fact, heat loads higher than in those comparably sized radiation furnaces can be effected, that is, smaller chambers produce the same equivalent heat load. Moreover, fluidized systems can operate under substantial pressures, thereby allowing more efficient gas cleaning. Fluidized-bed combustion is a means for providing high heat transfer rates, controlling sulfur, and reducing nitrogen oxide emissions due to the low temperatures in the combustion zone. However, there is the suggestion that NO_x abatement and control technologies may increase the emissions of N_2O from coal utilization (Takeshita and Sloss, 1993).

A fluidized-bed is an excellent medium for bringing gases in contact with solids. This can be exploited in a combustor since sulfur dioxide emissions can be reduced by adding limestone ($CaCO_3$) or dolomite ($CaCO_3.MgCO_3$) to the bed. The sulfur oxides react to form calcium sulfate, which leaves the bed as a solid with the ash:

$$2SO_2 + O_2 = 2SO_3$$
$$SO_3 + CaCO_3 = CaSO_4 + CO_2$$

or

$$2SO_2 + O_2 + 2CaCO_3 = 2CaSO_4 + 2CO_2$$

The spent sorbent from fluidized-bed combustion may be taken directly to disposal and is much easier to dispose of than the disposal salts produced by wet limestone scrubbing. These latter species are contained in wet sludge having a high volume and a high content of salt-laden water. The mineral products of fluid-bed combustion, however, are quite dry and in a chemically refractory state, and therefore disposal is much easier and less likely to result in pollution.

The spent limestone from fluidized-bed combustion may be regenerated, thereby reducing the overall requirement for lime and decreasing the disposal problem. Regeneration is accomplished with a synthesis gas (consisting of a mixture of hydrogen and carbon monoxide) to produce a concentrated stream of sulfur dioxide:

$$CaSO_4 + H_2 = CaO + H_2O + SO_2$$
$$CaSO_4 + CO = CaO + CO_2 + SO_2$$

The calcium oxide product is supplemented with fresh limestone and returned to the fluidized bed. Two undesirable side reactions can occur in the regeneration of spent lime, leading to the production of calcium sulfide:

$$CaSO_4 + 4H_2 = CaS + 4H_2O$$
$$CaSO_4 + 4CO = CaS + 4CO_2$$

which results in the recirculation of sulfur to the fluidized bed.

In entrained beds, fine grinding and increased retention times intensify combustion, but the temperature of the carrier and degree of dispersion are also important. In practice, the coal is introduced at high velocities, which may be greater than 100 feet/sec (30.5 m/sec) and involve expansion from a jet to the combustion chamber.

Entrained systems include cyclone furnaces (which have been used for various coals) and other systems developed and utilized for the injection of coal-oil slurries into blast furnaces or for the burning of coal-water slurries.

An advantage of the cyclone furnace is the low dust burden in the secondary furnace and, hence, its lower emission of particulate matter from the stack. Most cyclone furnaces capture about 90% of the ash in the coal and convert it to molten slag. Cyclone furnaces have two major shortcomings: (1) the ash of the coal must be convertible to molten slag at furnace temperatures and (2) the nitrogen oxide emissions are excessive (about 1000 ppm) because of the high furnace temperature. Cyclone furnaces have been widely used in areas where the coal contains ash with a low fusion temperature. For successful removal of slag, the slag viscosity cannot exceed 250 poise (m^2/sec) at 1420°C (2600°F) and many cyclone furnaces may not meet this requirement. Addition of iron ore, limestone, or dolomite makes it possible to flux the coal ash, thereby decreasing the viscosity at furnace temperatures.

Carbonization is the process by which coal is converted into solid (coke), liquid (tar and distillate), and gaseous products. The prime motive for carbonization varies between the need to produce one, or more, of the products as the prime product.

$$[C]_{organic} = [C]_{coke/char/carbon} + \text{liquids} + \text{gases}$$

Documented efforts at coke manufacture date from the late sixteenth century (1584, to be precise) (Fess, 1957), and it has seen various adaptations of conventional wood-charring methods with the eventual evolution of the self-descriptive beehive oven, which by the mid-nineteenth century had become the most common vessel for the coking of coal (Speight, 1994). The proportions of gas, coke, tar, and other liquid vary according to the particular method used for the carbonization (especially on the retort configuration), process temperature, as well as on the nature (rank) of the coal used. Purification of the gas is necessary (Speight, 1993). In more general terms, coal carbonization processes are regarded as low temperature when the temperature of the process does not exceed 700°C (1290°F) or high temperature if the temperature of the process is at, or in excess of, 900°C (1650°F); the temperature of a medium temperature carbonization process, of course, lies in between these values.

Low-temperature carbonization was mainly developed as a process to supply *town gas* for lighting purposes as well as to provide a smokeless (devolatilized) solid fuel for domestic consumption (Seglin and Bresler, 1981). However, the process by-products (tars) were also found to be valuable insofar as they served as feedstocks for an emerging chemical industry and were also converted to gasolines, heating oils, and lubricants (Aristoff et al., 1981).

On a commercial scale, the low-temperature carbonization of coal was employed extensively in the industrialized nations of Europe but suffered a major decline after 1945 as oil and natural gas became more widely available. The subsequent rapid escalation in oil prices as well as recent environmental regulations have stimulated (and reactivated) interest in the recovery of hydrocarbon liquids from coal by low-temperature thermal processing.

Coal gasification is, essentially the conversion of coal (by any one of a variety of processes) to produce combustible gases, which may be of low, medium, or high Btu content depending upon the defined use (Bodle and Huebler, 1981; Rath and Longanbach, 1991). Briefly, gases are defined on the basis of heat content expressed as British thermal units (Btu). Pure methane has a Btu content of ~1000-1100 (Btu;

1 Btu = 1.055 joule). A low Btu gas produced by some coal gasification processes may have a heat content as low as ~100-150 Btu.

High-Btu gas consists predominantly of methane with a heating value of approximately 1000 Btu/ft^3 (37.3 x 10^3 kJ/m^3) and is compatible with natural gas insofar as it may be mixed with, or substituted for, natural gas. Medium-Btu gas consists of a mixture of methane, carbon monoxide, hydrogen, and various other gases. The heating value of medium-Btu gas usually falls in the range 300-700 Btu/ft^3 (11-26 x 10^3 kJ/m^3), and it is suitable as a fuel for industrial consumers. Finally, low-Btu gas consists of a mixture of carbon monoxide and hydrogen and has a heating value of less than 300 Btu/ft^3 (11 kJ/m^3). This gas is of interest to industry as a fuel gas or even, on occasion, as a raw material from which ammonia, methanol, and other compounds may be synthesized.

The importance of coal gasification as a means of producing fuel gas(s) for industry use cannot be underplayed, but coal gasification systems also produce a range of undesirable products that must be removed before the products are used to provide fuel and/or to generate electric power (Alpert and Gluckman, 1986).

Coal gasification involves the thermal decomposition of coal and the reaction of the carbon in the coal, and other pyrolysis products with oxygen, water, and hydrogen to produce fuel gases such as methane:

$$[C]_{coal} + [H]_{coal} = CH_4$$

or with added hydrogen:

$$[C]_{coal} + 2H_2 = CH_4$$

If a high-Btu gas is required, efforts must be made to increase the methane content of the gas:

$$CO + H_2O = CO_2 + H_2$$
$$CO + 3H_2 = CH_4 + H_2O$$
$$2CO + 2H_2 = CH_4 + CO_2$$
$$CO + 4H_2 = CH_4 + 2H_2O$$

In summary, coal gasification produces a mixture of gases, some of which can be used but others of which must be removed from the product mix before use if the environment is to be protected from their harmful effects.

2.2 Petroleum

Liquid fossil fuels, generally termed oil, crude oil, petroleum, or heavy oil, and all are capable of producing pollutant materials (Loehr, 1992; Olschewsky and Megna, 1992). The near-solid bitumen (which is also referred to as natural asphalt although somewhat incorrectly since asphalt is a product of refinery processing) occurs in various locations throughout the world and is often classified as a heavy oil although various subdivisions of this classification are possible (Speight, 1990, 1991).

Refining petroleum follows a somewhat different scenario from that of coal conversion, and refinery gas streams often contain substantial amounts of acid gases such as hydrogen sulfide and carbon dioxide (Chapter 2). More particularly hydrogen sulfide arises from the hydrodesulfurization of feedstocks that contain organic sulfur (Bland and Davidson, 1967; Speight, 1981, 1991):

$$[S]_{feedstock} + H_2 = H_2S + \text{hydrocarbons}$$

Acid gases corrode refining equipment, harm catalysts, pollute the atmosphere, and prevent the use of hydrocarbon components in petrochemical manufacture. When the amount of hydrogen sulfide is high, it may be removed from a gas stream and converted to sulfur or sulfuric acid. Some natural gases contain sufficient carbon dioxide to warrant recovery as dry ice (Bartoo, 1985; Kumar, 1987).

The terms *refinery gas* and *process gas* are also often used to include all of the gaseous products and by-products that emanate from a variety of refinery processes (Gary and Handwerk; 1975; Speight, 1991). There are also components of the gaseous products that must be removed prior to release of the gases to the atmosphere or prior to use of the gas in another part of the refinery, i.e., as a fuel gas or as a process feedstock.

Processing crude petroleum, with the exception of some of the more viscous crude oils, involves a primary distillation of the hydrogen mixture, which results in its separation into fractions differing in carbon number, volatility, specific gravity, and other characteristics. The most volatile fraction, that contains most of the gases which are generally dissolved in the crude, is referred to as pipestill gas or pipestill light ends and consists essentially of hydrocarbon gases ranging from methane to butane(s), or sometimes pentane(s). The gas varies in composition and volume, depending on crude origin and on any additions to the crude made at the loading point. It is not uncommon to reinject light hydrocarbons such as propane and butane into the crude before dispatch by tanker or pipeline. This results in a higher vapor pressure of the crude, but it allows one to increase the quantity of light products obtained at the refinery. Since light ends in most petroleum markets command a premium, while in the oil field itself propane and butane may have to be reinjected or flared, the practice of "spiking" crude with liquefied petroleum gas is becoming fairly common.

In addition to the gases obtained by distillation of crude petroleum, more highly volatile products result from the subsequent processing of naphtha and middle distillate to produce gasoline, from desulfurization processes involving hydrogen treatment of naphtha, distillate, and residual fuel; and from the coking or similar thermal treatments of vacuum gas oils and residual fuels. The most common processing step in the production of gasoline is the catalytic reforming of hydrocarbon fractions in the heptane (C_7) to decane (C_{10}) range.

In a series of processes commercialized under Platforming, Powerforming, Catforming, and Ultraforming, paraffinic and naphthenic (cyclic nonaromatic) hydrocarbons, in the presence of hydrogen and a catalyst are converted into aromatics, or isomerized to more highly branched hydrocarbons. Catalytic reforming processes thus not only result in the formation of a liquid product of higher octane number, but also produce substantial quantities of gases. The latter are rich in hydrogen, but also contain hydrocarbons from methane to butanes, with a preponderance of propane ($CH_3.CH_2.CH_3$), n-butane ($CH_3.CH_2.CH_2.CH_3$) and isobutane [$(CH_3)_3CH$]. Their composition will vary in accordance with reforming severity and reformer feedstock. Since all catalytic reforming processes require substantial recycling of a hydrogen stream, it is normal to separate reformer gas into a propane ($CH_3.CH_2.CH_3$) and/or a butane [$CH_3.CH_2.CH_2.CH_3/(CH_3)_3CH$] stream, which becomes part of the refinery liquefied petroleum gas production, and a lighter gas fraction, part of which is recycled. In view of the excess of hydrogen in the gas, all products of catalytic reforming are saturated, and there are usually no olefinic gases present in either gas stream.

A second group of refining operations that contributes to gas production is that of the catalytic cracking processes. These consists of fluid-bed catalytic cracking, Thermofor catalytic cracking, and other variants in which heavy gas oils are converted into cracked gas, liquefied petroleum gas, catalytic naphtha, fuel oil, and coke by contacting the heavy hydrocarbon with the hot catalyst. Both catalytic and thermal cracking processes, the latter being now largely used for the production of chemical raw materials, result in the formation of unsaturated hydrocarbons, particularly ethylene ($CH_2=CH_2$), but also propylene (propene, $CH_3.CH=CH_2$), isobutylene [isobutene, $(CH_3)_2C=CH_2$] and the n-butenes ($CH_3.CH_2.CH=CH_2$, and $CH_3.CH=CH.CH_3$) in addition to hydrogen (H_2), methane (CH_4) and smaller quantities of ethane ($CH_3.CH_3$), propane ($CH_3.CH_2.CH_3$), and butanes [$CH_3.CH_2.CH_2.CH_3$, $(CH_3)_3CH$]. Diolefins such as butadiene ($CH_2=CH.CH=CH_2$) and are also present.

Additional gases are produced in refineries with coking or visbreaking facilities for the processing of their heaviest crude fractions. In the visbreaking process, fuel oil is passed through externally fired tubes and undergoes liquid phase cracking reactions, which result in the formation of lighter fuel oil components. Oil viscosity is thereby reduced, and some gases, mainly hydrogen, methane, and ethane, are formed. Substantial quantities of both gas and carbon are also formed in coking (both fluid coking and delayed coking) in addition to the middle distillate and naphtha. When coking a residual fuel oil or heavy gas oil, the feedstock is preheated and contacted with hot carbon (coke) which causes extensive cracking of the feedstock constituents of higher molecular weight to produce lower molecular weight products ranging from methane, via liquefied petroleum gas(es) and naphtha, to gas oil and heating oil. Products from coking processes tend to be unsaturated and olefinic components predominate in the tail gases from coking processes.

A further source of refinery gas is hydrocracking, a catalytic high-pressure pyrolysis process in the presence of fresh and recycled hydrogen. The feedstock is again heavy gas oil or residual fuel oil, and the process is mainly directed at the production of additional middle distillates and gasoline. Since hydrogen is to be recycled, the gases produced in this process again have to be separated into lighter and heavier streams; any surplus recycle gas and the liquefied petroleum gas from the hydrocracking process are both saturated.

Both hydrocracker and catalytic reformer tail gases are commonly used in catalytic desulfurization processes. In the latter, feedstocks ranging from light to vacuum gas oils are passed at pressures of 500-1000 psi (3.5-7.0 x 10^3 kPa) with hydrogen over a hydrofining catalyst. This results mainly in the conversion of organic sulfur compounds to hydrogen sulfide,

$$[S]_{feedstock} + H_2 = H_2S + \text{hydrocarbons}$$

but also produces some light hydrocarbons by hydrocracking.

Thus refinery streams, while ostensibly being hydrocarbon in nature, may contain large amounts of acid gases such as hydrogen sulfide and carbon dioxide. Most commercial plants employ hydrogenation to convert organic sulfur compounds into hydrogen sulfide. Hydrogenation is effected by means of recycled hydrogen-containing gases or external hydrogen over a nickel molybdate or cobalt molybdate catalyst.

In summary, refinery process gas, in addition to hydrocarbons, may contain other contaminants, such as carbon oxides (CO_x, where x = 1 and/or 2), sulfur oxides (So_x, where x = 2 and/or 3), as well as ammonia (NH_3), mercaptans (R-SH), and carbonyl sulfide (COS).

The presence of these impurities may eliminate some of the sweetening processes, since some processes remove large amounts of acid gas but not to a sufficiently low concentration. On the other hand, there are those processes not designed to remove (or incapable of removing) large amounts of acid gases whereas they are capable of removing the acid gas impurities to very low levels when the acid gases are present only in low-to-medium concentration in the gas.

The processes that have been developed to accomplish gas purification vary from a simple once-through wash operation to complex multistep recycling systems. In many cases, the process complexities arise because of the need for recovery of the materials used to remove the contaminants or even recovery of the contaminants in the original, or altered, form (Kohl and Riesenfeld, 1979; Newman, 1985).

From an environmental viewpoint, it is not means by which these gases can be utilized but it is the effects of these gases on the environment when they are introduced into the atmosphere.

In addition to the corrosion of equipment of acid gases, the escape into the atmosphere of sulfur-containing gases can eventually lead to the formation of the constituents of acid rain, i.e., the oxides of sulfur (SO_2 and SO_3). Similarly, the nitrogen-containing gases can also lead to nitrous and nitric acids (through the formation of the oxides NO_x, where x = 1 or 2) which are the other major contributors to acid rain. The release of carbon dioxide and hydrocarbons as constituents of refinery effluents can also influence the behavior and integrity of the ozone layer.

In summary, and from an environmental viewpoint, petroleum processing can result in similar, if not the same, gaseous emissions as coal. It is a question of degree insofar as the composition of the gaseous emissions may vary from coal to petroleum but the constituents are, in the majority of cases, the same.

2.3 Natural Gas

The gaseous component that often occurs in reservoirs with petroleum is natural gas. It is usually processed in a petroleum refinery but there are sources of natural gas that occur without the associated presence of petroleum. Even though the gas may, to all intents and purposes, be characterized as methane, there are those constituents of natural gas that present the potential for pollution and must be removed.

While the major constituent of natural gas is methane, there are components such as carbon dioxide (CO), hydrogen sulfide (H_2S), and mercaptans (thiols; R-SH), as well as trace amounts of sundry other emissions. The fact that methane has a foreseen and valuable end-use makes it a desirable product, but in several other situations it is considered a pollutant, having been identified as one of several greenhouse gases (Graedel and Crutzen, 1989; Smith et al., 1994).

A sulfur removal process must be very precise, since natural gas contains only a small quantity of sulfur-containing compounds that must be reduced several orders of magnitude. Most consumers of natural gas require less than 4 ppm in the gas.

A characteristic feature of natural gas that contains hydrogen sulfide is the presence of carbon dioxide (generally in the range of 1 to 4% v/v). In cases where the natural gas does not contain hydrogen sulfide, there may also be a relative lack of carbon dioxide.

3.0 Other Sources

Other industrial gas streams will vary depending upon the nature of the industry and the starting materials. However, there will always be the need to clean up the effluent gas before even the remotest chance of being allowed to discharge the gas to the atmosphere becomes a reality.

Emission sources may be characterized in a number of ways. First, a distinction may be made between natural and anthropogenic sources. Another frequent classification is in terms of *stationary sources* (power plants, incinerators, industrial operations, and space heating) and *mobile sources* (motor vehicles, ships, aircraft, and rockets). Another classification describes sources as *point sources* (a single stack), *line sources* (a line of stacks), or *area sources* (city).

3.1 Industrial Stationary Emission Sources

Stationary industrial sources are often classified as either point sources or fugitive sources. The term *industrial* in this context includes manufacturing, mining, power generation, commercial operations, and institutional sources. Stationary source emissions at industrial facilities can result from fuel combustion and from various process operations.

3.2 Nonindustrial Stationary Emission Sources

Although industrial sources are given the most attention in this chapter, recognition should also be made of the many nonindustrial sources such as human activities and natural sources. Some of the emission sources related to human activities are fireplaces, wood-burning stoves, architectural coatings (for example, painting of structures), intentional burning of woodlands for forest management or land-clearing purposes, and agricultural tilling. Examples of natural emission sources are wildfires (started by natural means), volcanoes and other geothermal sources (Pickering and Owen, 1994), oil seeps, biological decay, and windblown fugitive dust from barren areas.

3.3 Mobile Emission Sources

Mobile source emissions refer to the emissions generated by the combustion of fuels in various types of engine-driven vehicles (automobiles, trucks, buses, motorcycles, aircraft, water vessels, etc.). The principal emissions from operation of vehicles are carbon monoxide, oxides of nitrogen (primarily nitric oxide and nitrogen dioxide), various species of hydrocarbons, particulate matter, and depending on the fuel, sulfur compounds.

Vehicle emissions are of greatest concern in urban areas where the volume of traffic is highest and where vehicles often operate under conditions (such as stop-and-start driving) that result in higher emission rates. Vehicle emissions typically are the predominant cause of smog.

4.0 Effects

The major gaseous emissions produced in many processes are water vapor and carbon dioxide. The former is presumed to have little effect on the environment while the latter's effect is still open to considerable debate. In any event, it must be assumed that the effects of releasing unlimited amounts of carbon dioxide to the atmosphere will cause adverse, if not severe, perturbations to the chemistry of the atmosphere (Birks et al., 1993). It follows therefore that perturbations to the chemistry of the atmosphere can have consequences for the flora and fauna. In fact, the need for monitoring of personal exposure to air pollutants is a relatively new concept but is now recognized as being justified (Lioy, 1993).

Other gaseous emissions from processes include hydrogen sulfide (H_2S), as well as such sulfur compounds as carbonyl sulfide (COS), carbon disulfide (CS_2), and mercaptans (R-SH). Nitrogen compounds, particularly the oxides (NO_x, where x = 1 or 2) will also be evident, as will nitriles (-C=N) and miscellaneous organic nitrogen species.

The direct effects of these gases on life forms is well established, and standards have been set for the release of these gases to the atmosphere. However, the indirect effects of these gases are the subject of much concern and are still open to speculation. These effects stem from the formation of secondary pollutants such as (1) acid rain (acid deposition); (2) the onset of global warming, often referred to as the greenhouse effect; and (3) the formation of smog, often referred to as photochemical smog because of the chemistry of its formation.

4.1 Acid Rain

Acid rain occurs when the oxides of nitrogen that are released to the atmosphere during the combustion of fossil fuels (Gould, 1984) are deposited (as soluble acids) with rainfall, usually at some location remote from the source of the emissions. Emissions released from both stationary and mobile sources (Chapter 11) are dispersed in the atmosphere with the potential for transformation and ultimate depletion (Figure 10.1). Dispersion, transformation, and depletion are the terms that describe the initiation, formation, and deposition of acid rain (Longhurst et al., 1993a, 1993b; Smoot et al., 1993; Pickering and Owen, 1994).

It is generally believed (the chemical thermodynamics are favorable) that acidic compounds are formed when sulfur dioxide and nitrogen oxide emissions are released from tall industrial stacks. Gases such as sulfur oxides (usually sulfur dioxide, SO_2) as well as the nitrogen oxides (NO_x) react with the water in the atmosphere to form acids:

$$SO_2 + H_2O = H_2SO_3$$
$$2SO_2 + O_2 = 2SO_3$$
$$SO_3 + H_2O = H_2SO_4$$
$$2NO + H_2O = 2HNO_2$$
$$2NO + O_2 = 2NO_2$$
$$NO_2 + H_2O = HNO_3$$

Acid rain has a pH less than 5.0 and predominantly consists of sulfuric acid (H_2SO_4) and nitric acid (HNO_3). As a point of reference, in the absence of anthropogenic pollution sources the average pH of rain is ~6.0 (slightly acidic; neutral pH = 7.0). In summary, the sulfur dioxide that is produced during a variety of processes will react with oxygen and water in the atmosphere to yield environmentally detrimental sulfuric acid (Chapter 2). Similarly, nitrogen oxides will also react to produce nitric acid.

Another acid gas, hydrogen chloride (HCl), although not usually considered to be a major emission, is produced from mineral matter and other inorganic materials and is gaining increasing recognition as a contributor to acid rain (Gilleland and

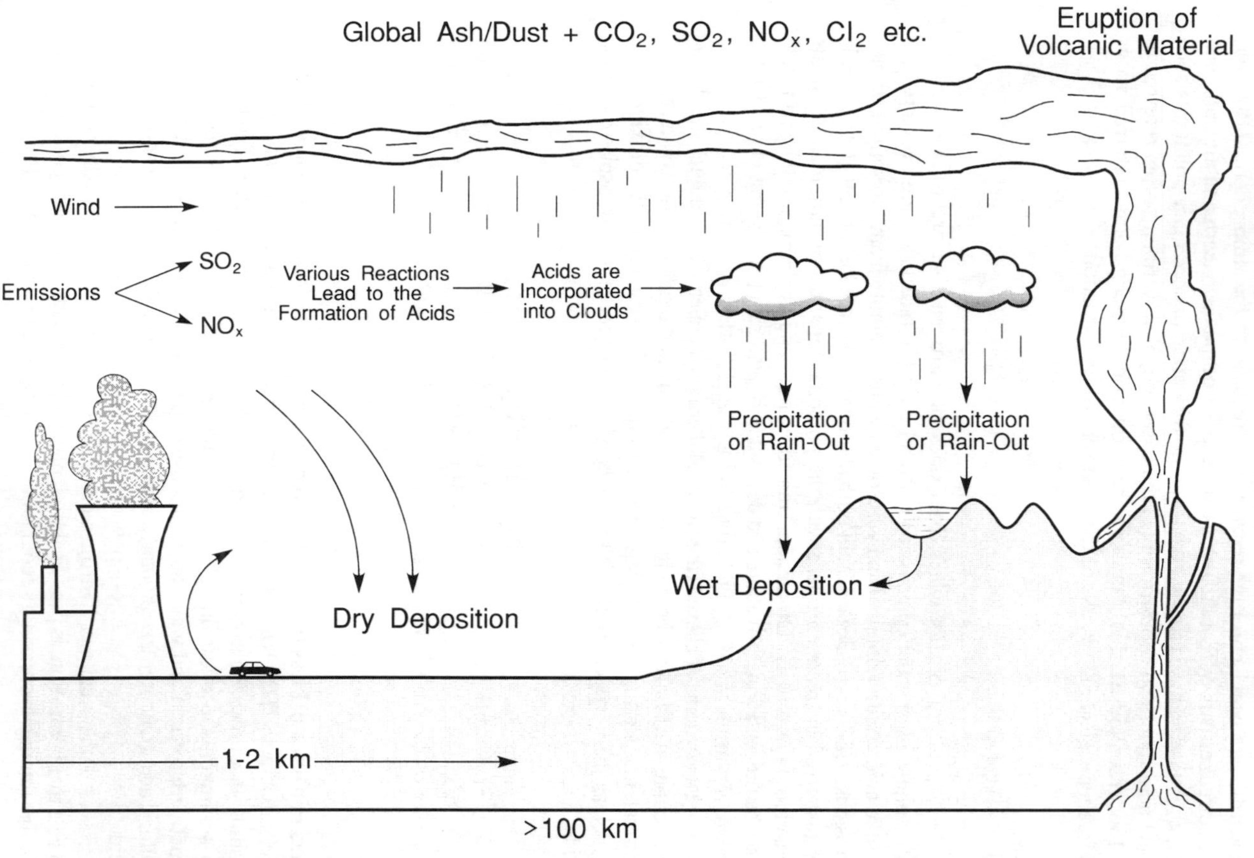

Figure 10.1: Events leading to acid rain formation and deposition

Swisher, 1986). However, hydrogen chloride may exert severe local effects because it does not need to participate in any further chemical reaction to become an acid. Under atmospheric conditions that favor a buildup of stack emissions in the areas where hydrogen chloride is produced, the amount of hydrochloric acid in rain water could be quite high.

In terms of coal usage, very little can be done during the pretreatment of coal to eliminate nitrogen since the nitrogen is part of the organic structure of coal. In addition, both organic chlorine and inorganic chloride salts contribute to the formation of hydrogen chloride during combustion. Coal-cleaning processes can reduce the mineral matter content of coal, but pretreatment processes do not remove organically bound chlorine, which is more likely to be the precursor to hydrogen chloride in a combustion process.

Pretreatment washing processes are successful methods for removing inorganic sulfur but they do not affect the organic sulfur content. Commercial methods crush the coal and separate the resulting particles on the basis of density, thereby removing up to 30% of the sulfur. But while pretreatment washing may remove up to 90% of the pyritic sulfur, up to 20% of the combustible coal may also be removed, and a balance must be struck between the value of the sulfur removed and coal lost to the cleaning process.

Although emissions from industrial operations and fossil fuel combustion are the major sources of acid-forming gases, acid rain has also been encountered in areas far from such sources. This is due in part to the fact that acid-forming gases are oxidized to acidic constituents and deposited over several days, during which time the air mass containing the gas may have moved as much as several thousand miles (1 mile = 1.6 km). It is likely that the burning of biomass, such as is employed in *slash-and-burn agriculture*, produces the gases that lead to the formation of acidic species in the atmosphere.

Ample evidence exists of the damaging effects of acid rain. The major such effects are as follows (Longhurst et al., 1993a, 1993b):

1. toxicity to plants from excessive acid concentrations.
2. toxicity from acid-forming gases that often accompany acid rain.
3. toxicity from metals, such as from aluminum (Al^{3+}), liberated from soil by the acid rain.
4. acidification of lake water with toxic effects to lake flora and fauna.
5. corrosion to exposed structures.
6. influence of sulfate aerosols on physical and optical properties of clouds (Charlson and Wigley, 1994).

In summary, acid rain has a significant effect (direct and indirect) on the various ecosystems. It is, however, the lasting effects of acidification of the land and the water systems that need attention. The obnoxious acid gases are removed as a result of rainfall, but not perhaps before some damage has been done to the ecosystem.

4.2 The Greenhouse Effect

The presence of infrared-absorbing trace gases (other than water vapor) in the atmosphere contributes to global warming (the *greenhouse effect*) by allowing incoming solar radiant energy to penetrate to the earth's surface while reabsorbing infrared radiation emanating from it. In brief, the differential in the behavior of the atmosphere toward outgoing/incoming radiation plays a similar role like that of the glass roof in a small horticultural ecosystem, i.e., a greenhouse.

For the past several years, there has been much concern about global warming (Keepin, 1986; Kellogg, 1987; Schlesinger and Mitchell, 1987; MacDonald, 1988; Smith, 1988; Hileman, 1989; Douglas, 1990; White, 1990; IEA Coal Research, 1992) whereby increased concentrations of carbon dioxide in the atmosphere, produced by the combustion of fossil fuels, are believed with much debate and conjecture, to lead to long-term global climatic changes (Stobaugh and Yergin, 1983; Michaels, 1991; Nordhaus, 1994).

But to maintain perspective, remembering that the magnitude and timing of the effect remain uncertain (Nordhaus, 1991), the possibility that changes in the amount of carbon dioxide in the atmosphere could lead to changes in world climate is not new and has been suggested since at least the end of the last century. Indeed, in this context it must be recognized that the earth's climate has undergone considerable changes in the past insofar as alternating greenhouse and icehouse effects have occurred (Pickering and Owen, 1994).

In a simplified illustration of the greenhouse effect (Figure 10.2) (Chapter 1) carbon dioxide and water vapor in the atmosphere absorb (or trap) part of the long wave (infrared) radiation from the earth's surface, while at the same time the atmosphere allows passage of the short-wave (visible) radiation from the sun. An increase in the concentration of greenhouse gases (such as carbon dioxide) in the atmosphere causes an increase in the absorption of the radiation from the earth, which ultimately causes an increase in the surface temperature.

Analyses of gases trapped in polar ice samples indicate that preindustrial levels of carbon dioxide and methane in the atmosphere were ~260 ppm and 0.70 ppm, respectively. Over the last 300 years, these levels have increased to current values of around 350 ppm and 1.7 ppm, respectively; most of the increase by far has taken place at an accelerating pace over the last 100 years.

A note of interest is the observation based upon analyses of gases trapped in ice cores, that the atmospheric level of carbon dioxide at the peak of the last ice age about 18,000 years past was 25% of preindustrial levels. About half of the increase in carbon dioxide in the last 300 years can be attributed to deforestation, which still accounts for approximately 20% of the annual increase in this gas. Carbon dioxide is increasing by about 1 ppm per year. Methane is going up at a rate of almost 0.02 ppm per year. The comparatively very rapid increase in methane levels is attributed to a number of factors resulting from human activities. Among these are direct leakage of natural gas, by-product emissions from coal mining and petroleum recovery, and release from the burning of savannas and tropical forests.

Biogenic sources resulting from human activities produce large amounts of atmospheric methane, including methane from bacteria during the degradation of organic matter, such as municipal refuse in landfills. Methane is also evolved from the anaerobic biodegradation of organic matter in rice paddies and as the result of bacterial action in the digestive tracts of ruminant animals.

However, the term *greenhouse effect* is deemed by some to be inappropriate because the actual events are much more complex. There are many other physical processes and climatic effects that must be included in the model than this simple absorption/radiation example would indicate. There are those who consider that the extent of the coupling of the effects is still sufficiently in doubt and that it cannot be stated, or estimated, with any degree of certainty whether the surface temperature of the earth will increase, is increasing, or may even be decreasing!

However, instincts alone tell us that the discharge of foreign species, or even the discharge of a large surplus of an indigenous species, must alter the delicate systems in an adverse fashion. Most models predict global warming of 1.5-5°C (2.7-9°F), about as much again as has occurred since the last ice age 18,000 years past. Such warming would have profound effects on rainfall, plant growth, and sea levels, which might rise as much as 1.5-5 feet (0.5-1.5 meters).

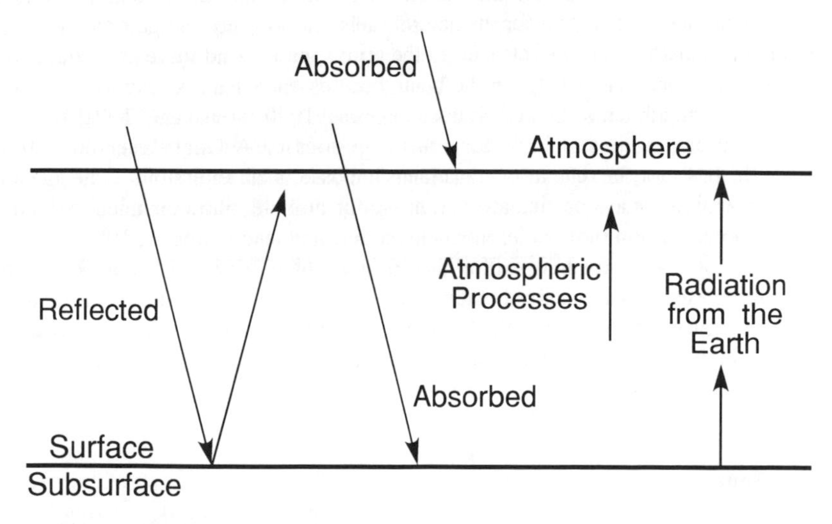

Figure 10.2: The greenhouse effect

There are several gases that contribute to the greenhouse effect (Energy Information Administration, 1993). Carbon dioxide is the gas most commonly thought of as a greenhouse gas; it is responsible for about half of the atmospheric heat retained by trace gases. It is produced primarily by burning of fossil fuels as well by burning and biodegradation of biomass. Deforestation programs essentially remove carbon dioxide sinks, thereby contributing to increased carbon dioxide in the atmosphere.

It is certain that atmospheric carbon dioxide levels will continue to increase significantly. The degree to which this occurs depends upon fixture levels of carbon dioxide production and the fraction of that production that remains in the atmosphere. Given plausible projections of carbon dioxide production and a reasonable estimate that half of that amount will remain in the atmosphere, indications are that sometime during the middle part of the next century the concentration of this gas will reach 600 ppm in the atmosphere. This is well over twice the levels estimated for preindustrial times. Much less certain are the effects that this change will have on climate. It is virtually impossible for the elaborate computer models to accurately take account of all variables, such as the degree and nature of cloud cover. Clouds both reflect incoming light radiation and absorb outgoing infrared radiation, the former effect tending to predominate. The magnitude of these effects depends upon the degree of cloud cover, brightness, altitude, and thickness. In the case of clouds, too, feedback phenomena occur; for example, warming induces formation of more clouds, which reflect more incoming energy.

It has been argued that the burning of hydrocarbon fuels with the highest hydrogen-to-carbon atomic ratio would minimize the carbon dioxide issue in that the minimum amount of carbon dioxide would be produced per unit of heat generated. This rationale seems to be true for the natural fuels but does not mitigate the need for cleaning gases before they are released to the atmosphere. And there are regulations to this effect (Majumdar, 1993). In the United States, the Clean Air Act of 1970 and the ensuing amendments (United States Congress, 1990; Stensvaag, 1991) to it, as well as the emissions standards for other countries (IEA Coal Research, 1991), provide the basis for the regulatory constraints imposed on air emissions. The specific regulated emissions are particulate matter, sulfur dioxide, photochemical oxidants, hydrocarbons, carbon monoxide, nitrogen oxides, and lead (Mintzer, 1990).

The acid-rain-forming sulfur dioxide may have a counteracting effect on greenhouse gases because sulfur dioxide is oxidized in the atmosphere to sulfuric acid, forming a light-reflecting haze. Furthermore, the sulfuric acid and resulting sulfates act as condensation nuclei that increase the extent, density, and brightness of this light-reflecting cloud cover.

4.3 Smog

Smog is the term used to denote a photochemically oxidizing atmosphere. The word originally was used to describe the unpleasant combination of *sm*oke and f*og* laced with sulfur dioxide, which was formerly prevalent in many of the heavily industrialized cities of Europe when high-sulfur coal was the primary fuel used in that city. Smog is the best known example of secondary pollutants formed by photochemical processes as a result of primary emissions of nitric oxide (NO) and reactive hydrocarbons from automobiles.

Smog is characterized by the presence of sulfur dioxide, a reducing compound, and therefore it is a reducing smog or sulfurous smog. In fact, readily oxidized sulfur dioxide has a short lifetime in an atmosphere where oxidizing smog is present.

A smoggy day is defined as one causing moderate to severe eye irritation or having visibility below 3 miles when the relative humidity is below 60%. The formation of oxidants in the air, particularly ozone, is indicative of smog formation. Serious levels of smog may be assumed to be present when the oxidant level exceeds 0.15 ppm for more than 1 hour.

A large number of specific reactions are involved in the overall scheme for the formation of smog (Figure 10.3). The formation of atomic oxygen by a primary photochemical reaction leads to several reactions involving oxygen and nitrogen oxide species.

Two major classes of inorganic products from smog are sulfates and nitrates. Inorganic sulfates and nitrates, along with sulfur and nitrogen oxides can contribute to acidic precipitation, corrosion, reduced visibility, and adverse health effects. Inorganic nitrates are formed by several reactions in smog. In fact, nitrates (and the resulting nitric acid) are among the more damaging end products of smog.

Aerosol particles that reduce visibility are formed by the polymerization of the smaller molecules produced in smog-forming reactions (Hidy et al., 1979). Since these reactions largely involve the oxidation of hydrocarbons, it is not surprising that oxygen-containing organic compounds make up the bulk of the particulate matter produced from smog. Among the specific kinds of compounds identified in organic smog aerosols are alcohols, aldehydes, ketones, organic acids, esters, and organic nitrates.

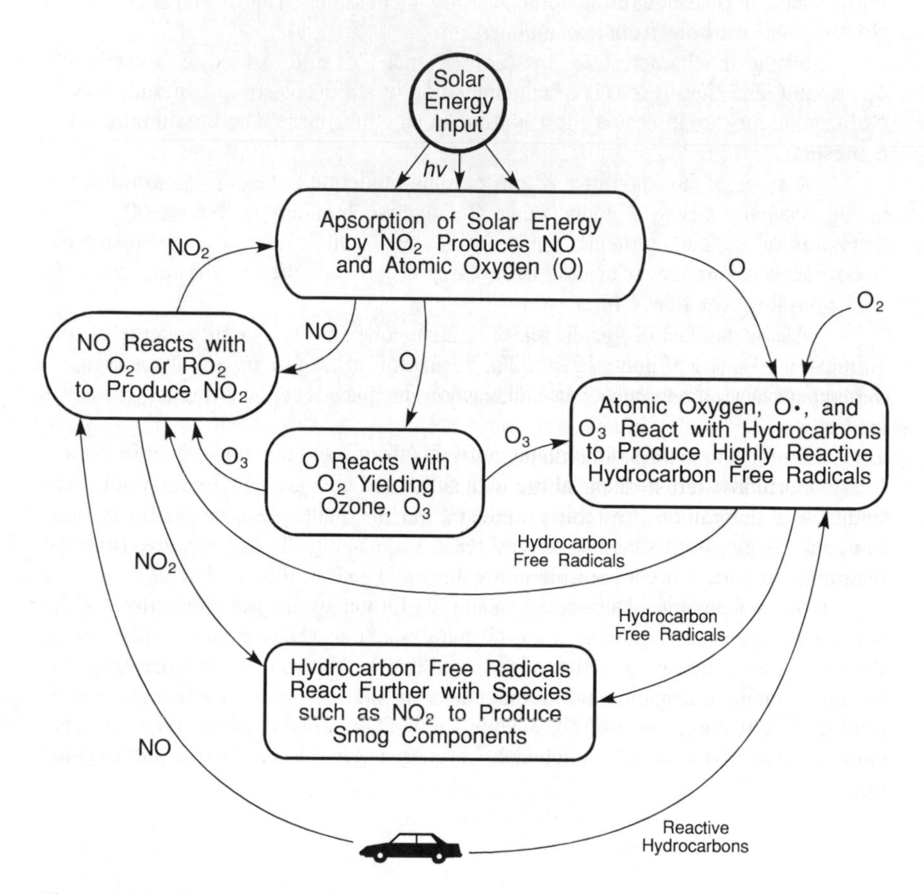

Figure 10.3: The events leading to smog formation

5.0 References

Alpert, S., and Gluckman, M.J. 1986. *Annual Reviews of Energy* 11: 315.

Aristoff, E., Rieve, R.W., and Shalit, H. 1981. Chap. 16 in *Chemistry of Coal Utilization*. Second Supplementary Volume. Edited by M.A. Elliott. John Wiley & Sons Inc., New York.

Austin, G.T. 1984. *Shreve's Chemical Process Industries*. McGraw-Hill, New York.

Bartoo, R.K. 1985. Chap. 13 in *Acid and Sour Gas Treating Processes*. Edited by S.A. Newman. Gulf Publishing Company, Houston, Texas.

Berkowitz, N. 1979. *Introduction to Coal Technology*. Academic Press Inc., New York.

Birks, J.W., Calvert, J.G., and Sievers, R.W. (eds.). 1993. *The Chemistry of the Atmosphere: Its Impact on Global Change*. American Chemical Society, Washington, D.C.

Bland, W.F., and Davidson, R.L. 1967. *Petroleum Processing Handbook*. McGraw-Hill, New York.

Bodle, W.W., and Huebler, J. 1981. Chap. 10 in *Coal Handbook*. Edited by R.A. Meyers. Marcel Dekker Inc., New York.

Charlson, R.J., and Wigley, T.M.L. 1994. *Scientific American* 270(2): 48.

Davidson, R. 1994. *Nitrogen in coal*. Report No. IEAPER/08. International Energy Agency Coal Research, London, England.

Dayton, D.C., and Milne, T.A. 1995. Preprints. *Division of Fuel Chemistry American Chemical Society* 40(3): 758.

Douglas, J. 1990. *EPRI Journal*. June, p.4. Electric Power Research Institute, Palo Alto, California.

Easterbrook, G. 1995. *A Moment on the Earth: The Coming Age of Environmental Optimism*. Viking Press, New York.

Elliott, M.A. (ed.). 1981. *Chemistry of Coal Utilization*. Second Supplementary Volume. John Wiley & Sons Inc., New York.

Energy Information Administration. 1993. *Emissions of Greenhouse Gases in the United States*. Report No. DOE/EIA-0573. Energy Information Administration. United States Department of Energy, Washington, D.C.

Fess, F.M. 1957. *History of Coke Oven Technology*. Essen, Gluckauf Verlag, Germany.

Frosch, R.A., and Gallopoulos, N.E. 1989. *Scientific American* 261(3): 144.

Funk, J.E. 1983. Chap. 3 in *Riegel's Handbook of Industrial Chemistry*. Edited by J.A. Kent. Van Nostrand Reinhold, New York.

Galloway, R.L. 1882. *A History of Coal Mining in Great Britain*. Macmillan, London, England.

Gary, J.H., and Handwerk, G.E. 1975. *Petroleum Refining: Technology and Economics*. Marcel Dekker Inc., New York.

Gilleland, D.S., and Swisher, J.H. (eds.). 1986. *Acid Rain Control II: The Promise of New Technology*. Southern Illinois University Press, Carbondale, Illinois.

Gould, R.R. 1984. *Annual Review of Energy* 9: 529.

Graedel, T.E., and Crutzen, P.J. 1989. *Scientific American* 261(3): 58.

Greenberg, M. 1993. *Hazmat World* 6(10): 43.
Hald, P. 1995. Preprints. *Division Fuel Chemistry American Chemical Society* 40(3): 753.
Hessley, R.K. 1990. In *Fuel science and technology handbook*. Edited by J.G. Speight. New York: Marcel Dekker.
Hessley, R.K., Reasoner, J.W., and Riley, J.T. 1986. *Coal Science*. John Wiley & Sons Inc., New York.
Hidy, G.M., Mueller, P.K., Grosjean, D., Appel, B.R., and Wesolowski, J.J. 1979. *The Character and Origins of Smog Aerosols*. John Wiley & Sons Inc., New York.
Hileman, B. 1989. *Chemical and Engineering News* March 13. p. 25.
Hoerning, J.M., Evans, M.A., Aerts, D.J., and Hagland, K.W. 1995. Preprints. *Division of Fuel Chemistry American Chemical Society* 40(3): 676.
IEA Coal Research. 1991. *Emissions Standard Data Base*. International Energy Agency Coal Research, London, England.
IEA Coal Research. 1992. *Greenhouse Gases Bulletin*. International Energy Agency Coal Research, London, England.
James, P., and Thorpe, N. 1994. *Ancient Inventions*. Ballantine Books, New York.
Keepin, B. 1986. *Annual Reviews of Energy* 11: 357.
Kellogg, W.W. 1987. *Climatic Change*. 10: 113.
Kohl, A.L., and Riesenfeld, F.C. 1979. *Gas Purification*. Gulf Publishing Company, Houston, Texas.
Kumar, S. 1987. *Gas Production Engineering*. Gulf Publishing Company, Houston, Texas.
Kyte, W.S. 1991. *Desulphurisation 2: Technologies and Strategies for Reducing Sulphur Emissions*. Institute of Chemical Engineers, Rugby, Warwickshire, England.
Lioy, P.J. 1993. Chap. 13 in *Measurement Challenges in Atmospheric Chemistry*. Edited by L. Newman. Advances in Chemistry Series No. 232. American Chemical Society, Washington, D.C.
Loehr, R.C. 1992. In *Petroleum Processing Handbook*. p. 190. Edited by J.J. McKetta. Marcel Dekker Inc., New York.
Longhurst, J.W.S., Raper, D.W., Lee, D.S., Heath, B.A., Conlan, B.,and King, H.J. 1993a. *Fuel* 72: 1261.
Longhurst, J.W.S., Raper, D.W., Lee, D.S., Heath, B.A., Conlan, B.,and King, H.J. 1993b. *Fuel* 72: 1363.
MacDonald, G.J. 1988. *Journal of Policy Analysis and Management* 7: 425.
Majumdar, S.B. 1993. *Regulatory Requirements for Hazardous Materials*. McGraw-Hill, New York.
Malin, C.B. 1990. *Oil and Gas Journal* 88(2): 20.
Meyers, R.A. 1981. Chap. 1 in *Coal Handbook* Edited by Robert A. Meyers. Marcel Dekker inc., New York.
Michaels, P.J. 1991. *Journal of Coal Quality* 10(1): 1.
Mintzer, I.M. 1990. *Annual Reviews of Energy* 15: 513.

Moilanen, A., and Kurkela, E. 1995. Preprints. *Division of Fuel Chemistry American Chemical Society*. 40(3): 688.

Newman, S.A. (ed.). 1985. *Acid and Sour Gas Treating Processes*. Gulf Publishing Company, Houston, Texas.

Nordhaus, W.D. 1991. In *Energy and Environment in the 21st Century*. p. 103. Edited by J.W. Tester, D.O. Wood, and N.A. Ferrari. MIT Press, Cambridge, Massachusetts.

Nordhaus, W.D. 1994. *American Scientist* 82(1): 45.

Olschewsky, D., and Megna, A. 1992. In *Petroleum Processing Handbook*. p. 179. Edited by J.J. McKetta. Marcel Dekker Inc., New York.

Pickering, K.T., and Owen, L.A. 1994. *Global Environmental Change*. Routledge Publishers, New York.

Probstein, R.F., and Hicks, R.E.. 1990. *Synthetic Fuels*. pH Press, Cambridge Massachusetts.

Rath, L.K., and Longanbach, J.R. 1991. *Energy Sources* 13: 443.

Schlesinger, M.E., and Mitchell, J.F.B. 1987. *Reviews of Geophysics* 25: 760.

Seglin, L., and Bresler, S.A.. 1981. Chap. 13 in *Chemistry of Coal Utilization*. Second Supplementary Volume. Edited by M.A. Elliott. John Wiley & Sons Inc., New York.

Skorupska, N.M. 1993. *Coal Specifications - Impact on Power Station Performance*. Report no. IEACR/52. International Energy Agency Coal Research, London, England.

Smith, I. 1988. *CO_2 and Climatic Change*. Report No. IEACR/07. International Energy Agency, London, England.

Smith, I., and Thambimuthu, K. 1991. *Greenhouse Gases, Abatement and Control: The Role of Coal*. Report No. IEACR/01. International Energy Agency Coal Research, London, England.

Smith, I.S., Nilsson, C., and Adams, D. 1994. *Grennhouse Gases - Perspectives on Coal*. Report No. IEAPER/12. International Energy Agency Coal Research, London, England.

Smoot, D.L., Boardman, R.D., Brewster, S.B., Hill, S.C., and Foli, A.K. 1993. *Energy & Fuels* 7: 786.

Soud, H., and Takeshita, M. 1994. *FGD Handbook*. Report no. IEACR/65. International Energy Agency Coal Research, London, England. .

Speight, J.G. 1981. *The Desulfurization of Heavy Oils and Residua*. Marcel Dekker Inc., New York.

Speight, J.G. (ed.). 1990. *Fuel Science and Technology Handbook*. Marcel Dekker Inc., New York.

Speight, J.G. 1991. *The Chemistry and Technology of Petroleum*. 2nd ed. Marcel Dekker Inc., New York.

Speight, J.G. 1993. *Gas Processing: Environmental Aspects and Methods*. Butterworth Heinemann, Oxford, England.

Speight, J.G. 1994. *The Chemistry and Technology of Coal*. 2nd ed. Marcel Dekker Inc., New York.

Stensvaag, J-M. 1991. *Clean Air Act Amendments: Law and Practice*. John Wiley & Sons Inc., New York.

Stobaugh, R., and Yergin, D. 1983. *Energy Future* and references cited therein.. Vintage Books/Random House, New York.

Takeshita, M., and Sloss, L. 1993. *N_2O Emissions from Coal Use*. Report No. IEAPER/06. International Energy Agency Coal Research, London, England.

United States Congress, 1990. Public Law 101-549. *An Act to Amend the Clean Air Act to Provide for Attainment and Maintenance of Health Protective National Ambient Air Quality Standards, and for Other Purposes*. Washington, D.C. November 15.

United States Department of Energy. 1995. *SO_2 Removal Using gas Suspension Absorption Technology*. Topical Report No. 4. Clean Coal Technology. United States Department of Energy, Washington, D.C

White, R.M. 1990. *Scientific American* 263(1): 36.

World Coal. 1995. *Markets Open for Western Coal*. June. p. 29.

Chapter 11

Control of Gaseous Emissions

1.0 Introduction

Emission control in the simplest sense is a means to protect the atmosphere. It is also the relationship of air quality to emission levels and the development of control standards. Finally, emission control is the availability of practical techniques to reduce emissions.

Gaseous emissions of most concern are the acid gases (mostly sulfur dioxide and hydrogen sulfide), volatile organic compounds, nitrogen oxides, hydrogen fluoride, and carbon monoxide (Argonne, 1990). Thus emission control must involve the use of a variety of physical and physical-chemical techniques.

Since the time when English kings recognized that the burning of coal could produce noxious fumes (Chapter 1), there have been attempts to mitigate the amounts of noxious gases entering the atmosphere (particularly sulfur dioxide). The most effective process modification in reducing emissions is the elimination of the source. When that is not possible, reduction of the emissions may be attempted.

Historically, the first method for removing sulfur dioxide from flue gases was used in London, England, during the 1930s. The method consisted of simple water scrubbing of the flue gas to absorb sulfur dioxide into solution (Plumley, 1971). Since then, various regulatory organizations in many countries have set standards for sulfur dioxide and other emissions, which must be met immediately or in the very near future (Chapters 2 and 12). A variety of techniques are involved (Table. 11.1).

Table 11.1: The chemistry of stack gas scrubbing systems.

Process	Chemical reactions
Lime slurry scrubbing	$Ca(OH)_2 + SO_2 = CaSO_3 + H_2O$
Limestone slurry scrubbing	$CaCO_3 + SO_2 = CaSO_3 + CO_2(g)$
Magnesium oxide scrubbing	$Mg(OH)_2(slurry) + SO_2 = MgSO_3 + 2H_2O$
Sodium base scrubbing	$Na_2SO_3 + H_2O + SO_2 = 2NaHSO_3$
	$2NaHSO_3 + heat = Na_2SO_3 + H_2O + SO_2$
Double alkali scrubbing	$2NaOH + SO_2 = Na_2SO_3 + 2H_2O$
	$Ca(OH)_2 + Na_2SO_3 = CaSO_3(s) + 2NaOH$

Gaseous emissions are characterized by chemical species identification, e.g., *inorganic gases* such as sulfur dioxide (SO_2), nitrogen oxides (NO_x), and carbon monoxide (CO) or *organic gases* such as chloroform ($CHCl_3$) and formaldehyde (HCHO). The rate of release or concentrating in the exhaust airstream (in parts per million or comparable units) along with the type of gaseous emission predetermines the applicable control technology.

On the other hand, particulate matter is classified as either *suspended particulate matter*, *total suspended particulate matter*, or simply *particulate matter*. For human health purposes, the fraction of particulate matter that has been shown to contribute to respiratory diseases is termed PM_{10} (i.e., particulate matter <10 microns). From a control standpoint, particulate matter can be characterized as follows: (1) by particle size distribution and (2) particulate matter concentration in the emission (mg/m^3).

Apart from the difference in type of emissions (i.e., gases and particulate matter), it is important to note that both mixing and phase transference between the gas and the solid phases can occur. Under such circumstances, the control technology should respond to the phase of the emission that is dictated by the requirements of the emissions reduction program. Most control systems are designed to be chemical-specific or phase-specific. As an example, a system for control of particulate matter will be designed to attain a specified efficiency for particulate matter removal.

Emissions are considered pollutants if they are being ejected into the atmosphere in greater than natural abundances or are not indigenous to the atmosphere. These pollutant emissions are characterized as criteria pollutants and hazardous air pollutants. *Criteria pollutants* are chemicals such as sulfur dioxide, carbon monoxide, nitrogen oxides, volatile organic compounds, and particulate matter. *Hazardous air pollutants* include, but are not limited to, many volatile organic compounds and particulate matter.

In terms of sulfur dioxide emissions, most control techniques involve alkaline scrubbing (Table 11.1). Both disposable (with the environmental regulations in mind) alkaline media such as magnesium carbonate ($MgCO_3$) and calcium carbonate ($CaCO_3$) and regenerable media such as the corresponding oxides (i.e., MgO and CaO) are in use.

Much attention has been given to coal pretreatment to reduce sulfur content, such as washing and solvent refining, although the use of low-sulfur coal for power generation is an extremely viable option (World Coal, 1995). Hydrogen sulfide from catalytic petroleum crackers (Chapter 2) is controlled by partial oxidation to sulfur (Claus method) (Speight, 1991). Tail gas alkaline scrubbers are used for a variety of acid gases (Speight, 1993).

Combustion modifications such as use of staged combustion, low excess air, and special burner and firebox designs have shown some success for control of carbon monoxide and nitrogen oxide(s), but efficiencies are usually on the order of 40-70%. Control techniques using ammonia injection (which reacts with nitric oxide to form nitrogen) with or without selective catalysts have received widespread attention (Wendt and Mereb, 1990; Sloss et al., 1992; Armor, 1994). In fact, catalytic technology has been refocused to address many of the enironmental issues such as,

Table 11.2: Use of catalysts to reduce emissions to the atmosphere

Event	Cause	Catalytic technology
Acid rain	NO_x, So_x emissions	hydrodesulfurization hydrodenitrogenation catalytic combustors
Ozone depletion	chlorofluorocarbons	demise of cholorfluorocarbon use; production of alternates
Smog	hydrocarbons in cities	NO_x removal catalytic combustors efficient fuel conversion
Odor	industrial processes	catalytic combustors catalytic processes

for example, the production of sulfur oxides and hydrogen sulfide in petroleum refineries and in other industries (Table 11.2) (van den Berg and de Jong, 1980; Speight, 1981, 1993). However, as with all emission control systems there are also issues that need to be taken into consideration (Table 11.3).

Volatile organic compounds originate from gasoline storage and transfer, chemical manufacturing, and solvent loss from surface coatings, printing inks, and materials compounding. Transfer losses of gasoline vapors are collected by return vapor systems where they are adsorbed, combusted, or condensed by refrigeration. Solvent losses can be controlled by solvent adsorption systems using activated carbon (Rook, 1994).

Table 11.3: Notes on the use of combustion systems

1. Simplicity of operation.
2. Capability of steam generation or heat recovery in other forms.
3. Capability for virtually complete destruction of organic contaminants.
4. Operating costs may be high depending upon fuel requirements.
5. Potential for catalyst poisoning (in the case of catalytic incineration).
6. Incomplete combustion could create potentially worse pollution problems.

2.0 Methods

2.1 Stationary Sources

Numerous methods exist for the reduction of emissions from stationary sources (Grano, 1995; Pereira and Amiridis, 1995). Strategies for reducing emissions have often concentrated on removing chemical pollutants from a stream of air after the pollutants are formed. Such control methods are referred to as *add-on control methods*, *flue gas control methods*, or *end-of-pipe control methods*. These methods entail the use of devices that remove or destroy pollutants after they are generated but before they are discharged to the atmosphere. Examples of such devices are wet scrubbers, carbon adsorption beds, incinerators, baghouse filters, and electrostatic precipitators.

An important economic consideration in planning an emission control program based on add-on pollution control devices is the recognition that control devices typically become progressively more expensive to install and operate as the required control efficiency increases. For example, the added cost of going from 90% control efficiency to 99% control efficiency could exceed the cost of achieving the initial 90% efficiency. This provides greater impetus for the use of pollution prevention techniques instead of pollution removal techniques (Noyes, 1993; Boubel et al., 1994). An additional drawback is that frequently a solid or liquid residue is created that must be disposed of in an environmentally protective manner. Thus, for stationary combustion emission sources, switching to cleaner fuels can be an effective emission control method when technically and economically feasible.

2.2 Mobile Sources

Because of the significant contribution of vehicular emissions to carbon monoxide, nitrogen oxides, and hydrocarbons in the atmosphere, it is convenient to consider vehicular or mobile sources apart from stationary sources. Mobile source emissions result from fuel combustion in engines and from evaporative fuel losses.

Although stationary emission sources may dominate air quality conditions in heavily industrialized areas or in remote areas around a single industrial source, urban air quality is greatly influenced by emissions from vehicles (Atkinson et al., 1991). Methods for reducing mobile source emissions consist of technological methods, such as the use of catalytic converters in automobiles (Heck and Farrauto, 1995), and transportation control methods.

Most controls on mobile sources are limited to motor vehicles. Evaporative emissions from fuel systems, crankcases, and exhaust are regulated by many countries. Carbon monoxide and hydrocarbons are controlled by (oxidation) catalysts and by special fuel-injection and combustion-chamber design. Nitrogen oxides are controlled by spark timing, exhaust gas recirculation, and (reduction) catalysts in the exhaust.

In addition to emission control devices such as catalytic converters and fuel vapor collectors, the use of reformulated fuels (as a means of controlling the emissions at the true sources) has recently received considerable attention (Kortum and Miller, 1994). Other potential emissions control concepts include the use of alternative fuels such as liquefied natural gas, the development of electric-powered vehicles, inspection and maintenance programs to identify and repair vehicles with excess emissions, and the development of mass transit systems.

The emphasis on the expanded use of reformulated gasoline warrants some comment here. A few years ago, the term *reformulated gasoline* did not exist. Reformulated gasoline is designed to mitigate smog production and to improve air quality through the reformulation of conventional gasoline by limiting the emission levels of certain chemical compounds such as benzene and other aromatic derivatives (Bobro et al., 1994). In addition, the vapor pressure (or volatility) of the gasoline is controlled or decreased (Erwin, 1994) and oxygenated compounds are added to improve combustion and other fuel properties.

The reformulated gasoline program in the United States, which contains both formula standards and performance standards, is a two-phase process. Phase I started on January 1, 1995 and mandates a maximum of 1% benzene, maximum of 25% total aromatic content, 2% minimum of oxygen, no metals, and detergent additives. Because reformulated gasoline can comply with regulations based on performance, many companies have unique additives and/or formulations to meet emission requirements. These additives are claimed to reduce tailpipe emissions, produce fewer engine deposits, and give the vehicle high performance and improved mileage.

In order to offer some predictability about the potential for air pollution to increase or to decrease, especially when new fuels are introduced, a variety of mathematical models have been developed and applied. There are various means by which models have been applied to air pollution (Zannetti, 1990), but for the present text it is essential to focus on two models in particular. These models were designed to predict emission performance from fuel composition without extensive direct testing, which is very costly and time consuming.

The simple model is available for use from 1995 to 1997. During the high-ozone season (summertime), the performance characteristics of the reformulated gasoline should reduce total car volatile organic compounds and total toxic emissions by 15% of 1990 baseline levels, or meet the equivalence of a formula fuel performance, whichever is more stringent. Baseline emission levels are from 1990 model year vehicles operated on a standard or baseline gasoline. The total volatile organic compounds from automobile use are broken down into evaporative emissions, which are made up of diurnal emissions (11%) and running loss (21%), exhaust emissions (64%), and refueling emissions (4%) (Kulakowski, 1994). The other constituents of the gasoline were given by a baseline fuel (derived from average survey data of the compositions of gasolines in the United States in 1989). This baseline fuel was defined in order to indicate a performance standard and not a formula composition standard. In order for a gasoline from a refinery or an importer to achieve equivalency certification, it gasoline must comply with the Clean Air Act Amendment requirements for emissions (Federal Register, 1994) or the composition requirements listed earlier. This includes meeting the requirements for detergent additives and lead content.

The simple model does not specify or limit any other gasoline properties, therefore allowing a wide variety of gasolines to comply with the regulations. For example, the formula states that oxygen content is to be at least 2% w/w, but it does not stipulate the form of the oxygen. This leaves the choice to the refiner, and as long as the fuel is blended to contain 2% w/w, the fuel complies. Also, the reformulated gasoline must produce no increase in nitrogen oxide emissions. Nitrogen oxides and hydrocarbons react in the lower atmosphere or troposphere to form ozone in the presence of sunlight (Chapter 5).

The complex model becomes mandatory in January 1998, with optional use prior to 1998. The complex model is a set of statistically derived equations that relate fuel properties to vehicle emissions. Thus the complex model relates oxygen, sulfur, Reid vapor pressure, the 93-149°C (200-300°F) distillation fraction of the target fuel in terms of volume percent (E200 and E300, respectively), aromatics, olefins, and benzene. The model can be divided into exhaust and nonexhaust portions, the nonexhaust referring to evaporative effects. The nonexhaust volatile organic compounds were derived directly from the simple model approach, and the nonexhaust benzene was modeled as a weight fraction of nonexhaust volatile organic compounds.

Finally, there are transportation control methods for reducing emissions to the atmosphere. These methods include use of high-occupancy vehicles (car pool) lanes on freeways, staggered work hours, and reversible lanes into and out of high vehicular areas. In some locations, more extreme measures have been implemented, such as restricting operation of vehicles by license tag number during episodes of high pollution levels.

3.0 Emission Control Technologies

Most human activities produce airborne emissions of gases or particulate matter. Therefore the technologies to control (by reducing) the emissions are essential to any industrial operation (Licht, 1988; Thambimuthu, 1993).

The technologies available for the control of industrial emission can be categorized as *integrated technologies* and *add-on* (or *end-of-pipe*) *technologies* (Figure 11.1). The former usually involve the control of emissions at the sources while the latter seek to control the emissions after their formation.

3.1 Integrated Technologies

With the evolution of technology and a better understanding of emission formation mechanisms (e.g., combustion elements such as burner and furnace configuration and design, flame temperature, and kinetic parameters) it is possible for many facilities to retard the formation of emissions in situ by means of oxidation and other chemical reactions (United States Department of Energy, 1993; Tullin et al., 1993; Ozkan et al., 1995).

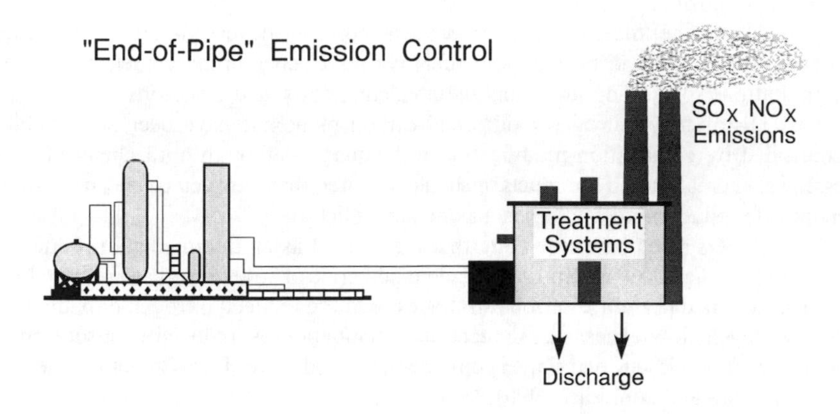

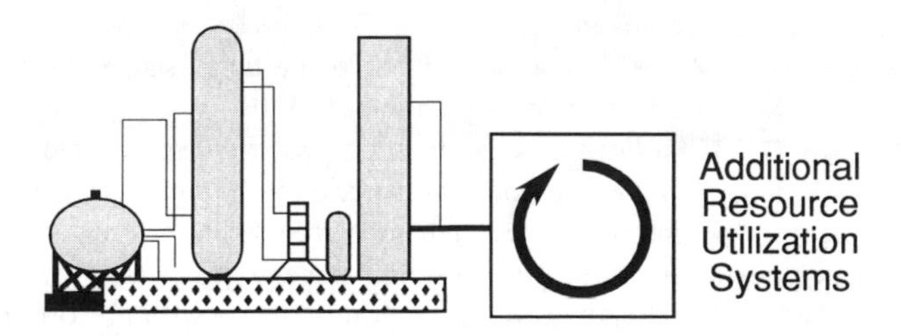

Figure 11.1: Representation of end-of-the-pipe and integrated emission control systems

However, integrated or in-process emissions control is severely hampered by its dependence on commercially available and proven technology and the fact that it is not readily amenable to retrofit situations. Further, the technology can be very process-specific and even plant-specific.

Most of the common fossil fuels fired for steam and/or power generation include coal, natural gas, and fuel oils. Other fuels may include bark, wood waste, sludge, pulp liquor, and fuels derived from refuse (Table 11.4). This use of various fuels for power generation excludes incinerators which burn a wide variety of waste materials and are mostly controlled through fuel preparation and postcombustion emission control.

Thus, the most rational in-process control, if feasible, involves fuel substitution, use of dirtier topping fuels with a cleaner primary fuel, and other permutations so as to optimize fuel requirements, costs, and emissions.

Historically, nitrogen oxides and carbon monoxide have been successfully controlled by combustion modification techniques. Although both chemicals are essentially combustion by-products, it should be noted that nitrogen oxide and carbon monoxide reduction technologies may be in conflict.

Levels of carbon monoxide that are formed as an intermediate product of combustion increase with most nitrogen oxide control strategies that are based on combustion modification. Various strategies used are reduced nitrogen dioxide and oxygen levels, low excess air, reduced peak temperatures, reduced exposure time, optimum burner design, and staged combustion (United States Department of Energy, 1993; Pereira and Amiridis, 1995).

Table 11.4: Types of refuse-derived fuels

Classification	Description
RDF-1	Wastes used as a fuel in as-discarded form; bulky wastes removed
RDF-2	Wastes processed to coarse particles size with or without ferrous metal separation
RDF-3	Combustible waste fraction processed to particle sizes, 95% weight passing 2-in.-square (5-cm-square) mesh screening
RDF-4	Combustible waste fraction processed into powder form, 95% weight passing 10-mesh screening
RDF-5	Combustible waste fraction (compressed into form of pellets or briquettes
RDF-6	Combustible waste fraction processed to liquid fuels
RDF-7	Combustible waste fraction processed to gaseous fuel

Various process reactions that are employed for the purpose of integrated control include feed substitution and/or reformulation, lower reaction temperatures, use of passive catalysts that interfere with pollutant formation, and solvent recovery and reuse.

Finally, the injection of lime or limestone into combustors with the feedstock (coal) (Figure 11.2) is another means of controlling emissions at the source (Seitzinger and Morrison, 1993).

3.2 Add-on Technologies

Add-on technologies are, as the name implies, a technological method (process) for controlling emissions from an emission-producing process. Emission control or gas processing is, generally, very simple in terms of chemical and/or physical principles and employs different process types (Speight, 1993). The processes (Table 11.5) that have been developed to accomplish emission control vary from a simple once-through wash operations to complex multistep systems with options for recycling the gases. In many cases, the process complexities arise because of the need for recovery of the materials used to remove the contaminants or even recovery of the contaminants in the original, or altered, form.

The selection of a particular process type (Speight, 1993) for an emission control operation is not a simple choice. Many factors have to be considered, not the least of which is the constitution of the gas stream that requires treatment. Indeed, process selectivity indicates the preference with which the process will remove one acid gas component relative to (or in preference to) another. For example, some processes remove both hydrogen sulfide and carbon dioxide while others are designed remove hydrogen sulfide only (Speight, 1993).

Table 11.5: Summary of gas cleaning processes.

Sorbent	Nature of interaction
Liquid	Absorption + chemical reaction
Liquid + solid	Adsorption + chemical reaction
Liquid	Physical adsorption
Solid	Physical adsorption
Solid	Chemical reaction

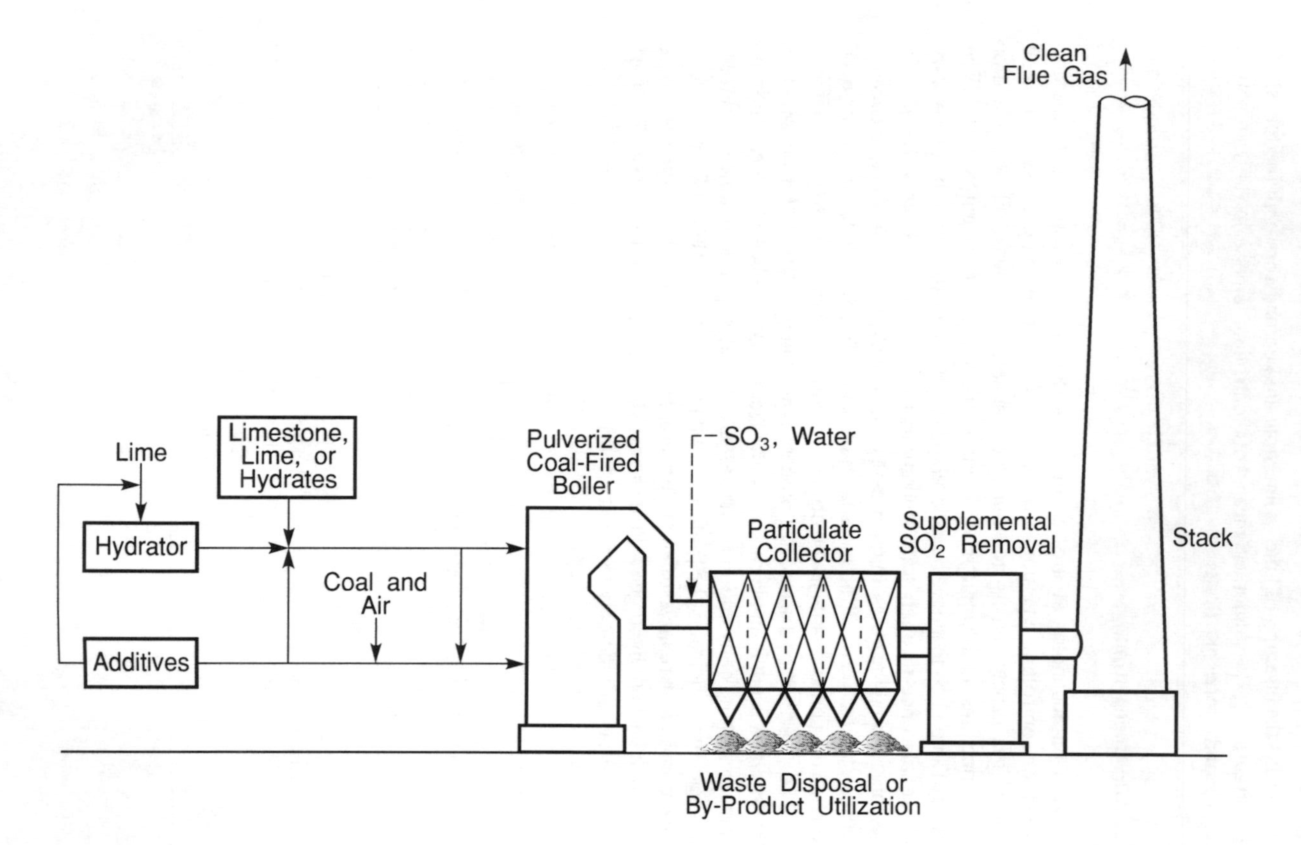

Figure 11.2: An integrated lime injection system

There are many variables in emission control, and the precise area of application of a given process is difficult to define, although there are several factors that need to be considered: (1) the types and concentrations of contaminants in the gas; (2) the degree of contaminant removal; (3) the selectivity of acid gas removal; (4) the temperature, pressure, volume, and composition of the emission stream; (5) the carbon dioxide to hydrogen sulfide ratio in the gas; and (6) the desirability of sulfur recovery due to process economics or environmental issues.

A variety of processes are commercially available for removal of acid gas from emission streams (Probstein and Hicks, 1990) and the processes generally fall into one of several categories. But, moreover, several factors control the choice of an acid gas removal process, namely gas flow rates, concentration of acid gases in the gas stream, and the necessity to remove carbon dioxide as well as hydrogen sulfide.

Emission control by some form of absorption by a liquid or adsorption by a solid is one of the most widely applied operations in the chemical and process industries (Speight, 1993). Some options have the potential for regeneration of the adsorbent, but in a few cases, the process is applied in a nonregenerative manner. The interaction between the chemical(s) to be adsorbed and the adsorbent may either be physical in nature or consist of physical sorption followed by chemical reaction. Other gas stream treatments use the principle of chemical conversion of the contaminants with the production of benign (noncontaminant) products to substances that can be removed much more readily than the impurities from which they are derived.

There are four general processes used for emission control (often referred to in another, more specific context as flue gas desulfurization: (1) adsorption; (2) absorption; (3) catalytic oxidation; and (4) thermal oxidation (Soud and Takeshita, 1994).

Adsorption is a physical-chemical phenomenon in which the gas is concentrated on the surface of a solid or liquid to remove impurities (Mantell, 1951). Usually, carbon is the adsorbing medium (Rook, 1994), which can be regenerated upon desorption (Fulker, 1972; Speight, 1993). The quantity of material adsorbed is proportional to the surface area of the solid and, consequently, adsorbents are usually granular solids with a large surface area per unit mass. Subsequently, the captured gas can be desorbed with hot air or steam either for recovery or for thermal destruction.

Adsorbers are widely used to increase a low gas concentration prior to incineration unless the gas concentration is very high in the inlet airstream. Adsorption also is employed to reduce problem odors from gases. There are several limitations to the use of adsorption systems, but it is generally felt that the major one is the requirement for minimization of particulate matter and/or condensation of liquids (e.g., water vapor) that could mask the adsorption surface and drastically reduce its efficiency (Table 11.6).

Absorption differs from adsorption, in that it is not a physical-chemical surface phenomenon, but an approach in which the absorbed gas is ultimately distributed throughout the absorbent (liquid). The process depends only on physical solubility and may include chemical reactions in the liquid phase (chemisorption). Common absorbing media used are water, aqueous amine solutions, caustic, sodium

Table 11.6: Notes on the use of adsorption systems

1.	Possibility of product recovery.
2.	Excellent control and response to process changes.
3.	No chemical-disposal problem when pollutant (product) recovered and returned to process.
4.	Capability to remove gaseous or vapor contaminants from process streams to extremely low levels.
5.	Product recovery may require distillation or extraction.
6.	Adsorbent may deteriorate as the number of cycles increase.
7.	Adsorbent regeneration.
8.	Prefiltering of gas stream possibly required to remove any particulate materials capable of plugging the adsorbent bed.

carbonate, and nonvolatile hydrocarbon oils, depending on the type of gas to be absorbed. Usually, the gas-liquid contactor designs which are employed are plate columns or packed beds.

Absorption is achieved by dissolution (a physical phenomenon) or by reaction (a chemical phenomenon) (Barbouteau and Dalaud, 1972; Ward, 1972). Chemical adsorption processes adsorb sulfur dioxide onto a carbon surface where it is oxidized (by oxygen in the flue gas) and absorbs moisture to give sulfuric acid impregnated into and on the adsorbent.

As currently practiced, acid gas removal processes (Figure 11.3) involve the selective absorption of the contaminants into a liquid, which is passed countercurrent to the gas. Then the absorbent is stripped of the gas components (regeneration) and recycled to the absorber. The process design will vary and, in practice, may employ multiple absorption columns and multiple regeneration columns (Table 11.7).

Table 11.7: Notes on the use of packed column and plate column absorption systems

1.	Relatively low pressure drop.
2.	Capable of achieving relatively high mass-transfer efficiencies.
3.	Increasing the height and/or type of packing or number of plates capable of improving mass transfer without purchasing a new piece of equipment.
4.	Ability to collect particulate materials as well as gases.
5.	May require water (or liquid) disposal.

Liquid absorption processes (which usually employ temperatures below 50°C (120°F) are classified either as *physical solvent processes* or *chemical solvent processes*. The former employ an organic solvent, and absorption is enhanced by low temperatures, or high pressure, or both. Regeneration of the solvent is often accomplished readily (Staton et al., 1985). In chemical solvent processes, absorption of the acid gases is achieved mainly by use of alkaline solutions such as amines or carbonates (Kohl and Riesenfeld, 1979). Regeneration (desorption) can be brought about by use of reduced pressures and/or high temperatures, whereby the acid gases are stripped from the solvent.

Solvents used for emission control processes should have a high capacity for acid gas, low tendencies to dissolve valuable components such as hydrogen and low-molecular weight hydrocarbons, low vapor pressure at operating temperatures to minimize solvent losses, low viscosity, thermal stability, absence of reactivity toward gas components, low tendencies toward fouling of equipment and corrosion, and be economically acceptable (Speight, 1991, 1994).

Amine washing (Figure 11.3) of gas emissions involves chemical reaction of the amine with any acid gases with the liberation of an appreciable amount of heat and it is necessary to compensate for the absorption of heat. Amine derivatives such as ethanolamine (monoethanolamine, MEA), diethanolamine (DEA), triethanolamine (TEA), methyldiethanolamine (MDEA), diisopropanolamine (DIPA), and diglycolamine (DGA) have been used in commercial applications (Kohl and Riesenfeld, 1979; Maddox et al., 1985; Polasek and Bullin, 1985; Jou et al., 1985; Pitsinigos and Lygeros, 1989; Speight, 1993).

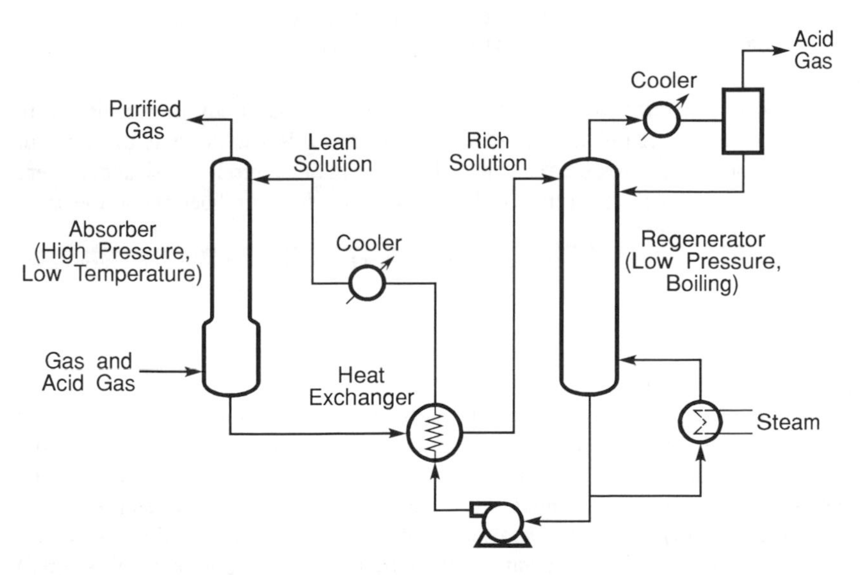

Figure 11.3: Gas cleaning by absorption

The chemistry can be represented by simple equations for low partial pressures of the acid gases:

$$2RNH_2 + H_2S = (RNH_3)_2S$$
$$2RHN_2 + CO_2 + H_2O = (RNH_3)_2CO_3$$

At high acid gas partial pressures, the reactions will lead to the formation of other products:

$$(RNH_3)_2S + H_2S = 2RNH_3HS$$
$$(RNH_3)_2CO_3 + H_2O = 2RNH_3HCO_3$$

The reaction is extremely fast, the absorption of hydrogen sulfide being limited only by mass transfer; this is not so for carbon dioxide.

Regeneration of the solution leads to near complete desorption of carbon dioxide and hydrogen sulfide. A comparison between monoethanolamine, diethanolamine, and diisopropanolamine shows that monoethanolamine is the cheapest of the three but shows the highest heat of reaction and corrosion; the reverse is true for diisopropanolamine.

Carbonate washing is a mild alkali process for emission control by the removal of acid gases (such as carbon dioxide and hydrogen sulfide) from gas streams (Speight, 1993) and uses the principle that the rate of absorption of carbon dioxide by potassium carbonate increases with temperature. It has been demonstrated that the process works best near the temperature of reversibility of the reactions:

$$K_2CO_3 + CO_2 + H_2O = 2KHCO_3$$
$$K_2CO_3 + H_2S = KHS + KHCO_3$$

Water washing, in terms of the outcome, is analogous to washing with potassium carbonate (Kohl and Riesenfeld, 1979), and it is also possible to carry out the desorption step by pressure reduction. The absorption is purely physical and there is also a relatively high absorption of hydrocarbons, which are liberated at the same time as the acid gases.

In *chemical conversion processes*, contaminants in gas emissions are converted to compounds that are not objectionable or that can be removed from the stream with greater ease than the original constituents. For example, a number of processes have been developed that remove hydrogen sulfide and sulfur dioxide from gas streams by absorption in an alkaline solution.

Catalytic oxidation is a chemical conversion process that is used predominantly for destruction of volatile organic compounds and carbon monoxide. These systems operate in a temperature regime of 205-595°C (400-1100°F) in the presence of a catalyst. Without the catalyst, the system would require higher temperatures. Typically, the catalysts used are a combination of noble metals deposited on a ceramic base in a variety of configurations (e.g., honeycomb-shaped) to enhance good surface contact.

Catalytic systems are usually classified on the basis of bed types such as *fixed bed* (or *packed bed*) and *fluid bed* (*fluidized bed*). These systems generally have very high destruction efficiencies for most volatile organic compounds, resulting in the formation of carbon dioxide, water, and varying amounts of hydrogen chloride (from halogenated hydrocarbons). The presence in emissions of chemicals such as heavy metals, phosphorus, sulfur, chlorine, and most halogens in the incoming airstream act as poison to the system and can foul up the catalyst.

Thermal oxidation systems, without the use of catalysts, also involve chemical conversion (more correctly, chemical destruction) and operate at temperatures in excess of 815°C (1500°F), or 220-610°C (395-1100°F) higher than catalytic systems.

Historically, particulate matter control (often referred to as dust control) (Mody and Jakhete, 1988) has been one of the primary concerns of industries, since the emission of particulate matter is readily observed through the deposition of fly ash and soot as well as in impairment of visibility. Differing ranges of control can be achieved by use of various types of equipment. Upon proper characterization of the particulate matter emitted by a specific process, the appropriate piece of equipment can be selected, sized, installed, and performance tested. The general classes of control devices for particulate matter are as follows:

1. Cyclones

Other than settling chambers for large particles, cyclone-type collectors are the most common of the inertial collector class. Cyclones are effective in removing coarser fractions of particulate matter. The particle-laden gas stream enters an upper cylindrical section tangentially and proceeds downward through a conical section. Particles migrate by centrifugal force generated by providing a path for the carrier gas to be subjected to a vortex-like spin. The particles are forced to the wall and are removed through a seal at the apex of the inverted cone. A reverse-direction vortex moves upward through the cyclone and discharges through a top center opening. Cyclones are often used as primary collectors because of their relatively low efficiency (50-90% is usual). Some small-diameter high-efficiency cyclones are utilized.

The equipment can be arranged either in parallel or in series to both increase efficiency and decrease pressure drop (Figure 11.4). However, there are disadvantages that must be recognized (Table 11.8).

Table 11.8: Notes on the use of cyclone collectors

1. Relatively low operating pressure drops (for degree of particulate removal obtained) in the range of approximately 2-inch to 6-inch water column.
2. Dry collection and disposal
3. Relatively low overall particulate collection efficiencies, especially on particulates below 10 mm in size.
4. Usually unable to process semisolid (tacky) materials

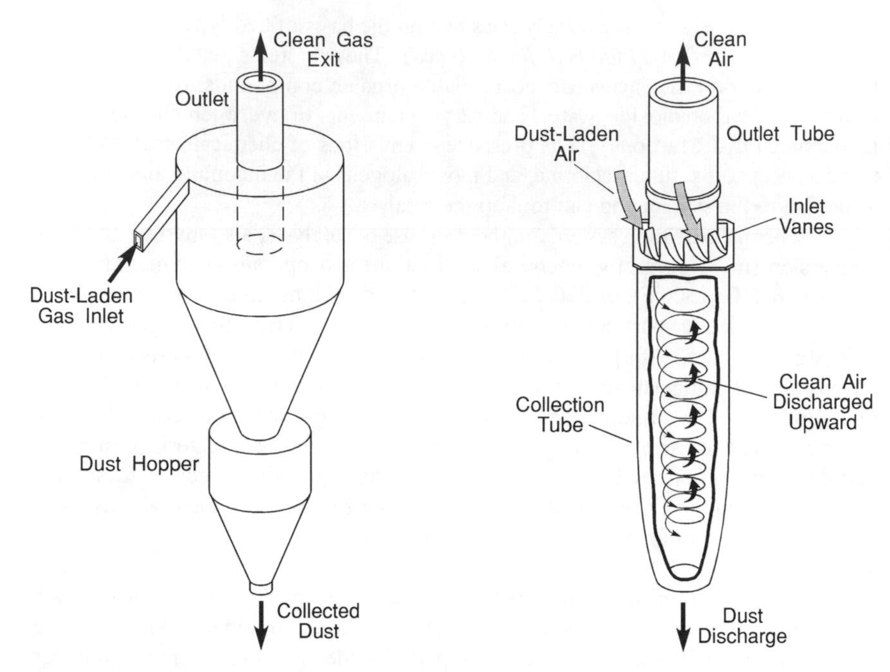

Figure 11.4: Cyclone systems for removal of particulate matter

These units for particulate matter operate by contacting the particles in the gas stream with a liquid. In principle the particles are incorporated in a liquid bath or in liquid particles which are much larger and therefore more easily collected.

2. Fabric filters

Fabric filters are typically designed with nondisposable filter bags. As the dusty emissions flow through the filter media (typically cotton, polypropylene, teflon, or fiberglass), particulate matter is collected on the bag surface as a dust cake. Fabric filters are generally classified on the basis of the filter bag cleaning mechanism employed (Figure 11.5).

Fabric filters operate with collection efficiencies up to 99.9% although other advantages are evident. There are several issues that arise during use of such equipment (Table 11.9).

3. Wet scrubbers

Wet scrubbers are devices in which a counter-current spray liquid is used to remove particles from an airstream. Device configurations include plate scrubbers, packed beds, orifice scrubbers, venturi scrubbers, and spray towers, individually or in various combinations. Wet scrubbers can achieve high collection efficiencies at the expense of prohibitive pressure drops (Table 11.10).

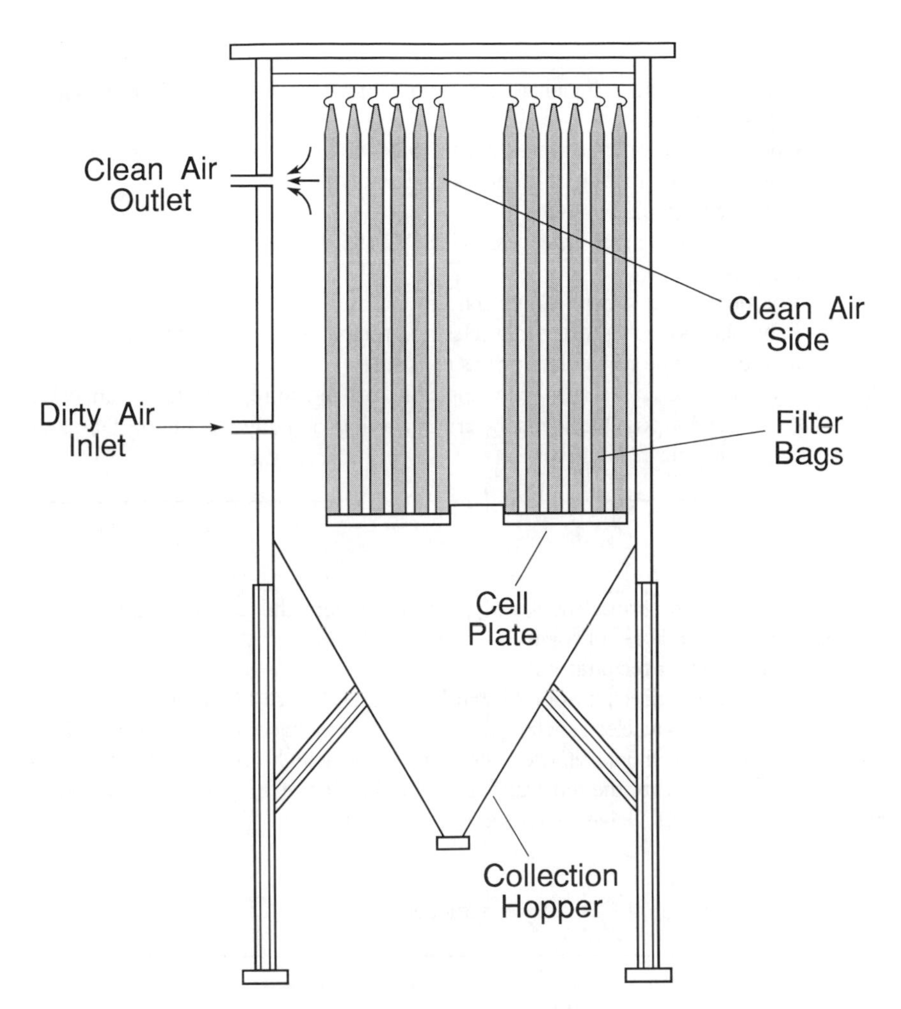

Figure 11.5: A filter (baghouse) system for removal of particulate matter

Other techniques include high-energy input venturi scrubbers (Figure 11.6), electrostatic scrubbers where particles or water droplets are charged, and flux force/condensation scrubbers where a hot humid gas is contacted with cooled liquid or where steam is injected into saturated gas. In the latter scrubber the movement of water vapor toward the cold water surface carries the particles with it (diffusiophoresis), while the condensation of water vapor on the particles causes the particle size to increase, thus facilitating collection of fine particles.

The foam scrubber is a modification of a wet scrubber in which the particle-

Table 11.9: Notes on the use of fabric filter systems

---

1. High collection efficiency on both coarse and fine (submicrometer) particulates.
2. Collected material recovered dry for subsequent processing of disposal.
3. Corrosion and rusting of components usually not major issues.
4. Relatively simple operation.
5. Temperatures much in excess of 288°C (550°F) require special refractory materials.
6. Potential for dust explosion hazard.
7. Fabric life possibly shortened at elevated temperatures and in the presence of acid or alkaline particulate or gas constituents.
8. Hygroscopic materials, condensation of moisture, or tarry adhesive components possible, causing crusty caking or plugging of the fabric or requiring special additives.

---

laden gas is passed through a foam generator, where the gas and particles are enclosed by small bubbles of foam.

4. Electrostatic precipitators.

Electrostatic precipitators (Table 11.11 and Figure 11.7) operate on the principle of imparting an electric charge to particles in the incoming airstream, which are then collected on an oppositely charged plate across a high-voltage field. Particles of high resistivity create the most difficulty in collection. Conditioning agents such as sulfur trioxide (SO_3) have been used to lower resistivity.

Table 11.10: Notes on the use of wet scrubbers

---

1. No secondary dust sources.
2. Ability to collect gases as well as particulates (especially "sticky" ones).
3. Ability to handle high-temperature, high-humidity gas streams.
4. Ability to achieve high collection efficiencies on fine particulates (however, at the expense of pressure drop).
5. May be necessary for water disposal.
6. Corrosion problems more severe than with dry systems.
7. Potential for solids buildup at the wet-dry interface.

Table 11.11: Notes on the use of electrostatic precipitators

1.	High particulate (coarse and fine) collection efficiencies.
2.	Dry collection and disposal.
3.	Operation under high pressure (to 150 lb/in^2) or vacuum conditions.
4.	Operation at high temperatures (up to 704°C, 1300°F) when necessary
5.	Relatively large gas flow rates capable of effective handling.
6.	Can be sensitive to fluctuations in gas-stream conditions.
7.	Explosion hazard when treating combustible gases/particulates.
8.	Ozone produced during gas ionization.

Important parameters include design of electrodes, spacing of collection plates, minimization of air channeling, and collection-electrode rapping techniques (used to dislodge particles). Techniques under study include the use of high-voltage pulse energy to enhance particle charging, electron-beam ionization, and wide plate spacing. Electrical precipitators are capable of efficiencies >99% under optimum conditions, but performance is still difficult to predict in new situations.

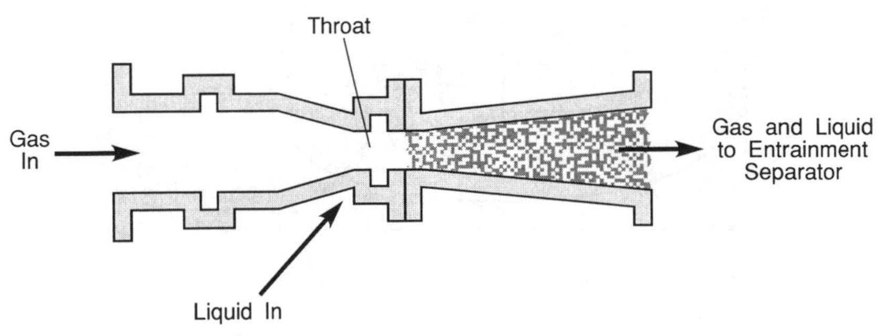

Figure 11.6: A venturi system

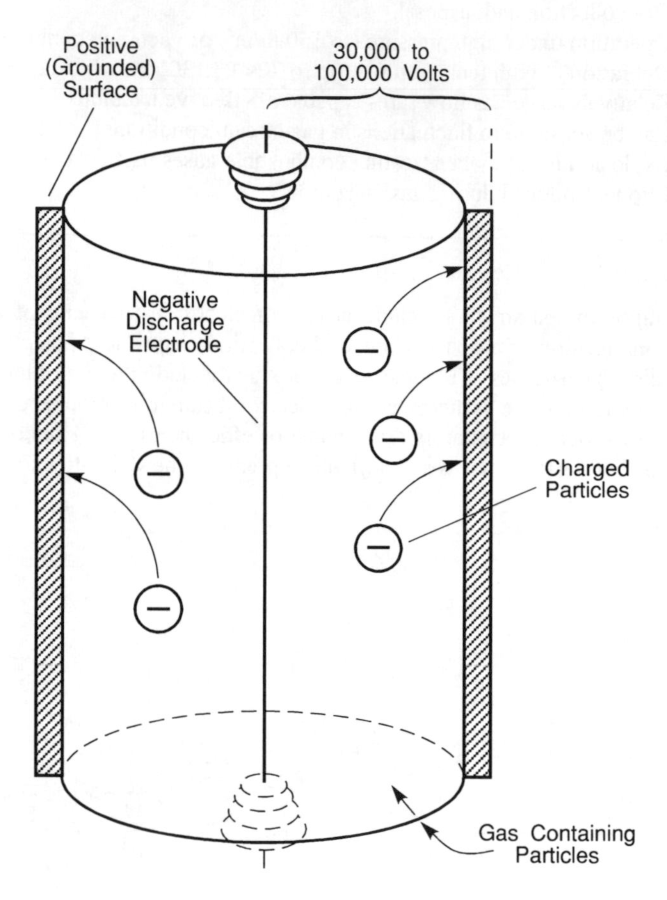

Figure 11.7: An electrostatic precipitator

4.0 References

Argonne. 1990. *Environmental Consequences of, and Control for, Energy Technologies*. Compiled by Argonne National Laboratory. Noyes Data Corp., Park Ridge, New Jersey.

Armor, J.N. (ed.).1994. *Environmental Catalysis*. Symposium Series No. 552. American Chemical Society, Washington, D.C.

Atkinson, D., Cristofaro, A., and Kolb, J. 1991. In *Energy and the Environment in the 21st Century*. p. 179. Edited by J.W. Tester, D.O. Wood, and N.A. Ferrari. MIT Press, Cambridge, Massachusetts.

Barbouteau, L., and Dalaud, R. 1972. Chapter 7 in *Gas Purification Processes for Air Pollution Control*. Edited by G. Nonhebel. Butterworth and Co., London.

Bobro, C.H., Karas, L.J., Leaseburge, C.D., and Skahan, D.J. 1994. Preprints. *American Chemical Society Division of Fuel Chemistry* 39(2): 305.

Boubel, R.W., Fox, D.L., Turner, D.B., and Stern, A.C. 1994. *Fundamentals of Air Pollution*. 3rd ed. Academic Press Inc., San Diego, California.

Erwin, J. 1994. Preprints. *American Chemical Society Division of Fuel Chemistry* 39(2): 310.

Federal Register. 1994. *Regulation of Fuels and Fuel Additives: Standards for Reformulated and Conventional Gasoline* 59(32): 7716.

Fulker, R.D. 1972. Chap. 9 in *Gas Purification Processes for Air Pollution Control*. Edited by G. Nonhebel. Butterworth and Co., London, England.

Grano, D. 1995. Chap. 2 in *Reduction of Nitrogen Oxide Emissions*. Edited by U.S. Ozkan, S.K. Agarwal, and G. Marcelin. Symposium Series No. 587. American Chemical Society, Washington, D.C.

Heck, R.M., and Farrauto, R.I. 1995. *Catalytic Air Pollution Control*. Van Nostrand Reinhold, New York.

Jou, F.Y., Otto, F.D., and Mather, A.E. 1985. Chap. 10 in *Acid and Sour Gas Treating Processes*. Edited by S.A. Newman. Gulf Publishing Company, Houston, Texas.

Kohl, A.L., and Riesenfeld, F.C. 1979. *Gas Purification*. 3rd ed. Gulf Publishing Company, Houston, Texas.

Kulakowski, M. 1994. Preprints. *American Chemical Society Division of Petroleum Chemistry* 39(4): 494.

Licht, W. 1988. *Air Pollution Control Engineering*. 2nd ed. Marcel Dekker Inc., New York.

Livo, K.B., Gallagher, J., and Boyd, M.W. 1994. Preprints. *American Chemical Society Division of Fuel Chem*istry 39(2): 282.

Maddox, R.N., Bhairi, A., Mains, G.J., and Shariat, A. 1985. Chap. 8 in *Acid and Sour Gas Treating Processes*. Edited by S.A. Newman. Gulf Publishing Company, Houston, Texas.

Mantell, C.L. 1951. *Adsorption*. McGraw-Hill, New York.

Mody, V., and Jakhete, R. 1988. *Dust Control Handbook*. Noyes Data Corp., Park Ridge, New Jersey, USA.

Noyes, R. (ed.). 1993. *Pollution Prevention Technology Handbook*. Noyes Data Corp., Park Ridge, New Jersey, USA.

Ozkan, U.S., Agarwal, S.K., and Marcelin, G. (eds.). 1995. *Reduction of Nitrogen Oxide Emissions*. Symposium Series No. 587. American Chemical Society, Washington, D.C.

Pereira, C.J., and Amiridis, M.D. 1995. Chap. 1 in *Reduction of Nitrogen Oxide Emissions*. Edited by U.S. Ozkan, S.K. Agarwal, and G. Marcelin. Symposium Series No. 587. American Chemical Society, Washington, D.C.

Pitsinigos, V.D., and Lygeros, A.I. 1989. *Hydrocarbon Processing* 58(4): 43.

Plumley, A.L. 1971. *Combustion* p. 36 (October).

Polasek, J., and Bullin, J. 1985. Chap. 7 in *Acid and Sour Gas Treating Processes*. Edited by S.A. Newman. Gulf Publishing Company, Houston, Texas.

Probstein, R.F., and Hicks, R.E. 1990. *Synthetic Fuels*. pH Press, Cambridge, Massachusetts.

Rook, R. 1994. *Chemical Processing* 57(11): 53.

Seitzinger, D.L., and Morrison, J.L 1993. Proceedings. *American Power Conference* 55: 806.

Sloss, L.L., Hjalmarsson, A-K., Soud, H.N., Campbell, L.M., Stone, D.K., Shareef, G.S., Livengood, C.D., Markussen, J.K., Emmel, T., and Maibodi, M. *Nitrogen Oxides Control Technology Fact Book*. Noyes Data Corp., Park Ridge, New Jersey, USA.

Soud, H., and Takeshita, M. 1994. *FGD Handbook*. Report No. IEACR/65. International Energy Agency Coal Research, London, England.

Speight, J.G. 1981. *The Desulfurization of Heavy Oils and Residua*. Marcel Dekker Inc., New York.

Speight, J.G. 1991. *The Chemistry and Technology of Petroleum*. 2nd ed. Marcel Dekker Inc., New York.

Speight, J.G. 1993. *Gas Processing: Environmental Aspects and Methods*. Butterworth Heinemann, Oxford, England.

Speight, J.G. 1994. *The Chemistry and Technology of Coal*. 2nd ed. Marcel Dekker Inc., New York.

Staton, J.S., Rousseau, R.W., and Ferrell, J.K. 1985. Chap. 5 in *Acid and Sour Gas Treating Processes*. Edited by S.A. Newman. Gulf Publishing Company, Houston, Texas.

Thambimuthu, K.V. 1993. *Gas Cleaning for Advanced Coal-based Power Generation*. Report No. IEACR/53. International Energy Agency Coal Research, London, England.

Tullin, C.J., Goel, S., Morihara, A., Sarofim, A.F., and Beer, J.M. 1993. *Energy & Fuels* 7: 796.

United States Department of Energy. 1993. *Reduction of NO_x and SO_x using Gas Reburning, Sorbent Injection, and Integrated Technologies*. Topical Report No. 3. Clean Coal Technology. United States Department of Energy, Washington, D.C.

van den Berg, P.J., and de Jong, W.A. 1980. Chap. VI in Introduction to Chemical Process Technology. Delft University Press, Delft, Netherlands.

Ward, E.R. 1972. Chap. 8 in *Gas Purification Processes for Air Pollution Control*. Edited by G. Nonhebel. Butterworth and Co., London.

Wendt, J.O.L., and Mereb, J.B. 1990. *Nitrogen Oxide Abatement by Distributed Fuel Addition*. Report No. DE92005212. Office of Scientific and Technical Information, Oak Ridge, Tennessee.

Winnick, J. *High-temperature Membranes for H_2S and SO_2 Separations*. Report No. DE92003115. Office of Scientific and Technical Information, Oak Ridge, Tennessee.

World Coal. 1995. *Markets Open for Western Coal*. June. p. 29.

Zannetti, P. 1990. *Air Pollution Modeling*. Van Nostrand Reinhold, New York.

Part IV Regulations and the Future

Chapter 12

Environmental Regulations

1.0 Introduction

With the advent of the industrial revolution, there was been a proliferation of waste materials of virtually every kind and complexity. Unfortunately, a significant portion of such waste is composed of toxic substances that are generally considered to be hazardous to various life forms either immediately or as a result of the formation of secondary pollutants (Zakrzewski, 1991). The *Hazardous Substances Act* defines a hazardous substance as:

> ...any substance or mixture which may be toxic, corrosive, an irritant, a strong sensitizer, flammable or combustible, or generating pressure through decomposition, heat, or other means and which may cause substantial personal injury or illness.

It is preferable that such materials be handled in the safest manner possible and sent for disposal as quickly as possible. However, any plan for the safe handling and disposal of hazardous materials (such as toxic chemicals) is complex because the status of hazardous (and potentially hazardous) materials must be periodically assessed on the basis of the most recent information.

Pollution prevention and control of hazardous materials are issues for many industries (Gray et al., 1991; Noyes, 1993). There are specific definitions for terms such as *hazardous substances*, *toxic substances*, and *hazardous waste* (Chapters 1 and 6) that must be fully understood in the context of the statutory and/or regulatory meanings. It is absolutely imperative from a legal sense that each statute or regulation promulgated be read in conjunction with terms defined in that specific statute or regulation (Majumdar, 1993).

Governments in a number of nations have passed legislation to deal with hazardous substances and wastes. In the United States such legislation has included a series of laws that deal with prevention of pollution to the land, water, and air (U.S. Congress, 1993; Pereira and Amiridis, 1995; Grano, 1995; Sullivan, 1995).

It is the purpose of this chapter to give a brief overview of the relevant legislation as it relates to the issues noted throughout this book.

2.0 Legislation

Many environmental issues have been brought to the courts on the basis of emotion rather than hard scientific fact. It is now time to shape policies on the basis of science (Fumento, 1993). Environmental watchdog groups and industry alike must prove scientifically that there will be no harm to the environment.

It is not possible in this text to deal with all of the legislation that is relevant to the environment. Selected pieces of legislation are chosen (Table 12.1) to give the reader a general flavor. Each piece has been in operation for some time, many since the 1970s, and in many cases, amendments have been added to the original legislation.

Table 12.1: Historical perspectives of environmental legislation in the United States

	First enacted	Amended
Atomic Energy Act	1954	
Clean Air Act	1970	1977
		1990
Clean Water Act	1948	1965[1]
(Water Pollution Control Act)		1972[2]
		1977
		1987[3]
Comprehensive Environmental Response, Compensation and Liability Act	1980	1986[4]
Hazardous Material Transportation Act	1974	1990
Low-level Radiation Waste Policy Act	1980	
National Environmental Policy Act	1970	1986
Occupational Safety and Health Act	1970	1987[5]
Oil Pollution Act	1924	1990[6]
Resource Conservation and Recovery Act	1976	1980[7]
Safe Drinking Water Act	1974	1986[8]
Superfund Amendments and Re-authorization Act (SARA)	1986	
Toxic Substances Control Act	1976	1984[9]

[1] Water Quality Act
[2] Water Pollution Control Act
[3] Water Quality Act
[4] SARA Amendments
[5] Several amendments during the 1980s
[6] Interactive with various water pollution acts
[7] Federal cancer policy initiated
[8] Several amendments during the 1970s and the 1980s
[9] Import rule enacted

2.1 The Clean Air Act Amendments

The first Clean Air Act of 1970 and the 1977 Amendments consisted of three titles. Title I dealt with stationary air emission sources, Title II with mobile air emission sources, and Title III with definitions of appropriate terms as well as applicable standards for judicial review.

The Clean Air Act Amendments of 1990 contain extensive provisions for control of the accidental release of toxic substances from storage or transportation (Thompson Publishing Group, 1995) as well as the formation of acid rain (acid deposition). They also provide new requirements for state implementation plans for attainment of the national ambient air-quality standards and permit requirements for the attainment and nonattainment areas (United States Environmental Protection Agency, 1993; Wyles, 1994).

The Title III (1990) relates to an expanded program for regulation of hazardous air pollutants (air toxics). As the first phase, the Environmental Protection Agency is required to establish technology-based (in place of the previous national emission standards) *maximum achievable control technology* (*MACT*) emission standards. These standards apply to those sources that emit at least 10 tons per year of any one air pollutant or 25 tons per year of any combination of the designated air pollutants. In addition, Title III provides for health-based standards to address the issue of residual risks due to toxic emissions from the sources equipped with maximum achievable control technology.

The requirement that the standards be technology based removes much of the emotional perception that all chemicals are hazardous (Wedin, 1994) as well as the guesswork from legal enforcement of the legislation. The requirement also dictates environmental and health protection with an ample margin of safety. In addition, the Clean Air Act Amendments of 1990 specify deadlines and detailed timetables for major events. All standards are required to be promulgated within 10 years of the passing of the 1990 Amendments, i.e., by November 15, 2000.

Implementation of these statutory requirements will be costly for many industries, particularly for smaller companies. The statute therefore offers a credit plan, whereby full implementation may be delayed by up to 6 years. This delay will be allowed if there is more than 90% reduction (for particulate matter, a 95% reduction) of hazardous air pollutants voluntarily achieved, using 1987 as the baseline.

2.2 The Water Pollution Control Act

Several acts are related to the protection of the waterways in the United States (U.S. Congress, 1993). Of particular interest in the present context is the Water Pollution Control Act (Clean Water Act). The objective of the Act is to restore and maintain the chemical, physical, and biological integrity of water systems.

The Water Pollution Control Act of 1948 and The Water Quality Act of 1965 were generally limited to control of pollution of interstate waters and the adoption of water-quality standards by the states for interstate water within their borders. The first comprehensive water-quality legislation in the United States came into being in 1972 as the Water Pollution Control Act. This Act was amended in 1977 and became the Clean Water Act. Further amendments in 1978 were enacted to deal more effectively with spills of petroleum or oil and hazardous substances. Other amendments followed in 1987 under the new name Water Quality Act and were aimed at improving water quality in those areas where there were insufficiencies in compliance with the discharge standards.

The statute distinguishes between conventional and toxic pollutants insofar as there are two standards of treatment that are required prior to discharge of an aqueous effluent into navigable waterways. Conventional pollutants generally include degradable nontoxic organic compounds and inorganic compounds, and the applicable treatment standard is *best conventional technology*. Toxic pollutants, on the other hand, require treatment by a higher standard, namely, *best available technology*.

The statutory provisions of the Clean Water Act have five major sections that deal with specific issues: (1) national water-quality standards; (2) effluent standards from specific industries; (3) permit programs for discharges into receiving water bodies based on the National Pollutant Discharge Elimination System; (4) discharge of toxic chemicals, including oil spills; and (5) construction grant programs for publicly owned treatment works.

In addition, Section 311 of the Clean Water Act includes elaborate provisions for regulating intentional or accidental discharges of petroleum and of hazardous substances. Included are response actions required for oil spills and the release or discharge of toxic and hazardous substances. As an example, the person in charge of a vessel or an onshore or offshore facility from which any designated hazardous substance is discharged, in quantities equal to or exceeding its reportable quantity, must notify the appropriate federal agency as soon as such knowledge is obtained. The *Exxon Valdez* is a well-known case.

2.3 The Safe Drinking Water Act

The Safe Drinking Water Act, first enacted in 1974, was amended several times in the 1970s and 1980s to set national drinking water standards. The Act calls for regulations that (1) apply to public water systems, (2) specify contaminants that may have any adverse effect on the health of persons, and (3) specify contaminant levels. In addition, the difference between primary and secondary drinking water regulations is defined, and a variety of analytical procedures are specified. Statutory provisions are included to cover underground injection control systems. The Act also requires maximum levels at which a contaminant must have no known or anticipated adverse effects on human health, thereby providing an *adequate margin of safety*.

The Superfund Amendments and Reauthorization Act (SARA) set standards the same for groundwater as for drinking water in terms of necessary cleanup and remediation of an inactive hazardous waste disposal site. Under the Act, all underground injection activities must comply with the drinking water standards as well as meet specific permit conditions that are in unison with the provisions of the

Clean Water Act.

In this respect, there are five broad classifications for underground wells.

1. *Class I wells* are used for industrial and municipal waste disposal, including storage and disposal of nuclear and hazardous waste.
2. *Class II wells* are used for injection of brine or reinjection for oil and gas recovery and for the storage of liquid hydrocarbons at standard temperature and pressure.
3. *Class III wells* are used for extraction of materials related to energy or mineral recovery.
4. *Class IV wells* are used for the injection of hazardous or radioactive wastes into or above underground sources of drinking water.
5. *Class V wells* are used for alternate energy development and in conjunction with other systems, such as septic systems and cesspools.

However, under the Resource Conservation and Recovery Act, class IV injection wells are no longer permitted and there are several restrictions on underground injection wells that may be used for storage and disposal of hazardous wastes.

2.4 The Resource Conservation and Recovery Act

Since its initial enactment in 1976, the Resource Conservation and Recovery Act (RCRA) continues to promote safer solid and hazardous waste management programs (Dennison, 1993). Besides the regulatory requirements for hazardous waste management, the Act specifies the mandatory obligations of generators, transporters, and disposers of hazardous waste as well as those of owners and/or operators of hazardous waste treatment, storage, or disposal facilities (United States Environmental Propection Agency, 1990a, 199b). The Act (Section 1004.27) also defines solid waste as:

> ...garbage, refuse, sludge from a waste treatment plant, from a water supply treatment plant, or air pollution control facility and other discarded material, including solid, liquid, semisolid, or contained gaseous material resulting from industrial, commercial, mining and agricultural operations and from community activities.

The Act also states that solid waste does not include solid, or dissolved, materials in domestic sewage, or solid or dissolved materials in irrigation return flows or industrial discharges, which are point-sources (Chapter 11) subject to permits under the Water Pollution Control Act, or source, special nuclear, or by-product material as defined by the Atomic Energy Act.

A solid waste becomes a hazardous waste if it exhibits any one of four specific characteristics: (1) ignitability, (2) reactivity, (3) corrosivity, or (4) toxicity (Chapter 6). Certain types of solid wastes (e.g., household waste) are not considered to be hazardous irrespective of their characteristics. Hazardous waste generated in a product or raw-material storage tank, transport vehicles, or manufacturing processes and samples collected for monitoring and testing purposes are exempt from the regulations.

Besides these four characteristics of hazardous wastes, the Environmental Protection Agency has established three hazardous waste lists: (1) hazardous wastes from nonspecific sources, such as spent non-halogenated solvents; (2) hazardous wastes from specific sources, such as bottom sediment/sludge from the treatment of wastewaters from wood preserving; and (3) discarded commercial chemical products and off-specification species, containers, and spill residues.

Hazardous waste management is based on a beginning-to-end concept so that all hazardous wastes can be traced and fully accounted for. All generators and transporters of hazardous wastes as well as owners and operators of related facilities in the United States must file a notification with the Environmental Protection Agency. The notification must state the location of the facility and a general description of the activities as well as the identified and listed hazardous wastes being handled. Thus all regulated hazardous waste facilities must exist and/or operate under valid, activity-specific permits.

Regulations pertaining to companies that generate and/or transport hazardous wastes require that detailed records be maintained to ensure proper tracking of hazardous wastes through transportation systems (Cheremisinoff, 1994). Approved containers and labels must be used, and wastes can only be delivered to facilities approved for treatment, storage, and disposal.

2.5 The Toxic Substances Control Act

The Toxic Substances Control Act was first enacted in 1976 and was designed to provide controls for those chemicals that may threaten human health or the environment (Sittig, 1991). Particularly hazardous are the cyclic nitrogen species (Figure 12.1) that may be produced when fossil fuels are processed (Chapter 2) and that often occur in tarry by-products. The objective of the Act is to provide the necessary control before a chemical is allowed to be mass produced and enter the environment. The list of the chemical substance inventory is fairly comprehensive (United States Environmental Protection Agency, 1990b).

The Act requires a *premanufacture notification* requirement by which any manufacturer must notify the Environmental Protection Agency at least 90 days prior to the production of a new chemical substance. Notification is also required even if there is a new use for the chemical which can increase the risk to the environment. No notification is required for chemicals that are manufactured in small quantities solely for scientific research and experimentation. A *new chemical substance* is defined as a chemical that is not listed in the Environmental Protection Agency Inventory of Chemical Substances or is an unlisted reaction product of two or more chemicals. In addition, the term *chemical substance* means any organic or inorganic substance of a particular molecular identity, including any combination of such substances occurring in whole or in part as a result of a chemical reaction or occurring

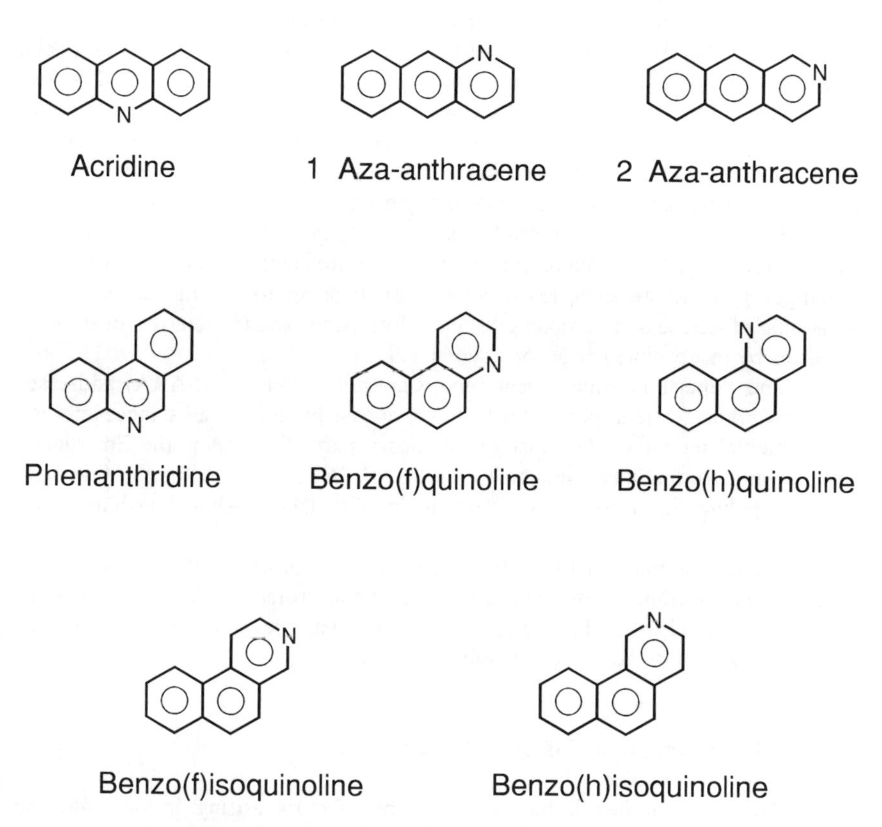

Figure 12.1: Polynuclear aromatic systems containing nitrogen

in nature, and any element or uncombined radical. The term *mixture* means any combination of two or more chemical substances if the combination does not occur in nature and is not, in whole or in part, the result of a chemical reaction.

2.6 The Comprehensive Environmental Response, Compensation, and Liability Act

The Comprehensive Environmental Response, Compensation, and Liability Act (CERCLA), popularly known as Superfund, was first signed into law in 1980 (U.S. Congress, 1986). The central purpose of this Act is to provide a response mechanism for cleanup of any hazardous substance released, such as an accidental spill, or of a threatened release of a hazardous substance (Nordin et al., 1995). While RCRA deals basically with the management of hazardous wastes that are generated, treated, stored, or disposed of, CERCLA provides a response to the environmental release of various pollutants or contaminants into the air, water, or land.

Under this Act, a hazardous substance is any substance requiring (1) special consideration due to its toxic nature under the Clean Air Act, the Clean Water Act, or the Toxic Substances Control Act and (2) defined as hazardous waste under RCRA. Additionally, a pollutant or contaminant can be any other substance not necessarily designated by or listed in the Act but that *will or may reasonably* be anticipated to cause any adverse effect in organisms and/or their offspring.

This Act requires responsible parties or the government to clean up waste sites. Among CERCLA's major purposes are: (1) site identification, (2) evaluation of danger from waste sites, (3) evaluation of damages to natural resources, (4) monitoring of release of hazardous substances from sites, and (5) removal or cleanup of wastes by responsible parties or government.

The Superfund Amendments and Reauthorization Act (SARA) addresses closed hazardous waste disposal sites that may release hazardous substances into any environmental medium. The most revolutionary part of SARA is the Emergency Planning and Community Right-to-Know Act (EPCRA), which for the first time mandated public disclosure. It is covered under Title III of SARA (Newburg-Rinn, 1992).

Perhaps a major part of the Superfund authorization is the technology development program (United States Environmental Protection Agency, 1994). In this program, promising technologies for site cleanup are developed with the goal of commercialization in the earliest possible time frame.

2.7 The Occupational Safety and Health Act

Occupational health hazards are those factors arising in or from the occupational environment that adversely impact health (Lipton and Lynch, 1994; Stricoff and Walters, 1995). Thus, the Occupational Safety and Health Administration (OSHA) came into being in 1970 and is responsible for administering the Occupational Safety and Health Act.

The goal of the Act is to ensure that employees do not suffer material impairment of health or functional capacity due to a lifetime occupational exposure to chemicals and hazardous substances. The statute imposes a duty on employers to provide employees with a safe workplace environment, free of known hazards that may cause death or serious bodily injury (Wang, 1994; Gilbert-Miller, 1995; Walter, 1995).

The Act is also responsible for the means by which chemicals are contained (Thompson Publishing Group, 1995). Workplaces are inspected to ensure compliance and enforcement of applicable standards under the Act. In keeping with the nature of the Act, there is also a series of standard tests relating to occupational health and safety (American Society for Testing and Materials, 1995) as well as the general recognition of health hazards in the workplace (Burgess, 1995). The Act is also the means by which guidelines have evolved for the management and disposition of chemicals used in chemical laboratories (American Chemical Society, 1994; Lunn and Sansone, 1994; Studt, 1995).

Under the Act, current standards impose certain record-keeping and reporting requirements on employers. Notices of certain specific materials must be posted at designated or prominent locations at the workplace for the purpose of keeping employees informed. The Act also provides employees with several rights, including the *right to information* (*right-to-know*) on hazards at the workplace.

2.8 The National Environmental Policy Act

The purpose of the National Environmental Policy Act (NEPA) is to provide a high priority to environmental concerns whenever the government or its contractors proposes any action that may affect the environment. For example, government contractors who generate chemical wastes must first assess any effects the wastes may have on the environment.

The centerpiece of the Act is the statutory requirement for an Environmental Impact Statement (EIS), in which the following specific matters must be addressed: (1) the environmental impact of the proposed action, (2) alternatives to the proposed action and their relative environmental impact, (3) all adverse environmental impacts due to the proposed action that cannot be avoided, and (4) all irreversible and irretrievable expenditure of natural resources in the event of implementation of the proposed action.

2.9 The Oil Pollution Act

The Oil Pollution Act of 1990 deals with pollution of waters by oil (Thompson Publishing Group, 1995). The Act specifically deals with petroleum vessels and onshore and offshore facilities and imposes strict liability for oil spills on their owners and operators.

2.10 The Atomic Energy Act and Other Nuclear Waste Statutes

The Atomic Energy Act of 1954 was truly the first comprehensive federal legislation to address the use of atomic energy for peaceful purposes. However, the Act does not provide any statutory guidance with regard to the handling and disposal of radioactive wastes.

Due to the prolific growth of both low-level and high-level radioactive wastes as a result of uranium mining, nuclear power plant operations, and production of nuclear weapons and bombs, the general provisions of the Act have very little impact today. Generally, high-level radioactive wastes are generated during reprocessing of spent nuclear fuel and use of fuel rods. Low-level radioactive wastes, however, are generated from various nuclear power plants and medical, commercial, and nonmilitary uses of radioactive isotopes as well as the contamination of items by radioactive materials.

The Low-Level Radioactive Waste Policy Act of 1980 specifies that each state is responsible for providing appropriate disposal of the low-level (radioactive) waste generated within its borders. In addition, the Low-Level Radioactive Waste Policy Act encourages a regional concept, popularly known as compacts, for establishing regional disposal sites for low-level radioactive waste.

In this context, it should be stressed that the RCRA mandate includes all types of radioactive waste except high-level. RCRA is also the federal statute with complete authority over *mixed waste* which is generated due to the contamination of hazardous waste by radioactive waste.

2.11 The Hazardous Materials Transportation Act

The Hazardous Materials Transportation Act authorizes the establishment and enforcement of hazardous material regulations for all modes of transportation by highway, water, and rail. The purpose of the Act is to ensure safe transportation of hazardous materials. The Act prevents any person from offering or accepting a hazardous material for transportation anywhere within this nation if that material is not properly classified, described, packaged, marked, labeled, and authorized for shipment pursuant to the regulatory requirements.

Under Department of Transportation regulations, a hazardous material is defined as:

> ...any substance or material, including a hazardous substance and hazardous waste, which is capable of posing an unreasonable risk to health, safety, and property when transported.

The Act also imposes restrictions on the packaging, handling, and shipping of hazardous materials. For shipping and receiving of hazardous chemicals, hazardous wastes, and radioactive materials, the appropriate documentation, markings, labels, and safety precautions are required.

3.0 Global Environmental Change

The environment is in a state of continual change, and not only as a result of modern activities. In fact, ever since life first appeared, the atmosphere has been influenced by the metabolic processes of living organisms. Of these, humans have probably caused the greatest changes even though their tenure on earth has been the shortest relative to the life terms of other species.

When the first primitive life molecules were formed approximately 3.5 billion

(3.5×10^9) years ago (Chapter 1), the atmosphere was very different from its present state. At that time, the atmosphere is thought to have consisted primarily of methane, ammonia, water vapor, and hydrogen. These simple molecular species were bombarded by ultraviolet radiation, which (along with, for example, lightning from the prevalent atmospheric disturbances) provided the energy to bring about chemical reactions that resulted in the production of relatively complicated molecules, including even amino acids and sugars:

$$CH_4 + NH_3 + H_2O + H_2 + \text{energy} = CH_3.CH(NH_2).COOH + CH_2OH.(CHOH)_n.CH_2OH$$

From this rich chemical mixture, it is postulated that life molecules evolved (Miller, 1987).

Although, there are alternate theories (Britannica, 1969; Krauss, 1992; Christian Bible, 1995). Initially, these very primitive life forms derived their energy from fermentation of organic matter formed by chemical and photochemical processes. Eventually they gained the capability to produce organic matter by photosynthesis and the stage was set for the massive biochemical transformation that resulted in the production of the majority of the oxygen in the atmosphere.

Because of the anaerobic conditions of the primitive atmosphere, the oxygen initially produced by photosynthesis was probably quite toxic to the prevailing life forms. Eventually, enzyme systems developed that enabled organisms to survive in an oxygen-rich atmosphere and to mediate the reaction of waste-product oxygen with oxidizable organic matter. Later, this mode of waste-product disposal evolved as a means of producing energy for respiration, which is now the mechanism by which nonphotosynthetic organisms obtain energy.

As oxygen accumulated in the atmosphere there was the (now widely accepted) additional benefit of the formation of ozone which became a shield against harmful solar ultraviolet radiation. With this shield in place, life evolved on the earth in the form that we now know it insofar as the current environment is much more hospitable to the current flora and fauna (including humans).

There is always the question about the nature of the life forms that would have (or may have) evolved without the protection of the ozone layer. Would the equivalent human form have had a skin like a dinosaur or perhaps the appearance of a dried prune? Nevertheless, life forms as we now know them were able to venture from the protective surroundings of the sea to the more exposed environment of the land, and they were able to survive.

These changes to the primitive atmosphere are not usually considered to be environmentally harmful, even though extinct organisms yet to be discovered may have suffered as a result of the change. The goal of humankind is to protect the present environment. Yet, at an ever accelerating pace during the last 200 years, humankind has engaged in a number of activities that are altering the state of the atmosphere. Industrial activities emit a variety of atmospheric contaminants including sulfur dioxide, particulate matter, photochemically reactive hydrocarbons, chlorofluorocarbons, and inorganic substances (such as toxic heavy metals). These activities include the use (combustion) of fossil fuels which can introduce carbon dioxide, carbon monoxide, sulfur dioxide, nitrogen oxides, hydrocarbons (including methane), particulate soot, polynuclear aromatic hydrocarbons, and fly ash into the atmosphere.

Man, a bipedal animal, found that self-transport was slow, ponderous, and energy consuming and developed a variety of mechanical transportation practices. However, these practices result in the release of carbon dioxide, carbon monoxide, nitrogen oxides, photochemically reactive (smog forming) hydrocarbons (including methane), and polynuclear aromatic hydrocarbons.

Humans are also adept at altering the surrounding environment by alteration of land surfaces, including deforestation. The burning of vegetation, including tropical and subtropical forests and savanna grasses, produces atmospheric carbon dioxide, carbon monoxide, nitrogen oxides, particulate soot and polynuclear aromatic hydrocarbons. Agricultural practices produce methane (from the digestive tracts of domestic animals and from the cultivation of rice in waterlogged anaerobic soils) and dinitrogen oxide from bacterial denitrification of nitrate-fertilized soils.

Human activities that are too numerous to note here have significantly altered the atmosphere, particularly in regard to its composition of minor constituents and trace gases. The major effects have been (1) an increased acidity in the atmosphere that can result in the deposition of acid rain; (2) production of pollutant oxidants in localized areas of the lower troposphere (Chapters 5 and 10); and (3) elevated levels of infrared-absorbing gases (greenhouse gases) that can contribute to threats to the ultraviolet-filtering ozone layer in the stratosphere.

Of all environmental hazards, there is little doubt that major disruptions in the atmosphere and climate have the greatest potential for environmental damage. If levels of greenhouse gases and reactive trace gases continue to increase at the projected rates, potential environmental harm can become real. However, the bulk of these emissions is being curtailed by the environmentally conscious industrialized nations, and there are serious efforts to substantially reduce harmful emissions.

It is important to keep in mind that the atmosphere has the ability to cleanse itself of pollutant species. Water-soluble gases, including greenhouse gases, acid gases, and fine particulate matter are removed with precipitation. For most gaseous contaminants, oxidation precedes or accompanies the removal processes.

The measures to be taken in dealing with the emissions issue can be partly overcome by switching to alternate energy sources, increasing energy conservation, and reversing deforestation. It is especially sensible to use measures that have major effects in addition to reduction of greenhouse warming. However, such a shift is not the complete answer. Indeed, shifting to nuclear-based energy sources just to prevent possible greenhouse warming may give rise to a host of other environmental disadvantages (Pickering and Owen, 1994).

Definite economic and political benefits would also accrue from lessened dependence on uncertain, volatile petroleum supplies. Increased energy efficiency would diminish both greenhouse gas and acid rain production, while lowering costs of production and reducing the need for expensive and environmentally disruptive new power plants.

Perhaps the most pertinent response at this time is the introduction of relevant legislation. However, there must be a balance between the extent of the legislation and the evolution of industrial practices. Old and outdated laws must be taken to task (*U.S. News & World Report*, 1995).

4.0 References

American Chemical Society. 1994. *Task Force on Laboratory Waste Management*. American Chemical Society, Washington, D.C.

American Society for Testing and Materials. 1995. *Annual Book of ASTM Standards*. American Society for Testing and Materials, Philadelphia, Pennsylvania, USA. Volume 11.03.

Britannica. 1969. Creation, Myths of. *Encyclopedia Britannica*. William Benton, Chicago.

Burgess, W.A. 1995. *Recognition of Health Hazards in Industry: A Review of Materials and Processes*. 2nd ed. John Wiley & Sons Inc., New York.

Cheremisinoff, N.P. 1994. *Transportation of Hazardous Materials: A Guide to Compliance*. Noyes Data Corp., Park Ridge, New Jersey, USA.

Christian Bible. 1995. New International Version. Book of Genesis. Chapters 1 and 2. Zondervan Publishing House, Grand Rapids, Michigan.

Dennison, M.S. 1993. *RCRA Regulatory Compliance Guide*. Noyes Data Corp., Park Ridge, New Jersey, USA.

Fumento, M. 1993. *Science Under Siege: Balancing Technology and the Environment*. p. 372. William Morrow and Company Inc., New York.

Gilbert-Miller, S. 1995. *Environmental Solutions* 8(5): 25.

Grano, D. 1995. Chap. 2 in *Reduction of Nitrogen Oxide Emissions*. Edited by U.S. Ozkan, S.K. Agarwal, and G. Marcelin. Symposium Series No. 587. American Chemical Society, Washington, D.C.

Gray, P.E., Tester, J.W., and Wood, D.O. 1991. In *Energy and the Environment in the 21st Century*. p. 120. Edited by J.W. Tester, D.O. Wood, and N.A. Ferrari. MIT Press, Cambridge, Massachusetts.

Krauss, L.M. 1992. In *Mysteries of Life and the Universe*. p. 47. Edited by W.H. Shore. Harcourt Brace and Co., New York.

Lipton, S., and Lynch, J. 1994. *Handbook of Health Hazard Control in the Chemical Process Industry*. John Wiley & Sons Inc., New York.

Lunn, G., and Sansone, E.B. 1994. *Destruction of Hazardous Chemicals in the Laboratory. 2nd Edition*. McGraw-Hill, New York.

Majumdar, S.B. 1993. *Regulatory Requirements for Hazardous Materials*. McGraw-Hill, New York.

Miller, S.L. 1987. In *Encyclopedia of Science and Technology. 6th Edition*. Vol. 10. p. 45. Edited by S.P. Parker. McGraw-Hill, New York.

Newburg-Rinn, S.D. 1992. Chap. 3 in *Pollution Prevention in Industrial Processes: The Role of Process Analytical Chemistry*. Edited by J.J. Breen and M.J. DeMarco. Symposium Series No. 508. American Chemical Society, Washington, D.C.

Nordin, J.S., Sheesley, D.C., King, S.B., and Routh, T.K. 1995. *Environmental Solutions* 8(4): 49.

Noyes, R. (ed.). 1993. *Pollution Prevention Technology Handbook*. Noyes Data Corp, Park Ridge, New Jersey, USA.

Pereira, C.J., and Amiridis, M.D. 1995. Chap. 1 in *Reduction of Nitrogen Oxide Emissions.* Edited by U.S. Ozkan, S.K. Agarwal, and G. Marcelin. Symposium Series No. 587. American Chemical Society, Washington, D.C.

Pickering, K.T., and Owen, L.A. 1994. *Global Environmental Issues.* Routledge Publishers Inc., New York.

Sittig, M. 1991. *Handbook of Toxic and Hazardous Chemicals and Carcinogens.* 3rd ed. Noyes data Corp., Park Ridge, New Jersey, USA.

Stricoff, R.S. and Walters, D.B. (eds.). 1995. *Handbook of Laboratory Health and Safety.* John Wiley & Sons Inc., New York.

Studt. T. 1995. *R&D Magazine.* February: p. 69.

Sullivan, T.F.P. (ed.). 1995. *Environmental Law Handbook.* Government Institutes Inc., Rockville, Maryland.

Thompson Publishing Group. 1995. *Aboveground Storage Tank Guide.* Vols. 1 and 2. Thompson Publishing Group, Washington, D.C.

United States Congress. 1986. *The Comprehensive Environmental Response, Compensation, and Liability Act of 1980.* PL-99-499. United States Congress. US Government Printing Office, Washington, D.C.

United States Congress. 1993. *Compilation of Selected Water Resources and Environmental Laws.* Report No. 103-14. United States Congress. US Government Printing Office, Washington, D.C.

United States Environmental Protection Agency. 1990a. *RCRA Orientation Manual.* Report No. EPA/530-SW-90-036. Office of Solid Waste, Environmental Protection Agency, Washington, D.C.

United States Environmental Protection Agency. 1990b. *Toxic Substances Control Act Chemical Substance Inventory.* Report No. EPA560/7-90-003. Office of Pesticides and Toxic Substances, Environmental Protection Agency, Washington, D.C.

United States Environmental Protection Agency. 1993. *The Plain English Guide to the Clean Air Act.* Report No. EPA400-K-93-001. Office of Air and Radiation, Environmental Protection Agency, Washington, D.C.

United States Environmental Protection Agency. 1994. *Superfund Innovative Technology Evaluation Program. Technology Profiles.* Office of Research and Development, United States Environmental Protection Agency, Washington, D.C.

U.S. News & World Report. 1995. March 13, p.34.

Walter, M. 1995. *Environmental Solutions* 8(5): 34.

Wang, C.C.K. 1994. *OSHA Compliance and Management Handbook.* Noyes Data Corp., Park Ridge, New Jersey, USA.

Wedin, R.E. 1994. *Today's Chemist* 3(3): 12.

Wyles, T.R. 1994. *Hazmat World* 7(1): 24.

Zakrzewski, S.F. 1991. *Principles of Environmental Toxicology.* American Chemical Society, Washington, D.C.

Epilogue

The environment is changing. In some cases this change is for the better but in many cases it is for the worse as people develop resources for the advancement of the modern world. But it does not have to be so. Planned development can occur whereby resources can be obtained with minimal disturbance to the environment.

Environmental issues are numerous and often overlap. The macro-issues such as acid rain and global warming involve more than one country. The micro-issues occur on a more localized scale. Whether or not the environment was ever pristine is another issue. It must be recognized by those advocating a pristine environment that some disturbance of the environment will occur if meaningful progress is to be made. On the other hand, the environment must not be raped! A major factor overall is control of the emotionalism that is apt to be the driving force behind many environmental movements; logic should be the driver. Are species really becoming extinct by the dozens or are they becoming more able to survive in other locales and to avoid humankind?

Darwinism is essentially the survival of the fittest whereby adaptation to change is genetic in origin. The variations in a species that are beneficial to the survival of that species are preserved. *Lamarckism* is the beneficial adaptations that occur during the lifetime of an organism. The Lamarckian change is not usually genetic in character. The ability of the human species to adapt to the environment and the ability of the environment to adapt to the human species is perhaps a combination of both Darwinism and Lamarckism. One hopes that both theories will be in play during the next decades for the protection of both the human species and the environment.

The development of various resources is necessary for the continued emergence of humans. But resources must be developed through judicious planning. The formation of various types of emissions (gaseous, liquid, and solids) must be thoroughly understood in order that they may be controlled and even mitigated.

Tolerance is necessary on both sides of the clean environment equation. Although industry has taken great strides to ensure protection of the environment, further steps are still required. The environmentalists must realize that no one wishes to "freeze in the green darkness".

The ability to determine our environmental future is held firmly in our hands. The influence of human development on the environment continues to be a major point for debate. There is the realization that human development must move forward with some consideration for its effects on the flora and fauna of the region. Without such consideration, the environment may be doomed, to be followed by the doom of human lifestyles as we know them or as we would prefer to have them.

The overriding question remains: how much do we need and how much is enough? It is said that as Michelangelo painted the ceiling of the Sistine Chapel, his continued response to Pope Julius' question, "When will you finish?" was "When I am done!" Perhaps this is the case with the environment. Using the Sistine Chapel ceiling as the analogy, the base coat to seal the plaster may be likened to the cleanup of the so-called Superfund sites. That is, the initial effort is focused on correcting the flaws and errors of the past. Concurrent regulations are in place to make sure that there are no further mistakes in the color scheme. The ceiling is becoming prepared for the main color schemes to follow. Thus, as we progress through the task, we can see the figures and the colors develop and start to form a presentable picture.

Government, industry, and domestic consumers in various countries realize the need for protecting the environment. Like Michelangelo's ceiling, the task may never be complete, and we must be ever watchful that the overall picture is maintained and not spoiled. Just as it would be an error to focus on one small part of the picture, it would also be an error to focus on one small part of the environment.

The environment is changing continually, not because of resource development but also from the general activities associated with human existence. Therefore, before any decision is made to dictate what the environment shall be, the question about what the environment can be must also be asked!

It would be a tragedy for the environment to be polluted to such an extent that human existence became impossible, but policies to protect the environment should be based on hard scientific fact, not on rampant emotionalism! A balance is needed between what should be and what can be. A balance must be struck between human development and the necessary environmental considerations. To swing the pendulum too far to any one side could be a serious error, and perhaps even a disastrous mistake.

Glossary

Abiotic: nonliving factors or physical factors; the abiotic elements of an ecosystem constitute its climatic, geological, and pedologic (soil) components.
Acaricide: a chemical detrimental to mites.
Acid deposition: acid rain; a form of pollution depletion in which pollutants are transferred from the atmosphere to soil or water; often referred to as atmospheric self-cleaning.
Acid mine drainage: the seepage of sulfuric acid solutions (pH 2.0-4.5) from mines and their wastes which have been placed on the surface; these solutions result from the interaction of water/precipitation with sulfide minerals exposed by mining.
Acid mine water: water in used/disused mines and contains an appreciable concentration of free mineral acid.
Acid rain: the precipitation phenomenon that incorporates anthropogenic acids and other acidic chemicals from the atmosphere to the land and water (see **Acid deposition**).
Acidity: the capacity of an acid to neutralize a base such as a hydroxyl ion (OH-).
Actinomycetes: microorganisms that are morphologically similar to both bacteria and fungi.
Activated sludge: the biologically active sediment produced by the repeated aeration and settling of sewage and/or organic wastes.
Activated sludge process: production of a biologically active sludge that is usually brown in color and that destroys the polluting organic matter in sewage and waste.
Add-on control methods: the use of devices that remove process emissions after they are generated but before they are discharged to the atmosphere.
Adsorption: transfer of a substance from a solution to the surface of a solid.
Aerobic: in the presence of air.
Aerosol: a colloidal system in which a gas, frequently air, is the continuous medium and particles of solids (usually less than 100 microns in size) or liquids are dispersed in it. .
A-horizon (E-horizon): the uppermost layer in a soil, also known as topsoil; mineral and organic materials, including soil fauna and roots of vegetation, accumulate in the A-horizon, while soluble salts and clays are removed as a result of eluviation (hence the alternative term: E-horizon).
Air pollution: the discharge of toxic gases and particulate matter introduced into the atmosphere, principally as a result of human activity.
Algae: microscopic organisms that subsist on inorganic nutrients and produce organic matter from carbon dioxide by photosynthesis.
Alkalinity: the capacity of a base to neutralize the hydrogen ion (H+).
Alluvium: deposits of stream-borne sediments.

Amine washing: a method of gas cleaning whereby acidic impurities such as hydrogen sulfide and carbon dioxide are removed from the gas stream by washing with an amine (usually an alkanolamine).

Anaerobic: in the absence of air.

Anthropogenic: created by human activities.

Anthropogenic acids: those acids that are the result of human activities.

Anthropogenic stress: the effect of human activity on other organisms.

Aquasphere: water systems.

Aquiclude: a rock formation that is too impermeable or unfractured to yield groundwater.

Aquifer: a subsurface water-bearing formation that yields economically important amounts of water to wells.

Areic: regions that lack surface steams because of low rainfall or lithologic conditions.

Asbestos: a group of fibrous silicate minerals, typically those of the serpentine group, of which there are three main types: (1) blue asbestos, or crocidolite; (2) white asbestos; (3) brown asbestos.

Asbestosis: a medical condition caused by inhalation of asbestos fibers into the respiratory system.

Asphalt: a product of petroleum refining used for road construction.

Autotrophic organisms: utilize solar or chemical energy to fix elements from simple, nonliving inorganic materials into complex life molecules that compose living organisms.

B-horizon: layer of soil below the A-horizon, also referred to as the subsoil; receives material such as organic matter, salts, and clay particles leached from the top soil. Also known as subsoil and contains less organic matter (and less biological activity) than the horizon above and is also characterized by less biological activity.

Baghouse: a filter system for the removal of particulate matter from gas streams; so called because of the similarity of the filters to coal bags.

Becquerel (Bq): the derived SI unit of radioactivity equal to the number of atoms of a radioactive substance that disintegrate per second. Smaller than the Curie and can be used to measure much smaller doses of radioactivity (1 Bq = 3.7027×10^{-10} Ci).

Benthic zone: the sea bottom, which is divided into three areas: (1) the littoral zone, comprising the shore area between the limits of high tide and low tide; (2) the sublittoral zone, extending from the low-tide level to the edge of the continental shelf; (3) the deep-sea zone, divisible into the archibenthic zone extending from the lower edge of the continental shelf to a depth of about 1000 meters and the abyssal benthic zone, extending beyond 1000 meters.

Biochemical oxygen demand (BOD): the degree of oxygen consumption by microbial oxidation of contaminants in the water; also known as biological oxygen demand.

Bioconcentration: concentration of a chemical by ingestion into the organism followed by ingestion of the organism by a predator and without any discharge of the chemical.

Biodegradation: the conversion of waste materials by biological processes to simple inorganic molecules and, to a certain extent, to biological materials.

Biological half-life: the time required for half of a quantity of radioactive material absorbed by a living tissue or organism to be naturally eliminated.

Biological waste treatment: a generic term applied to processes that use microorganisms to decompose organic wastes either into water, carbon dioxide, and simple inorganic substances.

Biomagnification: see **Bioconcentration**.

Biomass: biological organic matter.

Bioreactors: reactors for the conversion of biomass.

Bioremediation: the use of living organisms (primarily microorganisms) to degrade pollutants previously introduced into the environment or to prevent pollution through treatment of waste streams before they enter the environment.

Biosphere: living organisms and their environments on the surface of the earth.

Biota: living organisms

Biotic: actions of other organisms.

Biotransformation: the conversion of a substance through metabolization thereby causing an alteration to the substance by biochemical processes in an organism.

Bitumen: a naturally occurring material that is petroleum-like in nature but that is immobile at room temperature; similar in properties to asphalt (q.v.)

Bog: a thick zone of vegetation floating on water and lacking a solid foundation.

Bottom ash: ash that occurs at the bottom of a (for example, coal) combustor.

Breccia: sedimentary rock consisting of angular fragments of rock debris within a matrix of finer material. Breccia is recognized according to its origin, e.g. volcanic breccia, glacial breccia.

Bronchogenic carcinoma: cancer originating in the air passages in the lungs.

Bronsted acid: a chemical species which can act as a source of protons.

Bronsted base: a chemical species which can accept protons.

Burden: the amount of a specific pollutant in a reservoir.

C- horizon: the layer of soil lying below the B-horizon and above the parent material in a well-developed mineral soil where the occurrence of soil fauna and flora is minimal; transitional between the weathered material of the A-horizon and B-horizon, and the underlying parent material.

Capping: process to cover the wastes, prevent infiltration of excessive amounts of surface water, and prevent release of waste to overlying soil and the atmosphere.

Carbonate washing: processing using a mild alkali (e.g., potassium carbonate) process for emission control by the removal of acid gases from gas streams.

Carbon cycle: the natural circulation of carbon in the biosphere.

Carbonization: a high-temperature process by which coal is converted into coke.

Carcinogen: a cancer-causing substance.

Carcinogenesis: development of cancer cells within an organism.

Carcinogenic: cancer-causing.

Catabolism: breaking down complex molecules.

Catalytic cracking : a process in which heavy gas oils are converted into products of higher volatility by contacting the higher molecular weight hydrocarbon with the hot catalyst.

Catalytic oxidation: a chemical conversion process used predominantly for destruction of volatile organic compounds and carbon monoxide.

Cellulase: a biological catalyst (enzyme).

Chelating agents: complex-forming agents having the ability to solubilize heavy metals.

Chemical disinfection: a process that inactivates bacteria, viruses, and protozoa in waste streams.

Chemical waste: any solid, liquid, or gaseous material discharged from a process and that may pose substantial hazards to human health and environment.

Chlorofluorocarbons: organic compounds containing chlorine and fluorine.

Clay: silicate minerals that also usually contain aluminum and have particle sizes less than 0.002 micron.

Co-incineration: burning different waste materials to produce fuel for energy recovery in furnaces and boilers.

Coal: an organic rock.

Coke: solid product produced by the carbonization (q.v.) of coal; also produced from petroleum during thermal processes.

Combustible liquid: a liquid with a flash point in excess of 37.8°C (100°F) but below 93.3°C (200°F).

Composting: the biodegradation of solid or solidified materials in a medium other than soil.

Connate water (fossil water): the water trapped in sedimentary rocks during their formation; an important source of groundwater.

Containment: emplacement of a physical, chemical, or hydraulic barrier to isolate contaminated areas.

Contaminant: a substance that causes deviation from the normal composition of an environment.

Corrosive wastes: wastes that are capable of corroding metal.

Corrosivity: a characteristic of substances that exhibit extremes of acidity or basicity or a tendency to corrode steel.

Crude oil: petroleum in its natural state before refining.

Cultural eutrophication: changes in species composition, population sizes, and productivity in groups of organisms throughout the aquatic ecosystem.

Curie (Ci): unit of measurement of radioactivity equal to 3.7×10^{10} disintegrations per gram per second. Recently replaced by the becquerel (q.v.).

Cyclone: a device for extracting dust from industrial waste gases which takes the form of an inverted cone into which the contaminated gas enters tangentially from the top; the gas is propelled down a helical pathway, and the dust particles are deposited by means of centrifugal force onto the wall of the scrubber.

Darwinism: the theory espoused by Charles Darwin that proposes that an organism's ability to survive is dependent upon hereditary factors.

Dehydrating agents: substances capable of removing water (drying, q.v.) or the elements of water from another substance.

Detoxification: the biological conversion of a toxic substance to a less toxic species, which may still be relatively complex, or biological conversion to an even more complex material.

Devolatilized fuel: smokeless fuel.

Dialysis: a process for separating components in a liquid stream by using a membrane.

Dispersion aerosols: formed from disintegration of larger particles and are usually above 1 micron in size, e.g., dust.

Disposal: the discharge, deposit, injection, or placing of a waste onto, or into, a land facility.

Dissolved air flotation: air is dissolved in the suspending medium under pressure and comes out of solution when the pressure is released as minute air bubbles attached to suspended particles, which causes the particles to float to the surface.

Distillation: a process for separating liquids with different boiling points.

Dry deposition: the removal of both particles and gases as they come into contact with the land surface.

Drying: removal of a solvent or water from a chemical substance; also referred to as the removal of solvent from a liquid or suspension.

Dust control: particulate matter control:

Dust explosions: explosions caused by the presence of dust particles in the atmosphere.

Dystrophic lakes: shallow, clogged with plant life, and normally contain colored water with a low pH.

Ecological cycles: cycles involving land systems, water systems, and the atmosphere which are important to life.

Ecology: the branch of science related to the study of the relationship of organisms and their environment.

Ecosystem: an ecological community, or living unit, considered together with nonliving factors of its environment as a unit.

Effective stack height: the combination of the physical stack height and plume rise above the stack.

Effluent: any contaminating substance, usually a liquid, that enters the environment via a domestic industrial, agricultural, or sewage plant outlet.

Electrodialysis: an extension of dialysis that is used to separate the components of an ionic solution by applying an electric current to the solution thereby causing ions to move in preferred directions through the dialysis membrane.

Electrolysis: a process in which ionic species in solution move to an electrode of opposite electric charge.

Electrostatic precipitators: devices used to trap fine dust particles (usually in the size range 30-60 microns) that operate on the principle of imparting an electric charge to particles in an incoming airstream and which are then collected on an oppositely charged plate across a high voltage field.

Emission control: the use of gas-cleaning processes to reduce emissions.

Emission standard: the maximum amount of a specific pollutant permitted to be discharged from a particular source in a given environment.

Emulsion breaking: the settling or aggregation of colloidal-sized emulsions from suspension in a liquid medium.

Encapsulation: a process used to coat waste with an impermeable material so that there is no contact between the waste constituents and the surroundings.

Endangered species: any plant or animal species that no longer can be relied on to reproduce itself in numbers ensuring its survival.

End-of-pipe emission control: the use of specific emission control processes to clean gases after production of the gases.

Endoreic: regions that drain to interior closed basins.

Energy: the capacity of a body or system to do work, measured in joules (SI units); also the output of fuel sources.

Energy from biomass: the production of energy from biomass (q.v.).

Ethanol: see Ethyl alcohol.

Environment: the combination of physical and chemical (abiotic) factors and living (biotic) factors) which influence the lives of individual organisms.

Environmental Impact Assessment (EIA): the interpretation of the significance of anticipated environmental changes related to a proposed project.

Environmental technology: the application of scientific and engineering principles to the study of the environment with the goal of improving the environment.

Ethyl alcohol (ethanol or grain alcohol): an inflammable organic compound (C_2H_5OH) formed during fermentation of sugars; used as an intoxicant and as a fuel.

Eutrophic lakes: lakes that contain high proportions of nutrients and in which abnormal growth of organisms occurs (see Eutrophication).

Eutrophication: the deterioration of the esthetic and life-sporting qualities of lakes and estuaries caused by excessive fertilization from effluent high in phosphorus, nitrogen, and organic growth substances.

Evaporation: a process for concentrating nonvolatile solids in a solution by boiling off the liquid portion of the waste stream.

Evaporites: soluble salts that precipitate from solution, e.g., the result of evaporation of seawater.

Exoenzyme: extracellular enzyme.

Exoreic: regions that drain to the sea.

Expanding clays: clays that expand or swell on contact with water, e.g., montmorillonite.

Explosives: chemicals which decompose spontaneously, or by initiation/stimulation, with a rapid release of a high amount of energy.

Extracellular enzyme: an enzyme that is capable of acting outside an organism.

F-type wastes: wastes from nonspecific sources.

Fabric filters: filters made from fabric materials and used for removing particulate matter from gas streams (see Baghouse).

Feldspar (felspar): a group of aluminum silicate minerals commonly found in many types of rock.

Felspar: see Feldspar

Filtration: the use of an impassable barrier to collect solids but which allows liquids to pass.

Fission: energy from nuclear sources which uses uranium ore or refined uranium-235 as the basic energy source and which splits into lower atomic weight nuclei.

Fission reactor: a nuclear reactor in which atoms of elements such as uranium-235 and plutonium-239 are split by neutrons to release energy

Fixation: a process that binds a waste in a less mobile and less toxic form and is generally included in the definition of stabilization; also the process by which aerial (molecular) nitrogen is converted to nitrates in the soil (see Nitrogen fixation).

Fixed bed: use of a stationary bed to accomplish a process (see Fluid bed).

Fixed hearth incinerators: incinerators with single or multiple (non-mobile) hearths which are used for the combustion of liquid or solid wastes.

Flammability range: the range of temperature over which a chemical is flammable.

Flammable: a substance that will burn readily.

Flammable liquid: a liquid having a flash point below 37.8°C (100°F).

Flammable solid: a solid that can ignite from friction or from heat remaining from its manufacture, or which may cause a serious hazard if ignited.

Flotation: a process for removing solids from liquids using air bubbles to carry the particles to the surface.

Fluid bed: use of an agitated bed of inert granular material to accomplish a process in which the agitated bed resembles the motion of a fluid.

Flushing: a cleanup process in which the soil is left in place and the water is pumped into and out of the soil in order to clean it.

Flux: the rate of transfer of a pollutant from one sphere or domain to another.

Fly ash: particulate matter produced from mineral matter in coal that is converted during combustion to finely divided inorganic material and which emerges from the combustor in the gases.

Fog: a dense, cloud-like mass of water droplets suspended in the lowest layers of the atmosphere that reduces visibility to less that 1 km.

Fossil fuel resources: coal, petroleum, and natural gas.

Fossil water: see Connate water.

Fugitive emissions: emissions that enter the atmosphere from an unconfined area.

Fulvic acids: a base-soluble fraction of humus (q.v.).

Fumigants: volatile substances used as soil pesticides and to control insects in stored products.

Fungi: nonphotosynthetic organisms which frequently possess a filamentous structure.

Fungicide: a chemical used to control fungal action.

Fusion: energy produced using the fusion of hydrogen nuclei.

Gaseous pollutants: gases released into the atmosphere that act as primary or secondary pollutants.

Gasohol: a term for motor vehicle fuel comprising between 80-90% unleaded gasoline and 10-20% ethanol (see also Ethyl alcohol).

Geosphere: the complex and variable mixture of minerals, organic matter, water, and air that make up soil; also known as land systems.

Glassification: encapsulation of a waste in a glass-like coat; vitrification (q.v.).

Grain alcohol: see Ethyl alcohol.

Greenhouse effect: warming of the earth due to entrapment of the sun's energy by the atmosphere.

Greenhouse gases: gases that contribute to the greenhouse effect (q.v.)

Groundwater: water located in aquifers and accessible from the surface.

Half-life: the time taken for half the atoms in a given amount of radioactive material to decay; the biological half-life is the time required for half of a quantity of radioactive material absorbed by a living tissue or organism to be naturally eliminated.

Hazardous waste (toxic waste): a solid waste, or combination of solid wastes, which because of the quantity, concentration, or physical, chemical, or infectious characteristics may (1) cause or significantly contribute to an increase in mortality or an increase in serious irreversible, or incapacitating reversible, illness or (2) pose a substantial present or potential hazard to human health or the environment when improperly treated, stored, transported, or disposed of or otherwise managed.

Herbicide: a chemical used to control plant growth.

Heterosphere: that part of the atmosphere above 60 mi (100 km) in altitude.

Heterotrophic organisms: organisms that utilize the organic substances produced by autotrophic organisms as energy sources and as the raw materials for the synthesis of their own biomass.

High-level waste: a type of radioactive waste which depends upon the amount of radioactivity (see also Low-level waste).

Homopause: the boundary between the homosphere (q.v.) and the heterosphere (q.v.); also known as the turbopause (q.v.).

Homosphere: the atmosphere below about 60 miles (100 km) in altitude.

Horizons: layers of soil that are more or less parallel to the surface and differ from those above and below in one or more properties such as color, texture, structure, consistency, porosity, and reaction.

Humic acids: a group of complex organic acids originating from the roots of plants and from rotting humus material; base-soluble fraction of humus (q.v.)

Humin: an insoluble fraction of humin which is the residue from the biodegradation of plant material.

Humus: a generic term for the water-insoluble material that makes up the bulk of soil organic matter; a brown or black amorphous mass of decayed organic material found in soil that is derived from natural biological and chemical decomposition or organic matter, such as leaf litter, dead animals and plants, and animal feces.

Hydrocracking: a catalytic high-pressure high-temperature process for the conversion of petroleum feedstocks in the presence of fresh and recycled hydrogen.

Hydrological cycle: the water cycle which involves the continuous and complex transfer of water through its gaseous, liquid, and solid states from the oceans to land and back again.

Hydrology: the study of water.

Hydrolysis: to dispose of chemicals that are reactive with water by allowing them to react with water under controlled conditions.

Hydrosphere: water in various forms; total mass of free water is solid or liquid state on the surface of the earth, i.e., oceans lakes, rivers, glaciers, and ice sheets.

Igneous rock: rock produced from the solidification of molten rock, e.g., granite, basalt, quartz, feldspar, and magnetite; igneous rocks may be either intrusive or extrusive in origin and may be classified in several ways: (1) by silica content, which determines whether igneous rocks are acid, intermediate, basic, or ultra basic in nature; (2) by grain size of the groundmass, which is dependent on the rate of cooling of the magma; an arbitrary scale used to indicate grain size: (i) very course, larger that 3 cm; (ii) coarse, 5 mm-3 cm; (iii) medium, 1-5 mm; (iv) fine, smaller than 1 mm; and (v) glassy, no apparent crystalline structure; (3) the texture of the rock, which results from its mineralogical and chemical characteristics; (4) the level of dark-colored minerals in the rock.

Ignitability: characteristic of liquids whose vapors are likely to ignite in the presence of an ignition source; also characteristic of non-liquids that may catch fire from friction or contact with water and that burn vigorously.

Impact assessment: the interpretation of the significance of anticipated changes related to the proposed project.

In situ immobilization: a process used to convert waste constituents to insoluble or immobile forms that will not leach from a disposal site.

In situ treatment: waste treatment processes that can be applied to wastes in a disposal site by direct application of treatment processes.

Incineration: the controlled combustion of materials in an enclosed area.

Insecticide: a chemical that interferes in the life cycle of certain insects to control insect populations; the chemical may also kill the insect.

Ion exchange: a means of removing cations or anions from solution onto a solid resin.

Ionizing radiation: cosmic rays from outer space.

Ionosphere: the layer of the atmosphere at altitudes of about 30 miles (50 km) and higher.

Kaolinite: a clay mineral formed by hydrothermal activity at the time of rock formation or by chemical weathering of rocks with high feldspar content; usually associated with intrusive granite rocks with high feldspar content.

K-type waste: waste from a specific source.

Land farming: solid phase bioremediation (q.v.)

Landfill: a site where waste is placed in or on land (often in separate trenches depending upon the waste and to prevent contact of reactive waste materials) and which may/should be lined to prevent leakage and runoff of the contaminated surface water.
Land systems: components that form the earth.
Leaching: washing chemicals out of the soil.
Lewis acid: a chemical species which can accept an electron pair from a base.
Lewis base: a chemical species which can donate an electron pair.
Light radiation: from the sun.
Lignin: a complex substance that is the major component of wood.
Limnology: the branch of science dealing with the characteristics of fresh water, including biological properties as well as chemical and physical properties.
Liquid particulate matter: mist, raindrops, fog, and other chemical mists.
Lithosphere: the minerals in the earth's crust.
Loess: a yellow-brown, fine-grained, wind-deposited loam composed chiefly of angular particles of quartz, feldspar, and calcite in a clay matrix and may b extensive and deep, up to 33 feet (10 meters); derived from wind erosion of fine surface material on glacial outwash plains, deltas, river flood plains, and deserts.
Low-level waste: a type of radioactive waste depending upon the amount of radioactivity (see also High-level waste).

Macronutrients: those elements that occur in standard levels in plant materials or in fluids in the plants.
Magma: molten rock, as occurs in volcanoes.
Marsh: an area of spongy waterlogged ground with large numbers of surface water pools; marshes usually result from: (1) an impermeable underlying bedrock; (2) surface deposits of glacial boulder clay; (3) a basin-like topography from which natural drainage is poor; (4) very heavy rainfall in conjunction with a correspondingly low evaporation rate; (5) low-lying land, particularly at estuarine sites at or below sea level.
Mesopause: the transitional zone of minimum temperature between the mesosphere and the thermosphere.
Mesophere: that part of the atmosphere immediately above the stratosphere at an altitude of 50 miles (80 km).
Mesothelioma: a tumor of the mesothelial tissue lining the chest cavity adjacent to the lungs.
Methanol: see Methyl alcohol.
Methyl alcohol (methanol; wood alcohol): a colorless, volatile, inflammable, and poisonous alcohol (CH_3OH) traditionally formed by destructive distillation of wood or, more recently, as a result of synthetic distillation in chemical plants.

Mica: a complex aluminum silicate mineral that is transparent, tough, flexible, and elastic.

Micronutrients: elements that are essential only at very low levels and that are generally required for the functioning of essential enzymes.

Microorganisms: algae, bacteria, and fungi.

Minerals: naturally occurring inorganic solids with well-defined crystalline structures.

Mitigation: identification, evaluation, and cessation of potential impacts on the environment.

Mobile emissions sources: anthropogenic air emissions produced by the various forms of transportation.

Mohs scale: measurement of hardness of a mineral ranging from 1 to 10 with 10 being the hardest.

Mutagenic: tending to cause mutations.

Naft: pre-Christian era (Greek) term for naphtha (q.v.).

Naphtha: the volatile fraction of petroleum which is used as a solvent or as a precursor to gasoline.

Natural gas: gaseous components that often occur in reservoirs with petroleum.

Neutralization: a process for reducing the acidity or alkalinity of a waste stream by mixing acids and bases to produce a neutral solution; also known as pH adjustment.

New chemical substance: a chemical that is not list in the Environmental Protection Agency's Inventory of Chemical Substances or is an unlisted reaction product of two or more chemicals.

Night soil: human excrement collected nightly in towns and cities from cesspools and latrines, and transported to nearby farmland to be used as fertilizer.

Nitrogen cycle: the natural circulation of nitrogen by living organisms via the atmosphere, oceans, and soil; although the atmosphere comprises about 79% nitrogen, most of this is unavailable to living organisms.

Nitrogen fixation: the process by which atmospheric nitrogen is converted to nitrogen compounds available to plants.

Nuclear fission: a process in which an atom of an element is struck by a neutron, causing the atomic nucleus to split apart and release other neutrons.

Nuclear fusion: a process in which two nuclei of light elements such a hydrogen are combined to form a heavier nucleus such as helium with a substantial release of energy.

Oceanography: the science of the ocean and its physical and chemical characteristics.

Oligotrophic lakes: deep, generally clear lakes, deficient in nutrients, and without much biological activity.
Open windrow: a type of composting.
Organic sedimentary rocks: rocks containing organic material such as residues of plant and animal remains/decay.
Overturn: disappearance of thermal stratification causing an entire body of water to behave as a hydrological unit.
Oxidation: a process which can be used for the treatment and removal of a variety of inorganic and organic wastes.
Ozone: a highly reactive gas comprising triatomic oxygen (O_3) formed by recombination of oxygen in the presence of ultraviolet radiation.

P-type waste: a hazardous waste; usually a specific chemical species.
Particulate matter: particles in the atmosphere or on a gas stream that may be organic or inorganic and originate from a wide variety of sources and processes.
Peat soils: soils containing as much as 95% organic material.
Permeability: the ease of flow of the water through the rock.
Petroleum (crude oil): a naturally occurring viscous mixture of gaseous, liquid, and solid hydrocarbon compounds usually found trapped deep underground beneath impermeable cap rock and above a lower dome of sedimentary rock such as shale; most petroleum reservoirs occur in sedimentary rocks of marine, deltaic, or estuarine origin.
Petroleum refining: a complex sequence of events that result in the production of a variety of products.
pH adjustment: neutralization.
Photochemical: light-induced chemical phenomena.
Photochemical smog: see Smog.
Phreatic zone: see water table.
Pipestill gas: the most volatile fraction that contains most of the gases that are generally dissolved in the crude. Also known as pipestill light ends.
Pitch: the nonvolatile product of the thermal decomposition of coal.
Plasma incinerators: incineration by use of an extremely hot plasma of ionized air injected through an electrical arc.
Podzol soil: soil formed in temperate-to-cold climates under relatively high rainfall conditions under coniferous or mixed forest or heath vegetation.
Point sources: sources that emit air emission through a confined vent or stack.
Polishing: the treatment of a waste product for safe discharge.
Pollution: the introduction into the land, water, and air systems of a chemical(s) that are not indigenous to these systems or the introduction into the land, water, and air systems of indigenous chemicals in greater-than-natural amounts.

Porosity: the percentage of rock volume available to contain water or other fluid.

Possible reserves: reserves where there is an even greater degree of uncertainty but about which there is some information.

Potential reserves: reserves based upon geological information about the types of sediments where such resources are likely to occur and they are considered to represent an educated guess.

Pozzolan: a naturally-occurring diatomaceous siliceous material which, when finely divided and reacted with lime in water, forms compounds possessing cementitious properties.

Premanufacture notification (PMN): requirement of any manufacturer to notify the Environmental Protection Agency at least 90 days prior to the production of a new chemical substance.

Primary emissions: pollutants that are emitted directly to the atmosphere.

Primary pollutants: pollutants that are emitted directly from the sources.

Primary waste treatment: preparation for further treatment, although it can result in the removal of by-products and reduction of the quantity and hazard of the waste.

Probable reserves: mineral reserves that are nearly certain but about which a slight doubt exists.

Proved reserves: mineral reserves that have been positively identified as recoverable with current technology.

Pump and treat: extraction of contaminated groundwater from the aquifer with subsequent treatment at the surface and disposal or reinjection.

Pyrethroids: nerve poisons, acting through interference of ion transport along the axonal membrane.

Pyrolysis: exposure of waste to high temperatures in an oxygen-poor environment.

Pyrophoric: substances that catch fire spontaneously in air without an ignition source.

Radiation: the mechanism by which heat is transported away from the earth, usually after conduction and convection effects have transported the heat to the atmosphere.

Radioactive waste: see High-level waste and Low-level waste.

Radioactive waste management: the treatment and containment of radioactive waste materials that arise from industry.

Radon: a gas product of radium decay.

Raw materials: minerals extracted from the earth prior to any refining or treating.

Reactive substances: chemicals that undergo rapid or violent reaction under certain conditions.

Reactive waste: waste unstable under ambient conditions.

Receptor: an object or location that is affected by a pollutant.

Recycling: the use or reuse of chemical waste as an effective substitute for commercial products or as an ingredient or feedstock in an industrial process.

Reformulated gasoline (RFG): gasoline designed to mitigate smog production and to improve air quality by limiting the emission levels of certain chemical compounds such as benzene and other aromatic derivatives.

Regulated pollutants: pollutants that have been singled out for regulatory control.

Renewable energy sources: solar, wind, and other nonfossil fuel energy sources.

Reserves: well-identified resources that can be profitably extracted and utilized with existing technology.

Reservoir: a domain where a pollutant may reside for an indeterminate time.

Resource: the total amount of a commodity (usually a mineral but can include non-minerals such as water and petroleum) that has been estimated to be ultimately available.

Reverse osmosis: separation of components in a liquid stream by applying external pressure to one side of the membrane so that solvent will flow in the opposite direction.

Rotary kiln: a versatile large refractory-lined cylinder capable of burning virtually any liquid or solid organic waste; the unit is rotated to improve turbulence in the combustion zone.

Rotenones: electron-transport inhibitors.

Run-of-the-river reservoirs: reservoirs with a large rate of flow-through compared to their volume.

Sand: a course granular mineral mainly comprising quartz grains that is derived from the chemical and physical weathering of rocks rich in quartz, notably sandstone and granite.

Sandstone: a sedimentary rock formed by compaction and cementation of sand grains; can be classified according to the mineral composition of the sand and cement.

Screening: a process for removing particles from waste streams and used to protect downstream pretreatment processes.

Secondary emissions: pollutants formed in the atmosphere as the result of the transformation of primary emissions.

Secondary pollutants: pollutants produced by interaction of primary pollutants with another chemical or by dissociation of a primary pollutant or other effects within a particular ecosystem.

Sedimentary strata: typically consist of mixtures of clay, silt, sand, organic matter, and various minerals.

Serpentine asbestos: the mineral chrysotile, a magnesium silicate.

Slurry phase reactors: tanks where wastes, nutrients and microorganisms are placed.

Smog: a description of the combination of smoke and fog; an atmospheric haze sometimes found above large industrial and urban areas. The result of (photochemical) reactions between pollutants.

Soil: finely divided rock-derived material containing organic matter and capable of supporting vegetation.

Soil heaping: piling wastes in heaps several feet high on an asphalt or concrete pad.

Solid phase bioremediation: treatment of wastes using conventional soil management practices to enhance the microbial degradation of the wastes; also known as land farming.

Solid waste: garbage refuse, sludge from a waste treatment plant, water supply treatment plant, or air pollution control facility and other discarded material, including solid, liquid, semisolid, or contained gaseous material resulting from industrial, commercial, mining, and agricultural operation, and from community activities.

Solvent extraction: a process for separating liquids by mixing the stream with a solvent that is immiscible with part of the waste but that will extract certain components of the waste stream.

Source reduction: the reduction or elimination of chemical waste at the source, usually within a process.

Spontaneous ignition: ignition of a fuel, such as coal, under normal atmospheric conditions; usually induced by climatic conditions.

Stabilization: the addition of chemicals or materials to a waste to ensure that the waste is maintained in their least soluble or least toxic form.

Static windrow: a type of composting

Strata: layers including the solid iron-rich inner core, molten outer core, mantle, and crust of the earth.

Stratopause: the transitional zone of maximum temperature between the stratosphere and the mesosphere.

Stratosphere: atmospheric layer directly above the troposphere.

Stripping: a means of separating volatile components from less volatile ones in a liquid mixture by the partitioning of the more volatile materials to a gas phase of air or steam.

Surface impoundment: use of natural depressions, engineered depressions, or diked areas for treatment, storage, or disposal of chemical waste.

Surface water: water in lakes, streams, and reservoirs.

Swamp: a wetland where trees and shrubs are an important part of the vegetative association.

Symbiosis: a relationship between members of two different species that results in mutual benefit; can often permit the survival of a species in a hostile environment.

Tar: a volatile liquid or semi-solid product from the thermal decomposition of coal (see also Pitch).
Teratogenic: tending to cause developmental malformations.
Thermocline: a permanent layer of water within lakes and oceans that warmer layer (the epilimnion) and a lower colder layer (the hypolimnion).
Thermosphere: the layer of the atmosphere that lies above the mesosphere.
Toxic wastes: chemicals discharged from domestic and industrial sources that are harmful or fatal when ingested or absorbed.
Toxicity: defined in terms of a standard extraction procedure followed by chemical analysis for specific substances.
Toxicity Characteristic Leaching Procedure (TCLP): a test designed to determine the mobility of both organic and inorganic contaminants present in liquid, solid, and multiphasic wastes.
Trace element: those elements that occur at very low levels in a given system.
Transpiration: the process by which water enters the plant's leaves from the atmosphere.
Treatment: any method, technique, or process that changes the physical, chemical, or biological character of any chemical waste so as to neutralize the effects of the constituents of the waste.
Tropopause: the transitional zone between the troposphere and the stratosphere.
Troposphere: the lowest layer of the atmosphere.
Turbopause: homopause (q.v.).
Turbulence: air disturbances that result from such factors as the friction of the land surface, physical obstacles to wind flow, and the vertical temperature profile of the lower atmosphere.

U-type waste: hazardous waste made up of specific compounds.
Unsaturated zone: the region in which water is held in the soil.
Unstable atmosphere: marked by a high degree of turbulence.

Vadose water: water present in the unsaturated zone.
Vitrification: a phase conversion process in which a chemical waste (or constituents of the waste) is melted at high temperature to form an impermeable capsule around the remainder of the waste; also known as glassification (q.v.).

Washout: the uptake of particles and gases by water droplets and snow and their removal from the atmosphere when rain and snow fall on the ground.
Waste piles: piles used to contain accumulation of solid waste and which may also be used for final disposal or for temporary storage.
Waste management: an organized system for waste handling leading to elimination or disposal in ways that protect the environment.

Water pollution: any change in natural waters that may impair their further use, caused by the introduction of organic or inorganic substances or a change in temperature of the water.
Water table (phreatic zone): the upper surface of the zone of permanent ground water saturation.
Wet flatlands: areas where mesophytic vegetation is more important than open water and which are commonly developed in filled lakes, glacial pits, and potholes, or in poorly drained coastal plains or flood plains.
Wetland: any area of low-lying land where the water table is at or near the surface most of the time, resulting in open-water habitats and waterlogged land areas; wetlands are typically found in estuaries, along rivers with little vertical descent, and in uplands where natural drainage is impeded due to extensive boulder clay deposits or by drainage channels disrupted by glacial watershed breaching; wetlands may also form in areas of very high rainfall and low evaporation; can be freshwater, brackish, or saltwater habitats.
Wet scrubbers: devices in which a counter-current spray liquid is used to remove impurities and particulate matter from a gas stream.
Wind: horizontally moving air.
Wood alcohol: see Methyl alcohol.

Zone of aeration: unsaturated zone.
Zone of incorporation: the top foot of soil.
Zone of saturation: at lower depths, in the presence of adequate amounts of water, all voids are filled.

Index